AF594958

Marine Fungus

Marine Fungus

Editor

Federico Baltar

MDPI • Basel • Beijing • Wuhan • Barcelona • Belgrade • Manchester • Tokyo • Cluj • Tianjin

Editor
Federico Baltar
Functional and Evolutionary
Ecology
University of Vienna
Vienna
Austria

Editorial Office
MDPI
St. Alban-Anlage 66
4052 Basel, Switzerland

This is a reprint of articles from the Special Issue published online in the open access journal *Journal of Fungi* (ISSN 2309-608X) (available at: www.mdpi.com/journal/jof/special_issues/Marine_Fungus).

For citation purposes, cite each article independently as indicated on the article page online and as indicated below:

LastName, A.A.; LastName, B.B.; LastName, C.C. Article Title. *Journal Name* **Year**, *Volume Number*, Page Range.

ISBN 978-3-0365-4718-3 (Hbk)
ISBN 978-3-0365-4717-6 (PDF)

Contents

About the Editor

Federico Baltar

Dr. Federico Baltar is an Associate Professor of the Department of Functional and Evolutionary Ecology at the University of Vienna, Austria. His research interests are in biological oceanography integrating marine microbial ecology and biogeochemistry. From coastal seas to open oceans, from Fjords to the seawater below the Antarctic Ice Shelfs, all are large seawater environments performing relevant ecosystem services and functions driven by tiny microbial engines, including fungi. Dr. Federico Baltar's research group studies the factors that control the functioning of these seawater microbes to get a mechanistic understanding of this microbial engine today, and predict how it might change in the future ocean. Dr. Federico Baltar received his Ph.D. in 2010 and has an outstanding publication record, comprising >70 publications in peer-reviewed international journals. Dr. Federico Baltar's Google Scholar Citations number >2500. He has received several research-related awards, including among others, the 2016 Outstanding Young Scientist in Biogeosciences Award (from the European Geosciences Union, EGU), the Rutherford Discovery Fellowship (from the Royal Society of New Zealand), and the 2021 JoF Youg Investigator Award.

Preface to "Marine Fungus"

Microbes are the central players in the ecology and biogeochemistry of marine systems. Yet, most of the research on microbial ecology in oceans has been performed on prokaryotes and eukaryotic phytoplankton, and fungi have been essentially neglected in open-ocean studies. One of the main reasons why the ecology of fungi in pelagic oceanic environments has been overlooked thus far is probably the preconception that fungi are outcompeted by prokaryotes when living in a resource-limited liquid environment such as the ocean. Most of the available studies on marine fungi are based on the isolation and identification of fungi from different surfaces (e.g., submerged wood, sediments, macrophytes), mostly in coastal benthic environments. However, recent evidence suggests that fungi are also present in the oceanic water column, mainly associated to particles, with the genomic potential to significantly contribute to marine biogeochemical cycles. Still, we lack even basic information on the ecology of the oceanic mycobiome, precluding us from determining the ecological role of this enigmatic kingdom in our oceans. The aim of this book and Special Issue is to focus on the ecology of marine fungi. Topics include, fungal abundance, distribution, activity, and phylogenetic and/or functional diversity in coastal to open ocean environments, including seawater column and sediments, derived both from laboratory and field studies.

Federico Baltar
Editor

Journal of *Fungi*

Article

Adapting an Ergosterol Extraction Method with Marine Yeasts for the Quantification of Oceanic Fungal Biomass

Katherine Salazar Alekseyeva [1,*], Barbara Mähnert [1], Franz Berthiller [2], Eva Breyer [1], Gerhard J. Herndl [1,3] and Federico Baltar [1,*]

1 Department of Functional and Evolutionary Ecology, Microbial Oceanography Working Group, University of Vienna, 1030 Vienna, Austria; barbara.maehnert@univie.ac.at (B.M.); eva.breyer@univie.ac.at (E.B.); gerhard.hernld@univie.ac.at (G.J.H.)

2 Department of Agrobiotechnology (IFA-Tulln), University of Natural Resources and Life Sciences, Vienna (BOKU), 3430 Tulln, Austria; franz.berthiller@boku.ac.at

3 Department of Marine Microbiology and Biogeochemistry, Royal Netherlands Institute for Sea Research (NIOZ), University of Utrecht, 1790 AB Den Burg, Texel, The Netherlands

* Correspondence: katherine.salazar@univie.ac.at (K.S.A.); federico.baltar@univie.ac.at (F.B.); Tel.: +43-(0)1-4277-76438 (K.S.A.); +43-(0)1-4277-76436 (F.B.)

Abstract: Ergosterol has traditionally been used as a proxy to estimate fungal biomass as it is almost exclusively found in fungal lipid membranes. Ergosterol determination has been mostly used for fungal samples from terrestrial, freshwater, salt marsh- and mangrove-dominated environments or to describe fungal degradation of plant matter. In the open ocean, however, the expected concentrations of ergosterol are orders of magnitude lower than in terrestrial or macrophyte-dominated coastal systems. Consequently, the fungal biomass in the open ocean remains largely unknown. Recent evidence based on microscopy and -omics techniques suggests, however, that fungi contribute substantially to the microbial biomass in the oceanic water column, highlighting the need to accurately determine fungal biomass in the open ocean. We performed ergosterol extractions of an oceanic fungal isolate (*Rhodotorula sphaerocarpa*) with biomass concentrations varying over nine orders of magnitude. While after the initial chloroform-methanol extraction ~87% of the ergosterol was recovered, a second extraction recovered an additional ~10%. Testing this extraction method on samples collected from the open Atlantic Ocean, we successfully determined ergosterol concentrations as low as 0.12 pM. Thus, this highly sensitive method is well suited for measuring fungal biomass from open ocean waters, including deep-sea environments.

Keywords: marine fungi; chloroform-methanol extraction; HPLC-UV; LC-MS/MS; ergosterol; pelagic fungal biomass

Citation: Salazar Alekseyeva, K.; Mähnert, B.; Berthiller, F.; Breyer, E.; Herndl, G.J.; Baltar, F. Adapting an Ergosterol Extraction Method with Marine Yeasts for the Quantification of Oceanic Fungal Biomass. *J. Fungi* **2021**, *7*, 690. https://doi.org/10.3390/jof7090690

Academic Editor: Wei Li

Received: 21 July 2021
Accepted: 24 August 2021
Published: 26 August 2021

1. Introduction

In marine environments, heterotrophic microorganisms are the main drivers of biogeochemical cycles due to their orders of magnitude higher biomass-specific metabolic rates than metazoans. The main heterotrophic organisms in terms of biomass are bacteria, followed by their grazers [1], the hetero-, and mixotrophic protists [2]. A recent study on deep-sea marine snow in the North Atlantic showed that although bacteria dominated in terms of abundance, fungi dominated the marine snow-associated microbial community in terms of biomass [3]. However, in contrast to terrestrial systems [4,5], freshwater [6–8], and saltmarsh studies [9], marine pelagic fungi have been poorly studied. It has been assumed that fungi only play a trivial role in the ocean, so their role in the ecology and biogeochemistry of the ocean remains largely enigmatic [10]. A metagenomic study, however, suggests that fungi are potentially involved in many biogeochemical cycles in the ocean [11]. Moreover, a recent study supports that fungi contribute to the cycling of carbohydrates in the ocean [12]. Nonetheless, most of the recent discoveries on pelagic fungal ecology are based on molecular methods that inform about the relative abundance and genetic potential and expression; however, the lack of an adequate methodology to reliably determine other basic ecological parameters such as biomass and production precludes a deeper understanding of the ecological role of fungi in the ocean.

For terrestrial and aquatic environments, the concentration of ergosterol has been used as a proxy for fungal biomass [4,5,13–15]. For the majority of eukaryotic organisms, sterols are essential lipids serving as structural support and precursors for hormones [16]. Ergosterol is an essential component of membranes in fungi [17,18], and its concentration is species-specific and dependent on the physiological state [8]. A property that makes ergosterol especially well-suited as a proxy for fungal biomass is that it contains a 5,7-double bond, which is rare among other sterols [9]. Additionally, it represents more than 80% of sterols in fungal strains [19]. Not all fungi are capable to synthesize ergosterol such as many fungi belonging to the division Chytridiomycota [20]. Additionally, some non-fungal organisms such as the green algae *Chlorella vulgaris* and some Protozoa can synthesize ergosterol [21,22]. However, since the division Chytridiomycota is usually either not present or a low contributor to pelagic fungal communities, and only attains relevant biomass sporadically [23], the determination of fungal biomass via ergosterol quantification appears to be suitable.

The extraction of ergosterol from a sample is the first step in the quantification of fungal biomass. The most commonly used extraction method for ergosterol is the reflux extraction [9,14] which has been used for saltmarsh material, and also the chloroform-methanol extraction [24] applied to fish tissue samples. It is noteworthy that these methods were developed and applied for environments with considerably higher fungal biomass than in the oligotrophic open ocean. Thus, it is unknown whether the currently applied ergosterol extraction and detection methods are sufficiently sensitive to determine pelagic fungal biomass.

We tested the chloroform-methanol extraction using an oceanic fungal isolate at biomass concentrations varying over nine orders of magnitude. We tested and fine-tuned the method to determine low concentrations of fungal biomass using HPLC-UV and LC-MS/MS. Finally, we applied this method to samples collected from the Atlantic Ocean throughout the water column down to bathypelagic waters.

2. Materials and Methods

2.1. Testing a Method to Allow Detecting Low Concentrations of Ergosterol Using a Cultured Fungal Strain

2.1.1. Fungal Culture and Dilutions Preparation

Rhodotorula sphaerocarpa (HB 738), a fungus originally isolated from coastal Antarctic waters close to Marguerite Bay, was obtained from the Austrian Center of Biological Resources (ACBR). Ten fungal concentrations were prepared through serial dilution to identify the relationship between the fungal biomass and ergosterol concentration. For this, an initial amount of *Rhodotorula sphaerocarpa* cultured on yeast malt extract Agar [25,26] was diluted in 100 mL of artificial seawater (30 g/L sea salts S9883 Sigma-Aldrich) to obtain an $OD_{660} \approx 1$ (i.e., 1×). The optical density (OD) was measured with a UV-1800 Shimadzu spectrophotometer. Thereafter, 100 mL of the initial fungal biomass were added to 100 mL of artificial seawater. This process was repeated sequentially to obtain dilutions of 2×, 4×, 8×, 16×, 32×, 64×, 125×, 250×, 500×, and 1000×, all in triplicate. Thirty mL from each triplicate were gently vacuum-filtered onto combusted (450 °C; 6 h) 25 mm diameter Whatman GF/F filters (WHA1825047 Sigma-Aldrich). Finally, filters were wrapped in a combusted aluminum foil and stored at −20 °C until further processing.

2.1.2. Ergosterol Extraction

The chloroform-methanol method to extract ergosterol was used as described by Bligh and Dyer [24] with minor modifications. Briefly, one filter was gently shaken in a glass scintillation vial with 3 mL of 2:1 chloroform:methanol (*v*/*v*) in the dark at room temperature for 24 h. Thereafter, the extract was transferred to a 15 mL polypropylene tube, and 0.660 mL of Milli-Q water were added, thoroughly mixed, and centrifuged at 3200× *g* at room temperature for 3 min. To quantify the ergosterol potentially lost in the discarded extraction phase, the upper phase (usually discarded according to the standard protocol) was transferred to an amber glass vial. Afterwards, the lower phase was transferred to another amber glass vial and evaporated to dryness under a fume hood. Finally, the extracts were resuspended in 0.50 mL of methanol and stored at −20 °C until analysis.

The effect of repeated extractions of the same filter using the chloroform-methanol method was also tested. Each filter containing the various dilutions of *R. sphaerocarpa* was extracted three times using the chloroform-methanol approach as described above. The lower phase of each extraction was evaporated separately to dryness under a fume hood and the extracts were resuspended in 0.50 mL of methanol and stored at −20 °C until HPLC analysis.

2.1.3. Determining the Extraction Efficiency of Ergosterol

To determine the extraction efficiency of the chloroform-methanol method, 7.5 mL of the biomass of $OD_{660} \approx 1$ were added to 472.5 mL of artificial seawater to achieve the dilution 64×. Thirty mL were gently vacuum-filtered onto combusted (450 °C; 6 h) 25 mm diameter Whatman GF/F filters (WHA1825047 Sigma-Aldrich). In total, 12 filters were prepared in this way and each filter was wrapped in a combusted aluminum foil and stored at −20 °C until further processing. Before the extraction, 0.5 mL of ergosterol standards in methanol (1400 nM, 700 nM, and 350 nM final concentration; HPLC grade, Sigma-Aldrich) were added to each filter (containing the biomass) in triplicate, and no ergosterol standard was added to three filters. Subsequently, the filters were placed in the oven at 40 °C overnight. Finally, they were extracted using chloroform-methanol extraction as described above, and the extracts were analyzed by HPLC analysis.

2.1.4. Ergosterol Quantification by High-Performance Liquid Chromatography (HPLC-UV) Analysis

Ergosterol concentrations were quantified by applying HPLC according to the protocol of Gulis and Bärlocher [27]. An Agilent 1260 Infinity Bioinert HPLC System equipped with an autosampler was used together with a Zorbax StableBond Aq Analytical Guard Column (4.6 × 12.5 mm; 5 µm particle size; 80 Å pore size, Agilent Technologies, Santa Clara, CA, USA) and a Zorbax Eclipse AAA Rapid resolution column (4.6 × 150 mm; 3.5 µm particle size, 80 Å pore size, Agilent Technologies). For detecting ergosterol, the wavelength of the UV detector was set at 282 nm. The mobile phase consisted of 100% methanol (HPLC grade, Sigma-Aldrich) with a flow rate of 0.8 mL/min. The injection volume was 400 µL and the samples were run for 20 min in total, including 5 min of purging at a column temperature of 25 °C. Ergosterol standards (HPLC grade, Sigma-Aldrich) were prepared in a range from 500 µM to 30 nM in methanol and run together in the same sequence batch with the fungal samples. Furthermore, the temperature of the autosampler was kept at 10 °C to protect them from potential degradation. Peak areas were used for calculating ergosterol concentrations in samples.

2.1.5. Limit of Detection (LOD) and Limit of Quantitation (LOQ) of Ergosterol by HPLC Analysis

The lowest ergosterol concentration that can be reliably measured by HPLC was determined according to the protocol of Wenzl et al. [28]. Six ergosterol standards (HPLC grade, Sigma-Aldrich) were prepared in five replicates ranging from 10 µM to 30 nM in methanol. Linear calibrations were performed and LOD and LOQ concentrations were calculated.

2.2. Measuring Ergosterol Concentrations in Samples Collected throughout the Water Column of the Atlantic Ocean

2.2.1. Sampling

Samples were collected from the water column of the Atlantic Ocean in March-April 2019 during the Poseidon expedition on board the R.V. *Sarmiento de Gamboa* from Punta Arenas, Chile to Santa Cruz de Tenerife, Spain. Seawater was collected with a CTD rosette sampler containing 12 L Niskin bottles from three different depths corresponding to epipelagic (5 m), mesopelagic (950 m), and bathypelagic (4000 m) waters. Ten liters of seawater were vacuum-filtered onto combusted 47 mm diameter Whatman GF/F filters (WHA1825047 Sigma-Aldrich). Filters were wrapped in combusted aluminum foil and stored at −20 °C until analyses.

2.2.2. Ergosterol Extraction and Concentration

The ergosterol extraction of the samples collected in the Atlantic Ocean was performed with the chloroform-methanol method as mentioned above with minor modifications. Each filter was extracted twice, so the lower phases from extraction 1 and 2 were transferred to the same amber glass vial and evaporated to dryness under a fume hood. Finally, the dried extract was resuspended in 0.1 mL of methanol and stored at −20 °C until analysis.

2.2.3. Ergosterol Quantification via Liquid Chromatography–Mass Spectrometry/Mass Spectrometry (LC-MS/MS) Analysis

To ensure the identity of the analyte applying the HPLC-UV method and to compare quantification, LC-MS/MS analysis was also performed. Ergosterol concentrations were quantified with LC-MS/MS similar to the protocol of Ory et al. [29]. Briefly, an Agilent (Waldbronn, Germany) 1290 UHPLC System was used together with a Sciex (Framingham, MA, USA) QTrap 6500+ mass spectrometer and a Phenomenex (Aschaffenburg, Germany) Gemini column (150 × 4.6 mm, 5 μm particle size). Isocratic elution at 35 °C was performed with a mobile phase consisting of acidified acetonitrile (HPLC grade, Sigma-Aldrich) with 0.1% formic acid at a flow rate of 2.5 mL/min. The ergosterol standards (HPLC grade, Sigma-Aldrich) were prepared in a range from 1 nM to 100 μM in methanol and an injection volume of 20 μL was used. The retention time was 3.78 min, and the eluent was sent to the atmospheric pressure chemical ionization (APCI) source between 3.0 to 4.5 min.

For the ionization of ergosterol in positive ion mode, a heated nebulizer APCI probe was used in the Turbo V source with the following settings: curtain gas 30 psi, nebulizer gas 40 psi, temperature 500 °C, nebulizer current 3 μA. The mass spectrometer was operated in selected reaction monitoring mode using both the $[M+H-H_2O]^+$ (declustering potential 100 V) and the unusual $[M+H-2H_2]^+$ (declustering potential 50 V) ions as precursors. The following six transitions were monitored for 50 ms each (Q1 mass > Q3 mass (collision energy, CE)): 379.3 > 295.2 (20 eV), 379.3 > 159.1 (30 eV), 379.3 > 145.1 (35 eV), 379.3 > 69.0 (55 eV), 393.3 > 268.2 (30 eV), 393.3 > 173.1 (35 eV). Analyst version 1.6.3 was used for data acquisition and evaluation.

2.3. Correlation between HPLC-UV and LC-MS/MS

For the correlation between the two detection methods, we performed a growth experiment. *Sakaguchia dacryoidea* (HB 877), a fungus originally isolated from the Antarctic waters, was obtained from the Austrian Center of Biological Resources (ACBR). An initial amount of *Sakaguchia dacryoidea* cultured on yeast malt extract Agar [25,26] was diluted in 10 mL of artificial seawater (30 g/L sea salts S9883 Sigma-Aldrich) to obtain an $OD_{660} \approx 1$. Afterwards, these 10 mL were added to 1000 mL of liquid media containing: 2 g glucose, 2 g peptone, 2 g yeast extract, 2 g malt extract, 30 g artificial sea salts, and 0.5 g chloramphenicol. The liquid cultures were divided into ten bottles, each containing 100 mL of liquid media + fungi and incubated at 140 rpm (Aggro Lab, Ski 4) and at room temperature. The cultures were sampled in the lag, exponential, and stationary phase as determined by their optical density. Thirty milliliters from each triplicate were gently vacuum-filtered onto combusted

(450 °C; 6 h) 47 mm diameter Whatman GF/F filters (WHA1825047 Sigma-Aldrich). Filters were wrapped in a combusted aluminum foil and stored at −20 °C until further processing. Finally, the filter was extracted using the chloroform-methanol approach as described above and the extracts were analyzed via HPLC-UV and LC-MS/MS analysis.

3. Results and Discussion

Ergosterol is the most abundant fungal sterol [30] and it is almost exclusively produced by fungi [13]. Hence, it has been used to determine fungal biomass in terrestrial and aquatic environments [19]. However, quantifying ergosterol in open ocean waters and the deep sea requires highly sensitive methods due to the expected low fungal abundance in these environments. Accordingly, we extracted and concentrated ergosterol, and compared two detection methods.

The chloroform-methanol approach allows ergosterol extraction and purification in a single operation. This method is also less time-consuming and no special equipment is required such as a rotavapor as compared to the reflux method [9,14]. According to Bligh and Dyer [24], the chloroform-methanol extraction method produces a biphasic system where the chloroform layer contains the lipids and the aqueous layer contains the non-lipids. Nonetheless, we determined the ergosterol concentration in both the lower and upper phase to test whether ergosterol is carried over and discarded with the upper phase. Most of the ergosterol concentration was recovered in the lower phase (chloroform layer), irrespective of the fungal dilution (Figure 1). Approximately 98% of the extracted ergosterol was contained in the lower phase, whereas less than 2% was found in the upper phase. Ergosterol has a logP value of 8.86, which reflects the molecule affinity to the organic portion, in this case, chloroform. Moreover, this extraction allowed a recovery of ~86% of ergosterol standards. For a toluene extraction, a similar value (90%) was obtained by Verma et al. [18].

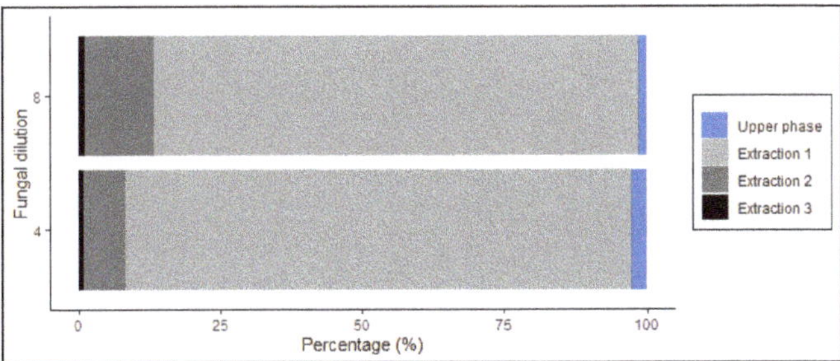

Figure 1. Percentages of ergosterol extracted with chloroform-methanol approach from two fungal dilutions (4×, and 8×) of the marine fungal isolate *Rhodotorula sphaerocarpa*. Extracts recovered from the upper (one extraction) and lower (three extractions) phases were analyzed.

The LOD for HPLC-UV was 37.2 nM, while the LOQ was 122.8 nM. As a result, we were able to determine ergosterol concentrations in 7 out of 9 fungal dilutions (4× to 250×) (Figures 2A and 3). Due to high biomass, 30 mL of the dilution 2× could not be filtered onto a 25 mm diameter Whatman GF/F filter, so it was not measured. The ergosterol concentration was linearly related to fungal biomass using *Rhodotorula sphaerocarpa* (Figure 4). Consequently, the chloroform-methanol extraction resulted in a good relationship between ergosterol concentration and fungal biomass (R^2= 0.9715).

We also tested the significance of a second and third extraction (Figure 2B,C). In the study of Bligh and Dyer [24] and Roose and Smedes [31], a second extraction yielded an additional lipid recovery of 6% and 8–10%, respectively. We found that a second extraction accounts for 5–16% of the total extracted ergosterol in the sample. For fungal dilutions 4×, 8×, 16×, 32×, 64×, and 125×, a second extraction allowed an additional recovery of 7.6%, 12.6%, 4.9%, 15.1%, 10.3%, and 15.5%, respectively. A third extraction led to an additional 1% of the total extracted ergosterol. Therefore, we recommend performing a second extraction, particularly when very low fungal biomass is expected as in oceanic samples.

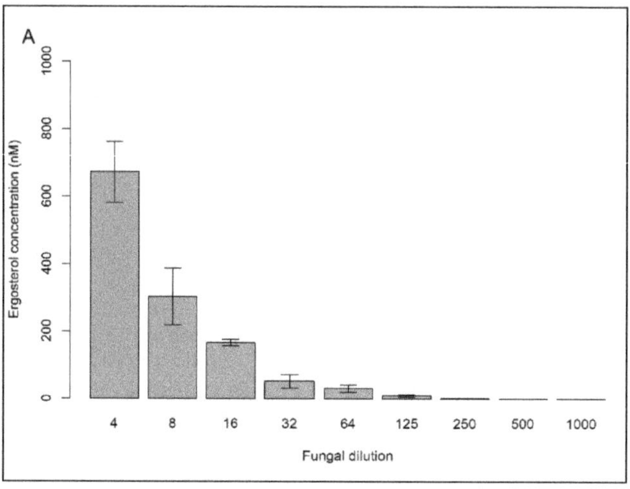

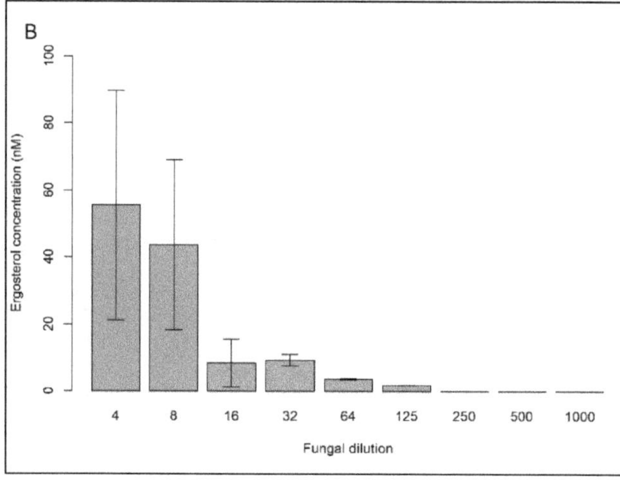

Figure 2. *Cont.*

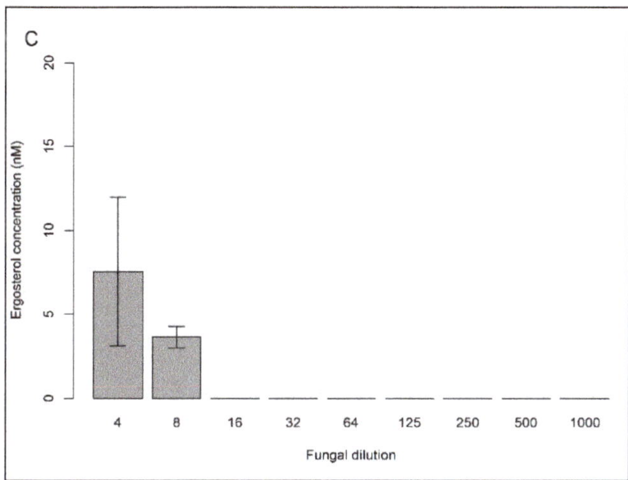

Figure 2. Ergosterol concentration in nM obtained with chloroform-methanol extraction from the marine fungal isolate *Rhodotorula sphaerocarpa*. Nine fungal dilutions in artificial seawater were used (4×, 8×, 16×, 32×, 64×, 125×, 250×, 500×, and 1000×). Extracts from the first (**A**), second (**B**), and third (**C**) extractions were analyzed.

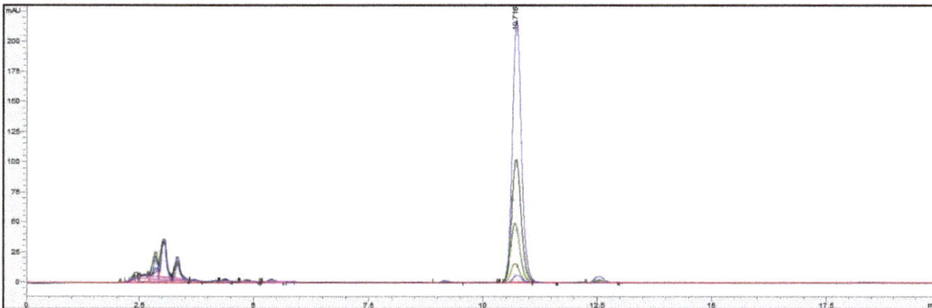

Figure 3. LC-UV-chromatogram of ergosterol obtained with chloroform-methanol extraction from the marine fungal isolate *Rhodotorula sphaerocarpa*. Five fungal dilutions in artificial seawater are shown (4×, 8×, 16×, 32×, and 64×), and the *x*-axis shows minutes of the run.

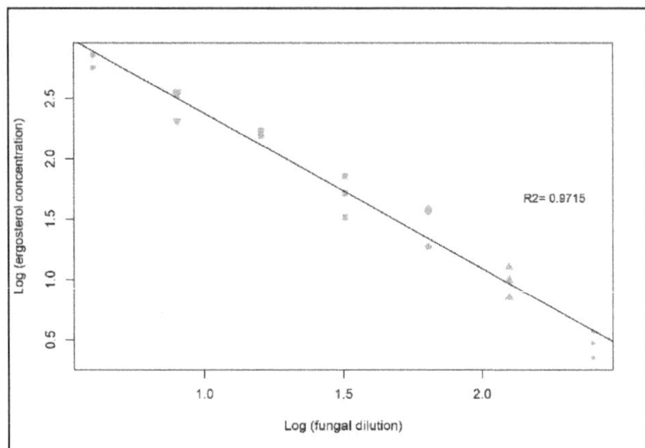

Figure 4. Relationship between ergosterol concentration and fungal biomass dilution.

The two detections, HPLC-UV and LCMS-MS were in good agreement (R^2 = 0.9985) (Figure 5). Only in the lag phase, where the lowest ergosterol concentrations were detected in the culture, both methods differed remarkably with 380.2 $\pm$ 59.0 nM for HPLC-UV and 97.6 $\pm$ 57.6 nM LC-MS/MS. In open ocean waters and the deep sea, highly sensitive methods are required to quantify ergosterol as a low fungal abundance is expected in these environments. Thus, for oceanic samples collected from the Atlantic Ocean over three depths, we determined ergosterol with LC-MS/MS.

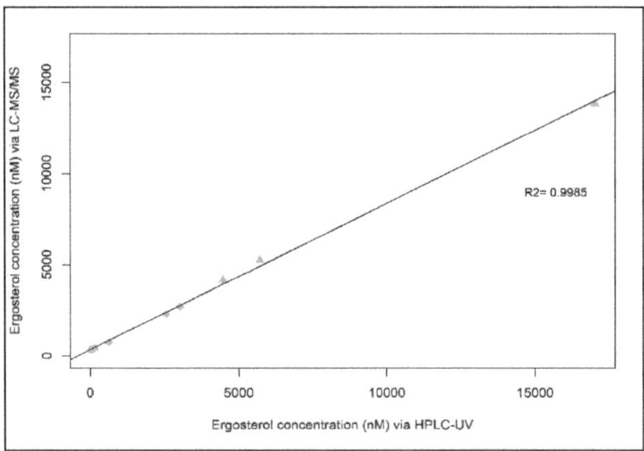

Figure 5. Ergosterol content in nM detected via HPLC-UV vs. LC-MS/MS from the marine fungal isolate *Sakaguchia dacryoidea* covering three growth phases (lag, exponential, and stationary).

For the developed LC-APCI-MS/MS method, the observed ionization was in agreement with results obtained from Ory et al. [29], with the $[M+H-H_2O]^+$ and the $[M+H-2H_2]^+$ ions being the most intense. An LOD of 1.5 nM (S/N = 3/1) and an LOQ of 5.0 nM (2 ng/mL) at S/N = 10/1 could be obtained, thus being about 25 times more sensitive than with the HPLC-UV method. A chromatogram of an ergosterol standard (1 μM) with all six SRM transitions is shown in Figure 6. We were able to successfully concentrate ergosterol and detect it in the pM range (Table 1). Marine samples were concentrated 10,000-fold (10 L water filtered and finally taken up in 0.1 mL); thus, the LOQ in seawater was about 0.05 pM (0.02 pg/mL). The highest concentration of ergosterol was 0.31 pM corresponding to epipelagic (5 m), whereas the lowest concentration was below the limit of detection corresponding to bathypelagic (4000 m). In the mesopelagic (950 m), the concentration was 0.12 pM. These numbers correspond to those of Hassett et al. [15], reporting an ergosterol concentration of 0.12 pM at a 246 m depth in the Arctic Ocean.

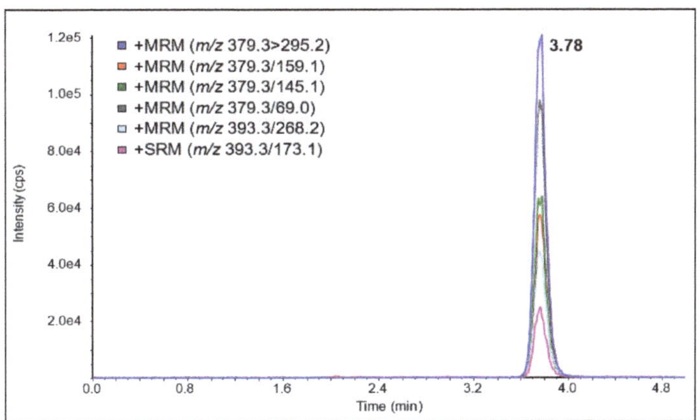

Figure 6. LC-APCI-MS/MS chromatogram of a 1000 nM ergosterol standard.

Table 1. Ergosterol content in the seawater of the Atlantic Ocean obtained with the chloroform-methanol approach (Bligh and Dyer 1959).

Coordinates		Pelagic Zone	Depth (m)	Ergosterol Concentration (pM)	Ergosterol Concentration (ng/L)
Latitude	Longitude				
		Epipelagic	5	0.306	0.121
13°35.748	−29°42.830	Mesopelagic	950	0.120	0.048
		Bathypelagic	4000	0.000	0.000

Overall, we successfully adapted a method originally developed for ergosterol extraction from fish tissue [24] to pelagic fungi. Based on our findings, we propose an ergosterol extraction method (detailed in Figure 7) and quantification method specially adapted for low fungal concentrations. We found three crucial steps in the determination of ergosterol which are (1) a re-extraction of the filter, (2) the volume of the extract, and (3) the filtration volume of the water sample. A second extraction of the filter accounts for an additional ~10% of the ergosterol recovered. Additionally, the final volume of the extraction was adjusted to 0.5 mL for culture samples and 0.1 mL for oceanic samples, which might also be essential for the ergosterol quantification. Nonetheless, as mentioned before, it is important to consider that the ergosterol concentration can vary between species, and also that it can be influenced by different environmental parameters [6–8]. Thus, it is now relevant to test this method with other fungal species from oceanic origin (including hyphae like), and to consider the importance of environmental conditions on the ergosterol concentration when estimating the biomass of oceanic fungi. In spite of these few limitations, the method characterized here represents an important step for future research on the ecological role of fungi in the ocean and an adequately sensitive method for environments with low fungal biomass.

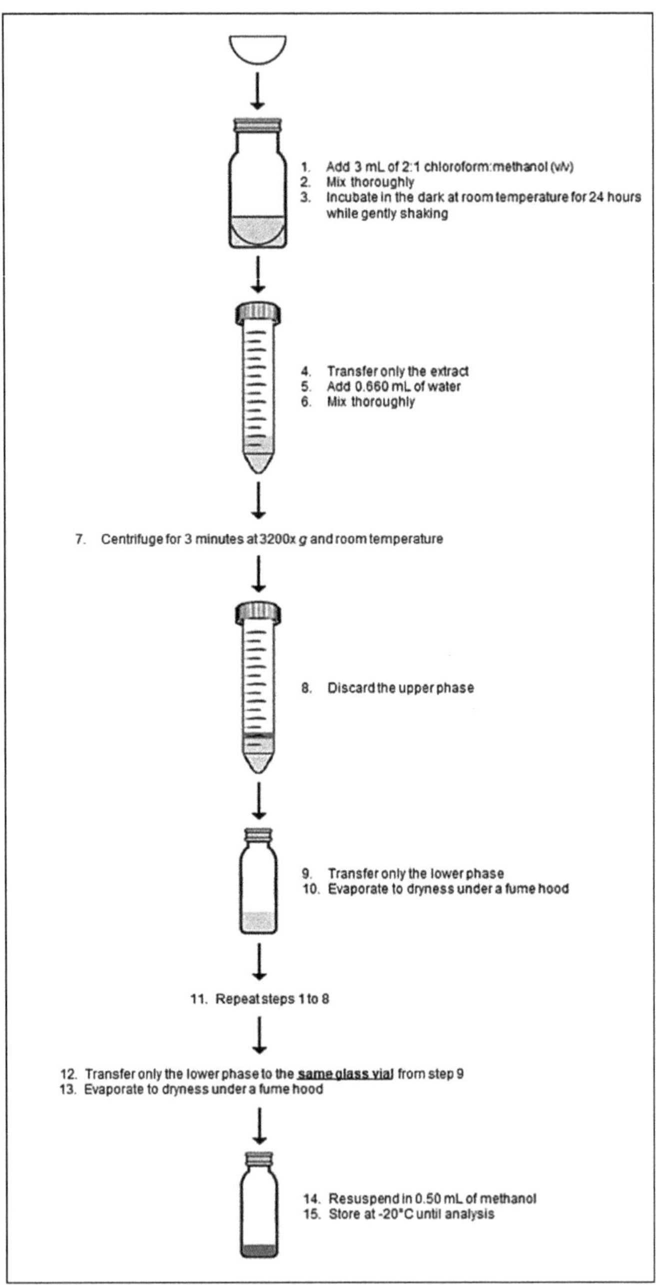

Figure 7. Ergosterol extraction method modified and adapted for pelagic fungi.

Author Contributions: Conceptualization, K.S.A., F.B. (Franz Berthiller), G.J.H. and F.B. (Federico Baltar); data curation, K.S.A. and F.B. (Franz Berthiller); formal analysis, K.S.A. and B.M.; funding acquisition, G.J.H. and F.B. (Federico Baltar); investigation, K.S.A. and E.B.; methodology, K.S.A., B.M., F.B. (Franz Berthiller), and E.B.; project administration, G.J.H. and F.B. (Federico Baltar); software, B.M. and F.B. (Franz Berthiller); supervision, G.J.H. and F.B. (Federico Baltar); visualization, K.S.A. and F.B. (Franz Berthiller); writing—original draft, K.S.A. and F.B. (Franz Berthiller); writing—review and editing, K.S.A., B.M., F.B. (Franz Berthiller), E.B., G.J.H. and F.B. (Federico Baltar). All authors have read and agreed to the published version of the manuscript.

Funding: This research was funded by Austrian Science Fund (FWF) project OCEANIDES (project number AP3430411/21) to Federico Baltar.

Institutional Review Board Statement: Not applicable.

Informed Consent Statement: Not applicable.

Data Availability Statement: The raw data supporting the conclusions of this article will be made available by the authors, without undue reservation to any qualified researcher.

Acknowledgments: Open Access Funding by the University of Vienna.

Conflicts of Interest: The authors declare no conflict of interest.

References

1. Thomson, P.G.; Davidson, A.T.; van den Enden, R.; Pearce, I.; Seuront, L.; Paterson, J.S.; Williams, G.D. Distribution and abundance of marine microbes in the Southern Ocean between 30 and 80 E. *Deep Sea Res. Part II Top. Stud. Oceanogr.* **2010**, *57*, 815–827. [CrossRef]
2. Bar-On, Y.M.; Milo, R. The biomass composition of the oceans: A blueprint of our blue planet. *Cell* **2019**, *179*, 1451–1454. [CrossRef] [PubMed]
3. Bochdansky, A.B.; Clouse, M.A.; Herndl, G.J. Eukaryotic microbes, principally fungi and labyrinthulomycetes, dominate biomass on bathypelagic marine snow. *ISME J.* **2017**, *11*, 362–373. [CrossRef]
4. Grant, W.; West, A. Measurement of ergosterol, diaminopimelic acid and glucosamine in soil: Evaluation as indicators of microbial biomass. *J. Microbiol. Methods* **1986**, *6*, 47–53. [CrossRef]
5. Stahl, P.D.; Parkin, T.B. Relationship of soil ergosterol concentration and fungal biomass. *Soil Biol. Biochem.* **1996**, *28*, 847–855. [CrossRef]
6. Grattan, R.; Suberkropp, K. Effects of nutrient enrichment on yellow poplar leaf decomposition and fungal activity in streams. *J. North Am. Benthol. Soc.* **2001**, *20*, 33–43. [CrossRef]
7. Baldy, V.; Gessner, M.O.; Chauvet, E. Bacteria, fungi and the breakdown of leaf litter in a large river. *Oikos* **1995**, *74*, 93–102. [CrossRef]
8. Gessner, M.O.; Chauvet, E. Ergosterol-to-biomass conversion factors for aquatic hyphomycetes. *Appl. Environ. Microbiol. J.* **1993**, *59*, 502–507. [CrossRef]
9. Newell, S.Y.; Arsuffi, T.L.; Fallon, R.D. Fundamental procedures for determining ergosterol content of decaying plant material by liquid chromatography. *Appl. Environ. Microbiol. J.* **1988**, *54*, 1876–1879. [CrossRef]
10. Grossart, H.P.; Massana, R.; McMahon, K.D.; Walsh, D.A. Linking metagenomics to aquatic microbial ecology and biogeochemical cycles. *Limnol. Oceanogr.* **2019**, *269*, 103–108. [CrossRef]
11. Morales, S.E.; Biswas, A.; Herndl, G.J.; Baltar, F. Global structuring of phylogenetic and functional diversity of pelagic fungi by depth and temperature. *Front. Mar. Sci.* **2019**, *6*, 1–12. [CrossRef]
12. Baltar, F.; Zhao, Z.; Herndl, G.J. Potential and expression of carbohydrate utilization by marine fungi in the global ocean. *Microbiome* **2021**, *9*, 1–10. [CrossRef]
13. Mille-Lindblom, C.; von Wachenfeldt, E.; Tranvik, L.J. Ergosterol as a measure of living fungal biomass: Persistence in environmental samples after fungal death. *J. Microbiol. Methods* **2004**, *59*, 253–262. [CrossRef]
14. Seitz, L.; Mohr, H.; Burroughs, R.; Sauer, D. Ergosterol as an indicator of fungal invasion in grains. *Cereal Chem.* **1977**, *54*, 1207–1217. [CrossRef]
15. Hassett, B.T.; Borrego, E.J.; Vonnahme, T.R.; Rama, T.; Kolomiets, M.V.; Gradinger, R. Arctic marine fungi: Biomass, functional genes, and putative ecological roles. *ISME J.* **2019**, *13*, 1484–1496. [CrossRef] [PubMed]
16. Dupont, S.; Lemetais, G.; Ferreira, T.; Cayot, P.; Gervais, P.; Beney, L. Ergosterol biosynthesis: A fungal pathway for life on land? *Evolution* **2012**, *66*, 2961–2968. [CrossRef] [PubMed]
17. Ravelet, C.; Grosset, C.; Alary, J. Quantitation of ergosterol in river sediment by liquid chromatography. *J. Chromatogr. Sci.* **2001**, *39*, 239–242. [CrossRef]
18. Verma, B.; Robarts, R.D.; Headley, J.V.; Peru, K.; Christofi, N. Extraction efficiencies and determination of ergosterol in a variety of environmental matrices. *Commun. Soil Sci. Plant Anal.* **2002**, *33*, 3261–3275. [CrossRef]

19. Gutiérrez, M.H.; Vera, J.; Srain, B.; Quiñones, R.A.; Wörmer, L.; Hinrichs, K.U.; Pantoja-Gutiérrez, S. Biochemical fingerprints of marine fungi: Implications for trophic and biogeochemical studies. *Aquat. Microb. Ecol.* **2020**, *84*, 75–90. [CrossRef]
20. Newell, S. Fungal biomass and productivity. *Methods Microbiol.* **2001**, *30*, 357–372. [CrossRef]
21. Raghukumar, S. *Fungi in Coastal and Oceanic Marine Ecosystems*; Springer: Berlin/Heidelberg, Germany, 2017.
22. Akihisa, T.; Hori, T.; Suzuki, H.; Sakoh, T.; Yokota, T.; Tamura, T. 24β-methyl-5α-cholest-9 (11)-en-3β-ol, two 24β-alkyl-Δ5, 7, 9 (11)-sterols and other 24β-alkylsterol from Chlorella vulgaris. *Phytochemistry* **1992**, *31*, 1769–1772. [CrossRef]
23. Taylor, J.D.; Cunliffe, M. Multi-year assessment of coastal planktonic fungi reveals environmental drivers of diversity and abundance. *ISME J.* **2016**, *10*, 2118–2128. [CrossRef]
24. Bligh, E.G.; Dyer, W.J. A rapid method of total lipid extraction and purification. *Can. J. Biochem. Physiol.* **1959**, *37*, 911–917. [CrossRef] [PubMed]
25. Wickerham, J. A taxonomic study of Monilia albicans with special emphasis on morphology and morphological variation. *J. Trop. Med. Hyg.* **1939**, *42*, 174–179.
26. Wickerham, L. *Taxonomy of Yeasts*; Technical Bulletin; US Department of Agriculture: Washington, DC, USA, 1951; Volume 1029, pp. 1–56.
27. Gulis, V.; Bärlocher, F. Fungi: Biomass, production, and community structure. *Methods Stream Ecol.* **2017**, *1*, 177–192. [CrossRef]
28. Wenzl, T.; Haedrich, J.; Schaechtele, A.; Robouch, P.; Stroka, J. *Guidance document on the estimation of LOD and LOQ for measurements in the field of contaminants in feed and food*; Publications Office of the European Union: Luxembourg, 2016; Volume 58. [CrossRef]
29. Ory, L.; Gentil, E.; Kumla, D.; Kijjoa, A.; Nazih, E.H.; Roullier, C. Detection of ergosterol using liquid chromatography/electrospray ionization mass spectrometry: Investigation of unusual in-source reactions. *Rapid Commun. Mass Spectrom.* **2020**, *34*, e8780. [CrossRef] [PubMed]
30. Rodrigues, M.L. The multifunctional fungal ergosterol. *mBio* **2018**, *9*, e01755-18. [CrossRef]
31. Roose, P.; Smedes, F. Evaluation of the results of the QUASIMEME lipid intercomparison: The Bligh & Dyer total lipid extraction method. *Mar. Pollut. Bull.* **1996**, *32*, 674–680. [CrossRef]

Journal of
Fungi

Communication

Autofluorescence Is a Common Trait in Different Oceanic Fungi

Eva Breyer *, Markus Böhm, Magdalena Reitbauer, Chie Amano , Marilena Heitger and Federico Baltar *

Department of Functional and Evolutionary Ecology, University of Vienna, 1030 Vienna, Austria; a00255710@unet.univie.ac.at (M.B.); a01441286@unet.univie.ac.at (M.R.); chie.amano@univie.ac.at (C.A.); a01540889@unet.univie.ac.at (M.H.)

* Correspondence: eva.breyer@univie.ac.at (E.B.); federico.baltar@univie.ac.at (F.B.)

Abstract: Natural autofluorescence is a widespread phenomenon observed in different types of tissues and organisms. Depending on the origin of the autofluorescence, its intensity can provide insights on the physiological state of an organism. Fungal autofluorescence has been reported in terrestrial and human-derived fungal samples. Yet, despite the recently reported ubiquitous presence and importance of marine fungi in the ocean, the autofluorescence of pelagic fungi has never been examined. Here, we investigated the existence and intensity of autofluorescence in five different pelagic fungal isolates. Preliminary experiments of fungal autofluorescence at different growth stages and nutrient conditions were conducted, reflecting contrasting physiological states of the fungi. In addition, we analysed the effect of natural autofluorescence on co-staining with DAPI. We found that all the marine pelagic fungi that were studied exhibited autofluorescence. The intensity of fungal autofluorescence changed depending on the species and the excitation wavelength used. Furthermore, fungal autofluorescence varied depending on the growth stage and on the concentration of available nutrients. Collectively, our results indicate that marine fungi can be auto-fluorescent, although its intensity depends on the species and growth condition. Hence, oceanic fungal autofluorescence should be considered in future studies when fungal samples are stained with fluorescent probes (i.e., fluorescence in situ hybridization) since this could lead to misinterpretation of results.

Keywords: marine fungi; fluorescence in situ hybridisation; mycoplankton; fungal cultures; pelagic; fluorescence

Citation: Breyer, E.; Böhm, M.; Reitbauer, M.; Amano, C.; Heitger, M.; Baltar, F. Autofluorescence Is a Common Trait in Different Oceanic Fungi. *J. Fungi* **2021**, *7*, 709. https://doi.org/10.3390/jof7090709

Academic Editor: Wei Li

Received: 19 July 2021
Accepted: 26 August 2021
Published: 29 August 2021

1. Introduction

Natural autofluorescence is a process in which endogenous cell compounds i.e., aromatic amino acids, become fluorescent when excited with light in a specific wavelength. Depending on the type and origin of the fluorescent compound, the intensity of the emitted light can change with the morphological or physiological state of the observed cells or tissue and, hence, the organism [1].

Autofluorescence in fungi was first observed when human tissue sections, used to diagnose mycotic infectious diseases, were exposed to UV-light. Diverse fungi including Candida, Aspergillus, Blastomyces, Cryptococcus and Coccidioides, emitted fluorescent light, allowing their easy detection in infected tissue samples without prior staining [2,3]. Moreover, experiments with terrestrial-derived fungal cultures investigating autofluorescence in spores revealed differences in fluorescence intensities between fungal species, which might be related to different degrees of cellular viability [4]. In contrast, it was found that arbuscular mycorrhizal fungal structures were auto-fluorescent under blue and green light excitation despite their viability [5]. Likewise, fungal autofluorescence persisted after cell death following sample fixation, further demonstrating its diagnostic value for direct analyses of histological samples [6].

The origin of fungal autofluorescence is still not clear. It has been hypothesized that chitin could be the cause for fungal autofluorescence due to the similarities observed when fungal cells were stained with Calcofluor White (that binds to chitin in fungal cell walls) [5]. Recently, ergosterol, a membrane lipid found in fungal cell walls in the sub-kingdom of Dikarya used to quantify fungal biomass was suggested as another source of autofluorescence [7].

Previous research only focused on terrestrial and human-associated fungi. Recently, the important contribution of marine fungi to oceanic microbial food webs [8–10], biomass [11,12] and functional diversity [13,14] has been highlighted. Despite the ubiquitous presence of pelagic fungi in the oceanic water column, there are no investigations studying the existence of autofluorescence of oceanic fungi, and on the factors affecting the presence and intensity of this autofluorescence. To fill this gap of knowledge, we studied the autofluorescence of five marine pelagic fungi and investigated their autofluorescence at different growth stages. Furthermore, we also investigated the effect of varying nutrient concentrations and DAPI co-staining (a common DNA-stain in microbiology) on this fungal autofluorescence. Based on previous research, we hypothesized that marine fungi would be auto-fluorescent, similarly to certain terrestrial and human-associated fungi [2,3,5]. We also hypothesized that the autofluorescence intensity changes with species and physiological state, since it is possible that the production and distribution of auto-fluorescent compounds changes in response to growth and physiological state [1].

2. Materials and Methods

2.1. Cultivation of Marine Fungal Cultures

Four marine fungal cultures (*Metschnikowia australis*, *Rhodotorula sphaerocarpa*, *Sakaguchia dacryoidea*, *Blastobotrys parvus*) obtained from the Austrian Centre of Biological Resources (ACBR), and one fungal species (*Rhodotorula* sp.) isolated during the 'Poseidon' research cruise in 2019, were used to study the existence and intensity of autofluorescence in growth experiments (Table 1). All the five fungal species used (four yeasts and one hyphae-morphotype) were originally isolated from open ocean waters. The fungi were grown in the dark at room temperature on solid agar media containing (g/L): 10 g glucose, 5 g peptone, 3 g yeast extract, 3 g malt extract, 35 g artificial sea salts, 20 g agar and 0.5 g chloramphenicol.

Table 1. Marine fungi used for the autofluorescence experiments.

Species	Division	Origin of Isolation	ACBR Code
Metschnikowia australis	Ascomycota	Antarctic Ocean	HA635
Rhodotorula sphaerocarpa	Basidiomycota	Antarctic Ocean, Marguerite Bay	HB738
Sakaguchia dacryoidea	Basidiomycota	Antarctic Ocean	HB877
Blastobotrys parvus	Ascomycota	Antarctic Ocean	HA1620
Rhodotorula sp.	Basidiomycota	Atlantic Ocean	-

Aliquot samples of the yeasts were transferred and grown in liquid media (Table 2) to obtain samples at specific growth stages. All yeasts were cultured in the oligotrophic medium, and *S. dacryoidea* was additionally cultured in the eutrophic medium. The liquid cultures were incubated on a shaker incubator (Argo Lab, Ski 4, 140 rpm, Carpi, Italy) at room temperature under normal day-night light regime. The exponential growth stage was sampled in all species according to the daily measured optical density (UV-1800 Shimadzu spectrophotometer, λ = 660 nm, Kyoto, Japan). Additionally, *S. dacryoidea* was sampled in the stationary growth stage. Sampling of the different growth stages were based on previous experiments and on the shape of the growth curve (Table 3, Figure 1).

Table 2. Growth media used for marine yeast liquid cultures.

Chemicals (g/L)	Eutrophic Medium	Oligotrophic Medium
Glucose	2	0.2
Peptone	2	0.2
yeast extract	2	0.2
sea salts	35	35
chloramphenicol	0.5	0.5

Table 3. Optical density (OD) of different fungal growth stages.

Growth Stage	OD for Eutrophic Medium	OD for Oligotrophic Medium
Adaptation phase	0.07–0.4	<0.15
Exponential phase	0.4–1.2	0.15–0.27
Stationary phase	>1.2	>0.27

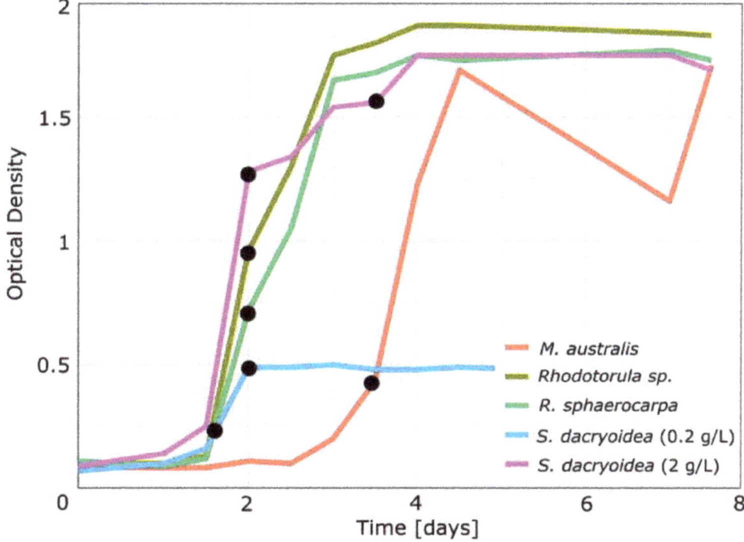

Figure 1. Growth of yeast cultures as indicated by optical density. Note: *S. dacryoidea* was cultured in two different nutrient concentrations (0.2 and 2 g/L glucose, blue and purple line, respectively). The other fungal species were grown with 2 g/L glucose. Black dots indicate the sampling times.

2.2. Sample Preparation to Investigate Fungal Autofluorescence

In the corresponding growth stages, 20 mL of the yeast cultures were sampled and fixed with 2% final conc. of formaldehyde (Sigma-Aldrich, 37%, St. Louis, MO, USA). Microscopic samples were prepared by filtering 250 μL of fixed culture, diluted in 5 mL MilliQ-water on GTTP filters (0.22 μm, 25 mm diameter, Merck Millipore, Burlington, MA, USA). Subsequently, filters were dried for 30 min and mounted with Vectashield (Vector Laboratories, H-1000, Burlingame, CA, USA) on a microscopic slide.

To further investigate the autofluorescence in hyphae-morphotype fungi, *Blastobotrys parvus* was diluted in artificial seawater (35 g/L sea salts), then filtered and mounted with Vectashield as described before.

To investigate fungal autofluorescence, samples were examined with a Zeiss Axio Imager 2 microscope (1250× magnification, Carl Zeiss, Jena, Germany) using four different channels and filter sets provided by Zeiss: DAPI (4′,6-diamidin-2-phenylindol, filter set 49); FITC (fluorescein isothiocyanate, filter set 44); DsRed (red fluorescent protein, filter set

43 HE); rhodamine (filter set 20 HE). For comparing potential species-specific differences in autofluorescence, each filter channel was analysed with fixed exposure times (2.6 s for DAPI, 5.2 s for FITC, 9.06 s for DsRed, 1.1 s for rhodamine) to obtain optimal results based on previous autofluorescence investigations. Here, we intentionally chose relatively long exposure times to examine autofluorescence. For usual fluorescence measurements, we used exposure times automatically calculated by the software (Axio Vision SE64-Re4.9, Carl Zeiss) of around 100–500 ms for Calcofluor-White (Sigma-Aldrich, St. Louis, MO, USA) and 150–500 ms for DAPI staining. Pictures were taken with an AxioCam MRm camera (Carl Zeiss).

3. Results and Discussion

Sakaguchia dacryoidea was grown in two different nutrient concentrations (0.2 and 2 g/L glucose) and sampled in the exponential and stationary growth stage (Figure 1). In both nutrient concentrations, *S. dacryoidea* entered the exponential growth phase after 1.5 d. In the media with less nutrients, the stationary phase was reached after 2 d, with a maximum OD of about 0.5. Conversely, at high nutrient concentrations, the *S. dacryoidea* yield was higher, reaching OD values in the stationary phase of about 1.8. It is noteworthy that all fungal species grown with 2 g glucose/L reached similar maximum biomass in the stationary phase.

When observed under the microscope, *S. dacryoidea* sampled in the exponential phase exhibited autofluorescence in all of the channels investigated (i.e., DAPI, FITC, DsRed and rhodamine). The strongest autofluorescence was detected in the DAPI channel as indicated by the shortest exposure time (in this case of 2.6 s) (Figure 2). The autofluorescence of *S. dacryoidea* was weaker in the FITC, DsRed and Rhodamine than in the DAPI channel.

To test whether the autofluorescence was affected by the growth stage we also examined it in the stationary phase of the same fungal culture (Figure 2). We found that the autofluorescence of *S.dacryoidea* became weaker in all observed channels compared to the exponential phase, indicating that nutrient limitation in the stationary phase affects marine fungal autofluorescence (Figure 2).

To gain more insight into the potential effect of nutrient concentrations on fungal physiology and autofluorescence, the autofluorescence of *S. dacryoidea* was analysed after culturing in a medium with reduced (10 times less) glucose concentration (i.e., 0.2 g/L) (Figure 3). In all of the analysed channels, the autofluorescence of *S. dacryoidea* in the exponential phase was slightly lower under lower nutrient conditions than at high nutrient concentrations (Figures 2 and 3). However, in the stationary phase, the autofluorescence of *S. dacryoidea* increased again under low nutrient conditions (Figure 3).

To determine whether autofluorescence is a general phenomenon in oceanic fungi or whether it is species-dependent, we compared the autofluorescence of four different marine yeasts in the DAPI channel (Figure 4). We found species-specific autofluorescence intensities with autofluorescence clearly visible in *S. dacryoidea*, *R. sphaerocarpa* and *Rhodotorula* sp. In contrast, *M. australis* showed only minor autofluorescence. These results suggest that autofluorescence is not restricted to a single species, but instead might be a widely distributed feature in marine fungi, albeit species-specific. Future research including a larger number of more diverse fungi would be help in confirming whether autofluorescence is a global characteristic of marine fungi.

To study the effect of additional staining with DAPI on the autofluorescence of *S.dacryoidea*, we stained the fungal cells sampled in both growth stages and nutrient conditions with DAPI (Figure 5).

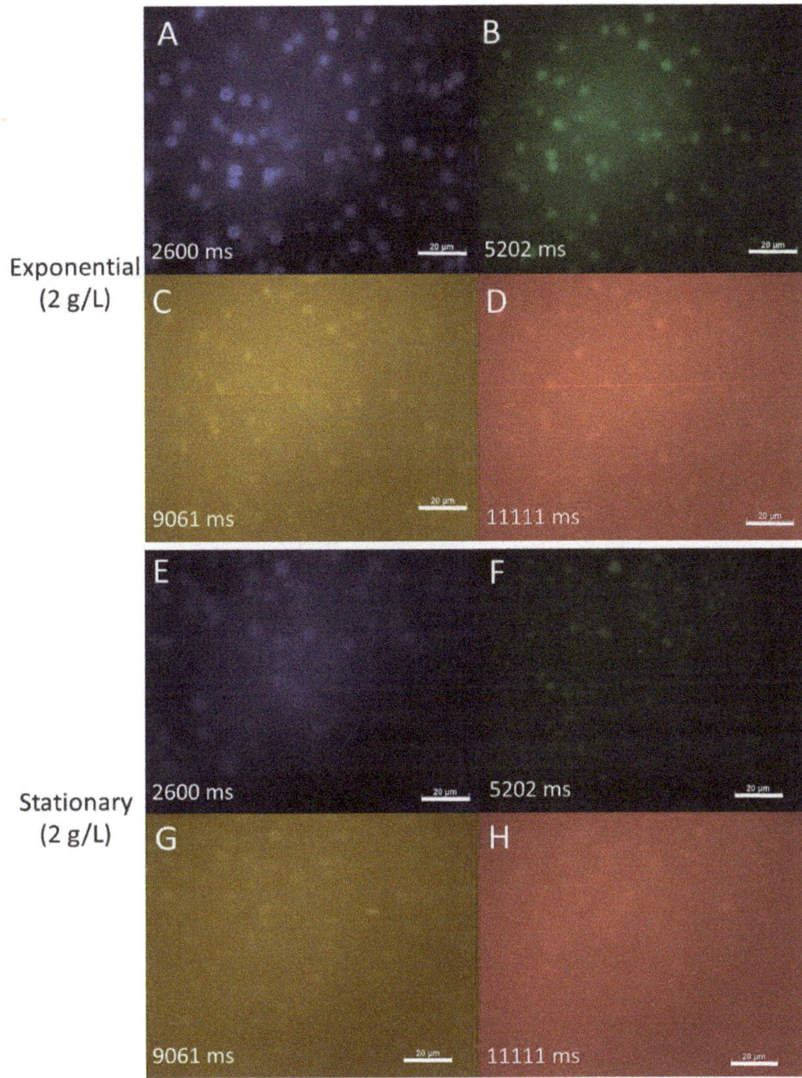

Figure 2. Autofluorescence of *S. dacryoidea* grown in the eutrophic medium and sampled in the exponential and stationary phase. (**A**,**E**): DAPI channel; (**B**,**F**): FITC channel; (**C**,**G**): DsRed channel; (**D**,**H**): rhodamine channel. Picture exposure times are depicted in milliseconds (ms). Scale bar = 20 μm.

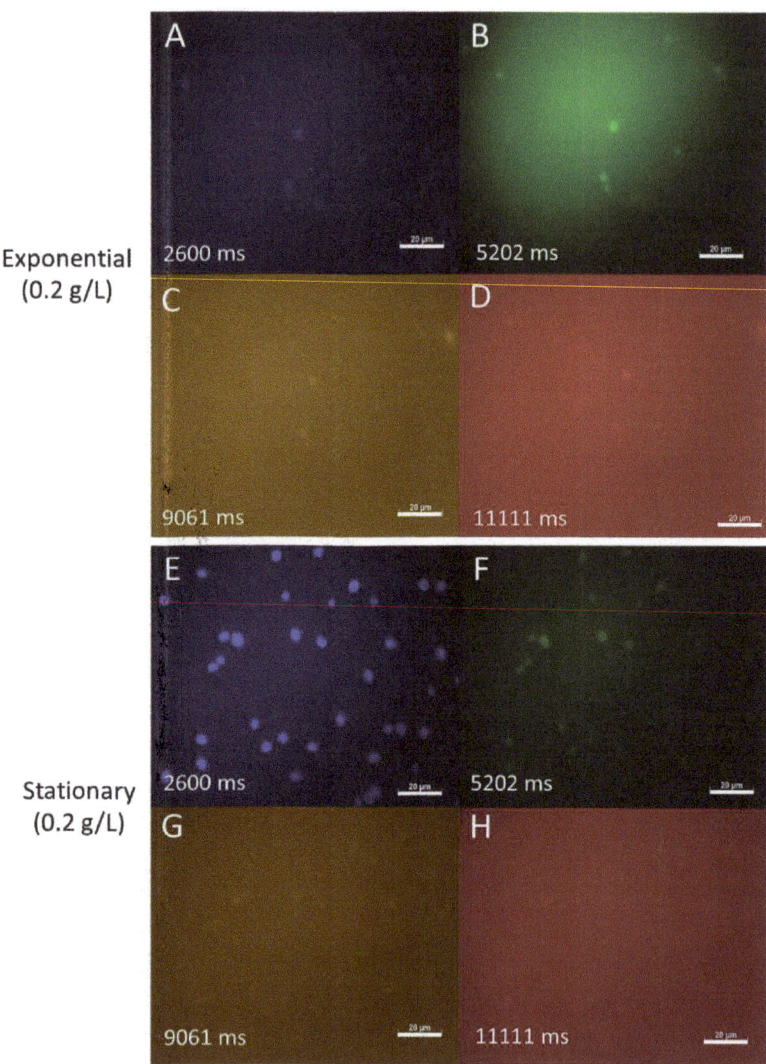

Figure 3. Autofluorescence of *S. dacryoidea* grown in the oligotrophic medium and sampled in the exponential and stationary phase. (**A**,**E**): DAPI channel; (**B**,**F**): FITC channel; (**C**,**G**): DsRed channel; (**D**,**H**): rhodamine channel. Picture exposure times are depicted in milliseconds (ms). Scale bar = 20 μm.

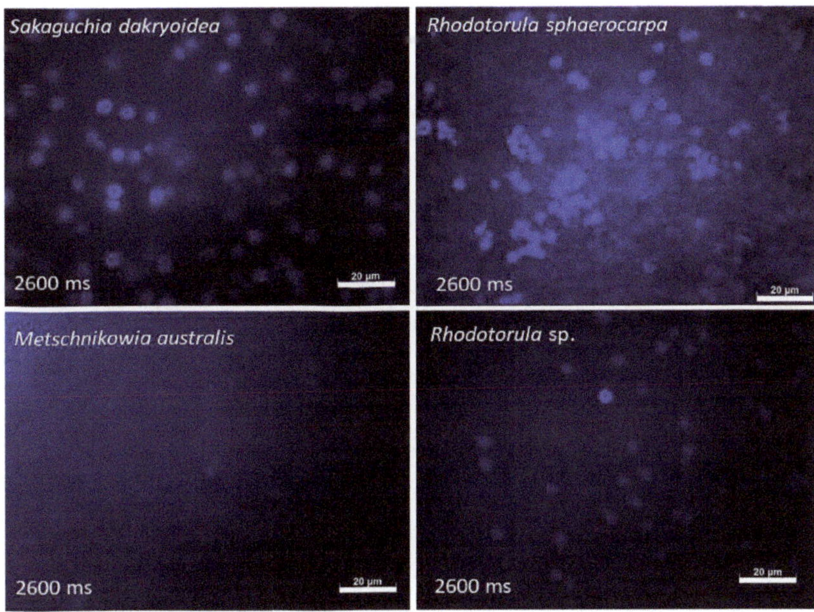

Figure 4. Autofluorescence of four marine yeasts grown in eutrophic medium and sampled in the exponential phase. All pictures were made in the DAPI channel, exposure times are depicted in milliseconds (ms). Scale bar = 20 µm.

DAPI-staining intensified fungal fluorescence in all conditions (note the shorter exposure time) with strongest fluorescence of the cell nucleus (Figure 5). Hence, although fungal cells are auto-fluorescent, the cell nucleus is still clearly visible and can be distinguished from the rest of the cell when stained with DAPI.

Finally, to study fungi with hyphae-morphotype, we investigated the autofluorescence of *Blastobotrys parvus* in four different channels (Figure 6).

When observed with the DAPI channel, *B. parvus* showed no autofluorescence. However, after changing to FITC, DsRed and Rhodamine channel, the autofluorescence of *B. parvus* was clearly visible, supporting that this phenomenon occurs in both fungal morphotypes.

Collectively, this is the first time that autofluorescence is shown in marine fungal species. These results are consistent with previous studies on terrestrial and human-derived fungal species [2–6]. Our results also suggest that the intensity of fungal autofluorescence changes with the excited wavelengths. We also provide a preliminary indication that the autofluorescence of marine fungi varies between fungal species and in relation to the nutrient availability and growth stage. This is consistent with a previous study where the autofluorescence of cells/tissues was suggested to be influenced by the physiological state of the organism [1].

The existence of autofluorescence in marine fungi is important to consider, particularly when dealing with fluorescence-based techniques for analysis or identification. This autofluorescence can be used as a methodological advantage without the need of prior staining as shown for human associated fungi [2,3]. It can be misleading, however, when working with fluorescence microscopy. For instance, a very common method in microbiological studies for identification of cells and estimating their relative abundance is fluorescence in situ hybridisation (FISH). Its principle relies on staining microbial cells with, for example, DAPI, which binds to the DNA in the cells, and also with some specific probe to target a specific taxon. As a result, cell compounds that contain DNA emit a fluorescent signal when exited in a specific wavelength [15,16] and the targeted taxa with the corresponding FISH-probe will also emit fluorescence in a different wavelength to DAPI [17]. Thus, the existence of autofluorescence in marine fungi might lead to potential interferences, potentially resulting in false "positive" results due to natural autofluorescence in marine fungal cells. Therefore, natural autofluorescence in marine fungi should be investigated before applying fluorescence-dependent analytical methods.

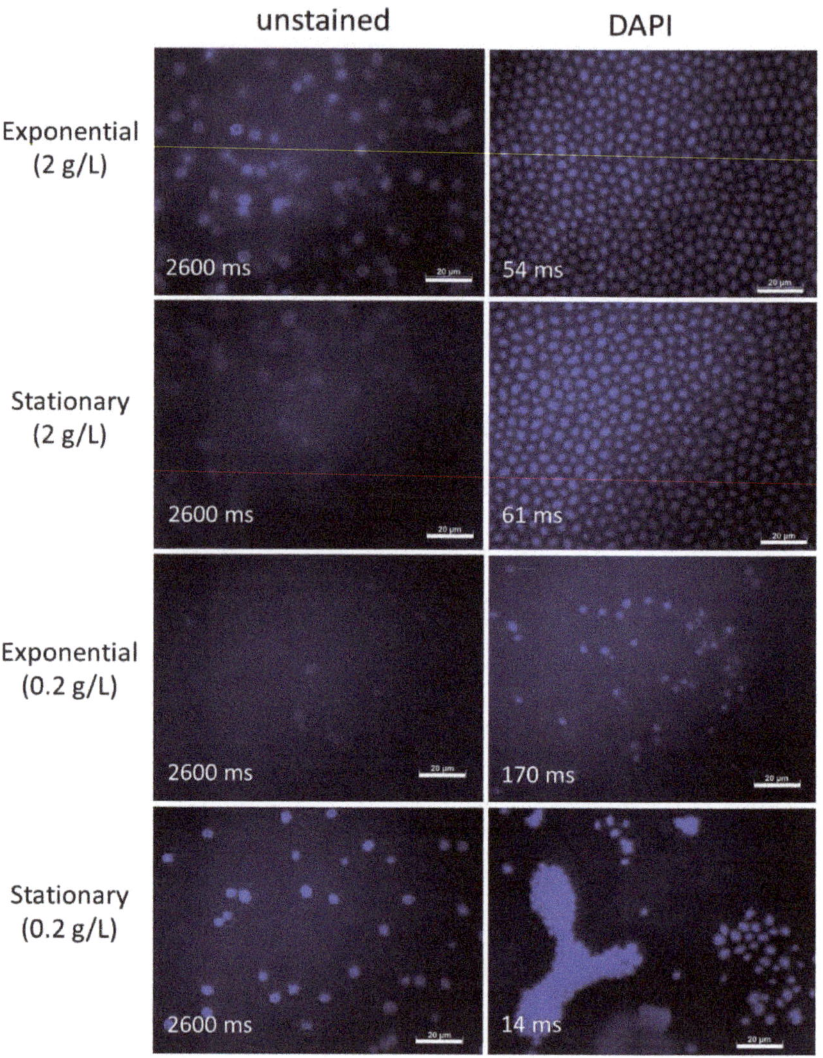

Figure 5. Autofluorescence of *S. dacryoidea* grown in two different nutrient concentrations and sampled in the exponential and stationary phase. Right panels show additional staining with DAPI. Exposure times are depicted in milliseconds (ms). Scale bar = 20 μm.

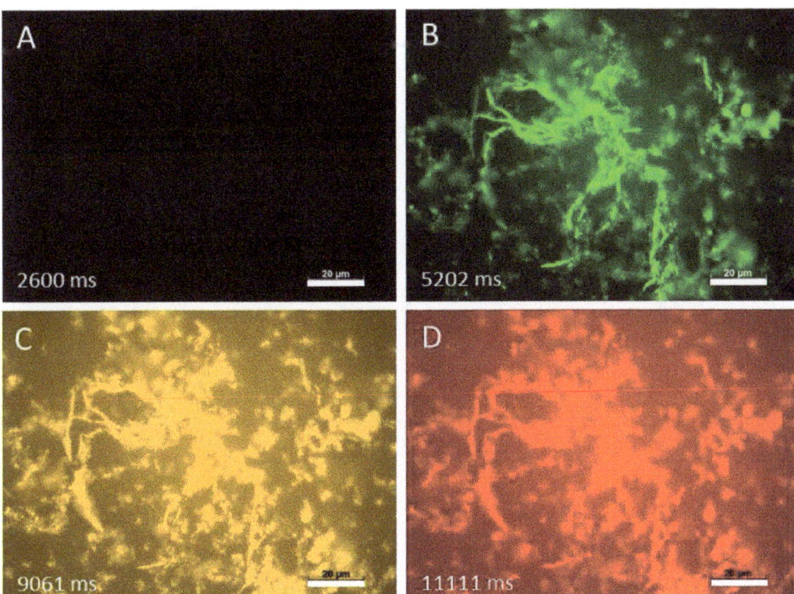

Figure 6. Autofluorescence of *B. parvus*. (**A**): DAPI channel; (**B**): FITC channel; (**C**): DsRed channel; (**D**): rhodamine channel. Picture exposure times are depicted in milliseconds (ms). Scale bar = 20 µm.

Author Contributions: Conceptualization, F.B. and E.B.; methodology, F.B., E.B. and C.A.; investigation, M.B. and M.R.; resources, M.H.; writing—original draft preparation, E.B. All authors have read and agreed to the published version of the manuscript.

Funding: This research was funded by the Austrian science fond (FWF) project OCEANIDES (project number AP3430411/21) to Federico Baltar.

Institutional Review Board Statement: Not applicable.

Informed Consent Statement: Not applicable.

Data Availability Statement: The raw data supporting the conclusions of this article will be made available by the authors, without undue reservation to any qualified researcher.

Acknowledgments: We would like to acknowledge Gerhard J. Herndl, for providing feedback on the drafting of the manuscript and also the Open Access Funding by the University of Vienna.

Conflicts of Interest: The authors declare no conflict of interest.

References

1. Monici, M. Cell and tissue autofluorescence research and diagnostic applications. *Biotechnol. Ann. Rev.* **2005**, *11*, 227–256.
2. Mann, J.L. Autofluorescence of Fungi: An Aid to Detection in Tissue Sections. *Am. J. Clin. Pathol.* **1983**, *79*, 587–590. [CrossRef] [PubMed]
3. Graham, A.R. Fungal Autofluorescence with Ultraviolet Illumination. *Am. J. Clin. Pathol.* **1983**, *79*, 231–234. [CrossRef] [PubMed]
4. Wu, C.H.; Warren, H.L. Natural Autofluorescence in Fungi, and its Correlation with Viability. *Mycologia* **1984**, *76*, 1049–1058. [CrossRef]
5. Dreyer, B.; Morte, A.; Pérez-Gilabert, M.; Honrubia, M. Autofluorescence detection of arbuscular mycorrhizal fungal structures in palm roots: An underestimated experimental method. *Mycol. Res.* **2006**, *110*, 887–897. [CrossRef] [PubMed]
6. Margo, C.E.; Bombardier, T. The diagnostic value of fungal autofluorescence. *Surv. Ophthalmol.* **1985**, *29*, 374–376. [CrossRef]
7. Mansoldo, F.R.P.; Firpo, R.; da Silva Cardoso, V.; Queiroz, G.N.; Cedrola, S.M.L.; de Godoy, M.G.; Vermelho, A.B. New method for rapid identification and quantification of fungal biomass using ergosterol autofluorescence. *Talanta* **2020**, *219*, 121238. [CrossRef] [PubMed]
8. Gutiérrez, M.H.; Pantoja, S.; Tejos, E.; Quiñones, R.A. The role of fungi in processing marine organic matter in the upwelling ecosystem off Chile. *Mar. Biol.* **2011**, *158*, 205–219. [CrossRef]

9. Hassett, B.T.; Gradinger, R. Chytrids dominate arctic marine fungal communities. *Environ. Microbiol.* **2016**, *18*, 2001–2009 [CrossRef] [PubMed]
10. Cleary, A.C.; Durbin, E.G.; Rynearson, T.A.; Bailey, J. Feeding by Pseudocalanus copepods in the Bering Sea: Trophic linkages and a potential mechanism of niche partitioning. *Deep Sea Res. Part. II Top. Stud. Oceanogr.* **2016**, *134*, 181–189. [CrossRef]
11. Bochdansky, A.B.; Clouse, M.A.; Herndl, G.J. Eukaryotic microbes, principally fungi and labyrinthulomycetes, dominate biomass on bathypelagic marine snow. *ISME J.* **2017**, *11*, 362–373. [CrossRef] [PubMed]
12. Hassett, B.T.; Borrego, E.J.; Vonnahme, T.R.; Rämä, T.; Kolomiets, M.V.; Gradinger, R. Arctic marine fungi: Biomass, functional genes, and putative ecological roles. *ISME J.* **2019**, *13*, 1484–1496. [CrossRef] [PubMed]
13. Morales, S.E.; Biswas, A.; Herndl, G.J.; Baltar, F. Global Structuring of Phylogenetic and Functional Diversity of Pelagic Fungi by Depth and Temperature. *Front. Mar. Sci.* **2019**, *6*, 131. [CrossRef]
14. Baltar, F.; Zhao, Z.; Herndl, G.J. Potential and expression of carbohydrate utilization by marine fungi in the global ocean. *Microbiome* **2021**, *9*, 106. [CrossRef] [PubMed]
15. Otto, F. DAPI staining of fixed cells for high-resolution flow cytometry of nuclear DNA. In *Methods in Cell Biology*; Academic Press: Cambridge, MA, USA, 1990; Volume 33, pp. 105–110.
16. Hamada, S.; Fujita, S. DAPI staining improved for quantitative cytofluorometry. *Histochemistry* **1983**, *79*, 219–226. [CrossRef] [PubMed]
17. Amann, R.; Fuchs, B.M. Single-cell identification in microbial communities by improved fluorescence in situ hybridization techniques. *Nat. Rev. Microbiol.* **2008**, *6*, 339–348. [CrossRef] [PubMed]

Journal of *Fungi*

Article

Extracellular Enzymatic Activities of Oceanic Pelagic Fungal Strains and the Influence of Temperature

Katherine Salazar Alekseyeva [1,*], Gerhard J. Herndl [1,2] and Federico Baltar [1,*]

1 Department of Functional and Evolutionary Ecology, University of Vienna, 1030 Vienna, Austria; gerhard.hernld@univie.ac.at
2 Department of Marine Microbiology and Biogeochemistry, Royal Netherlands Institute for Sea Research (NIOZ), University of Utrecht, 1790 Texel, The Netherlands
* Correspondence: katherine.salazar@univie.ac.at (K.S.A.); federico.baltar@univie.ac.at (F.B.); Tel.: +43-(0)1-4277-76438 (K.S.A.); +43-(0)1-4277-76436 (F.B.)

Abstract: Although terrestrial and aquatic fungi are well-known decomposers of organic matter, the role of marine fungi remains largely unknown. Recent studies based on omics suggest that marine fungi potentially play a major role in elemental cycles. However, there is very limited information on the diversity of extracellular enzymatic activities performed by pelagic fungi in the ocean and how these might be affected by community composition and/or critical environmental parameters such as temperature. In order to obtain information on the potential metabolic activity of marine fungi, extracellular enzymatic activities (EEA) were investigated. Five marine fungal species belonging to the most abundant pelagic phyla (Ascomycota and Basidiomycota) were grown at 5 °C and 20 °C, and fluorogenic enzymatic assays were performed using six substrate analogues for the hydrolysis of carbohydrates (β-glucosidase, β-xylosidase, and *N*-acetyl-β-D-glucosaminidase), amino acids (leucine aminopeptidase), and of organic phosphorus (alkaline phosphatase) and sulfur compounds (sulfatase). Remarkably, all fungal strains were capable of hydrolyzing all the offered substrates. However, the hydrolysis rate (V_{max}) and half-saturation constant (K_m) varied among the fungal strains depending on the enzyme type. Temperature had a strong impact on the EEAs, resulting in Q_{10} values of up to 6.1 and was species and substrate dependent. The observed impact of temperature on fungal EEA suggests that warming of the global ocean might alter the contribution of pelagic fungi in marine biogeochemical cycles.

Keywords: marine fungi; total extracellular enzymatic activity; kinetics; maximum velocity; half-saturation constant

Citation: Salazar Alekseyeva, K.; Herndl, G.J.; Baltar, F. Extracellular Enzymatic Activities of Oceanic Pelagic Fungal Strains and the Influence of Temperature. *J. Fungi* **2022**, *8*, 571. https://doi.org/10.3390/jof8060571

Academic Editor: Wei Li

Received: 27 April 2022
Accepted: 23 May 2022
Published: 26 May 2022

1. Introduction

Fungi are eukaryotic and osmoheterotrophic organisms depending on organic matter to grow and obtain energy [1]. Osmotrophy involves the secretion of different enzymes to break down complex biological polymers into smaller monomers that can then be taken up through the cell wall [2]. Due to this conversion of organic matter, osmoheterotrophs such as fungi should be major players in the recycling of organic matter. In the marine environment, most of the research on extracellular enzymatic activity (EEA) has been focused on prokaryotes [3,4]. Only a few studies have reported on fungal EEA related to phytoplankton blooms [5] or the degradation of plant-derived matter [6]. As a result, the ecological role of fungi in the biogeochemistry of the oceans, which represents the largest habitat in the Earth's biosphere, remains poorly known [7]. This contrasts with recent evidence of fungal biomass dominating the bathypelagic marine snow [8]. Additionally, recent studies based on omics suggest that pelagic fungi harbor genes indicative of an active role in marine biogeochemical cycles [9]. It has also been shown that a large variety of carbohydrate active enzymes (CAZymes) are expressed in marine pelagic fungi [10]. Still, there is very limited information on the diversity of EEA in pelagic fungi in the ocean, and how it might be affected by community composition and/or environmental parameters such as temperature.

Approximately 70% of the Earth's biosphere is composed of persistently cold environments, from the deep sea to polar regions [11,12]. Depending on the optimal growth temperature, organisms living there can be psychrophilic or psychrotrophic [13]. These organisms need to be well adapted to low temperatures, low nutrient availability, and light seasonality [14,15]. Moreover, as low temperatures influence the biochemical reaction rates, organisms must be prepared to overcome those challenges [11].

Here, we investigated the kinetic parameters (V_{max} and K_m) of the EEA of five oceanic fungal isolates using six different fluorogenic substrate analogues. β-glucosidase and β-xylosidase were used as a proxy for the degradation of plant-derived matter; *N*-acetyl-β-D-glucosaminidase were used for the utilization of animal and fungal chitinous compounds; alkaline phosphatase and leucine aminopeptidase were used for the cleavage of phosphate moieties and peptides, respectively; and sulfatase was used for the degradation of sulfate esters in macromolecules. We used five fungal strains isolated from the oceanic water column, with two strains belonging to the phylum Ascomycota and three to Basidiomycota. We selected these strains because Ascomycota and Basidiomycota are the most abundant pelagic fungal phyla [9,16,17]. Furthermore, the influence of temperature on fungal EEAs was determined.

2. Materials and Methods

2.1. Culture of Fungi Species

The fungal species *Blastobotrys parvus* (HA 1620), *Metschnikowia australis* (HA 635), *Rhodotorula sphaerocarpa* (HB 738), and *Sakaguchia dacryoidea* (HB 877) were obtained from the Austrian Center of Biological Resources (ACBR). All these species were isolated from Antarctic Ocean waters at temperatures ranging from -1.24 °C to 5.60 °C [18–21]. *M. australis* and *R. sphaerocarpa* were isolated close to the South Shetland Islands and Marguerite Bay, respectively. The fungus *Rhodotorula mucilaginosa* was isolated from the Atlantic Ocean at 21.03 °C during the Poseidon cruise on board of *RV* Sarmiento de Gamboa in March 2019. The maximum temperature growth reported is 25 °C for *B. parvus* [18] and *M. australis* [19], and 30 °C for *R. mucilaginosa* [22], *R. sphaerocarpa* [21], and *S. dacryoidea* [20]. In order to have fresh cultures, the pure isolates were cultured on yeast malt extract agar [23,24] for one week. Afterwards, an initial amount of each fungus was diluted in artificial seawater (30 g/L sea salts S9883 Sigma-Aldrich, Vienna, Austria) to obtain an $OD_{660} \approx 1$ [25]. The optical density (OD) was measured with a UV-1800 Shimadzu spectrophotometer. Then, 10 mL of this fungal culture was inoculated into an autoclaved growth medium containing 2 g/L of glucose, malt extract, peptone, and yeast extract; 35 g/L of artificial sea salts (S9883 Sigma-Aldrich); and 0.50 g/L of chloramphenicol. Afterwards, 150 mL of this medium containing fungi was filled in Schott bottles and to compare the effect of temperature, all strains were grown in triplicate at 5 °C and 20 °C on a rotary shaker (Jeio Tech ISS-7100 Incubated Shaker, Daejeon, South Chungcheong, Republic of Korea). The culture growth was tracked daily by OD. Once the exponential phase was reached, bottles with similar OD values were chosen in triplicate for further analysis (EEA and biomass).

2.2. Determining Extracellular Enzymatic Activity and Fungal Biomass

Fluorogenic substrate analogues such as 4-methylumbelliferyl β-D-glucopyranoside (M3633 Sigma-Aldrich), 4-methylumbelliferyl β-D-xylopyranoside (M7008 Sigma-Aldrich), and 4-methylumbelliferyl *N*-acetyl-β-D-glucosaminide (M2133 Sigma-Aldrich) were used to estimate the potential activity of the enzymes β-glucosidase (BGL), β-xylosidase (BXY), and *N*-acetyl-β-D-glucosaminidase (NAG), respectively (Table 1). These enzymes can hydrolyze cellulose [26,27], chitin, and xylan [28,29], respectively, thus mediating carbohydrate degradation by marine fungi. The hydrolysis of 4-methylumbelliferyl phosphate (M8883 Sigma-Aldrich) and N-succinyl-Ala-Ala-Pro-Phe-7-amido-4-methylcoumarin (L2145 Sigma-Aldrich) was used to estimate the potential enzymatic activity of alkaline phosphatase (ALP) and leucine aminopeptidase (LAP), respectively. ALP is indicative of the capability of microbes to acquire inorganic phosphorus from organic molecules, and LAP is involved in the hydrolysis of proteins and peptides [30]. Finally, 4-methylumbelliferyl sulfate potassium salt (M7133 Sigma-Aldrich) was used to determine the activity of sulfatase (SUL) degrading sulfate esters in macromolecules. The hydrolysis of these fluorogenic substrates analogues were standardized to the corresponding fluorophores. The fluorophores methylcoumaryl amide (MCA) (A9891 Sigma-Aldrich) and methylumbelliferone (MUF) (M1381 Sigma-Aldrich) were dissolved in 2-methoxyethanol to obtain a final concentration of 100, 50, 10, and 1 μM, and 2000, 1000, 100, and 50 μM.

Table 1. Targeted enzymes with an analogue fluorogenic substrate and their respective standards, methylumbelliferyl (MUF) and methylcoumaryl (MCA) amide.

Target	Code	Name	Standard
Carbohydrates	BGL	β-glucosidase	MUF
	BXY	β-xylosidase	MUF
	NAG	*N*-acetyl-β-D-glucosaminidase	MUF
Proteins, peptides	LAP	Leucine aminopeptidase	MCA
Phosphorus	ALP	Alkaline phosphatase	MUF
Sulfur	SUL	Sulfatase	MUF

Sterile microplates of 96 wells with an F bottom and low protein binding (XT64.1, Carl Roth, Karlsruhe, Baden-Wurtemberg, Germany) were used. The standards were distributed to each biological triplicate to establish a standard calibration curve, and each biological triplicate without any addition was used as a blank to determine the background fluorescence of the medium. Serial dilutions of the fluorogenic substrate were established resulting in 12 final concentrations ranging from 100 to 0.05 μM. The fluorescence was measured with a Tecan Infinite 200 PRO at an excitation wavelength of 365 nm and an emission wavelength of 445 nm. An initial measurement was performed, and then, every hour, a measurement was made over a total period of 3 h. Between measurements, the microplates were incubated in the dark at their respective temperature.

For fungal biomass determination, combusted (450 °C; for 6 h) Whatman GF/F filters (WHA1825047 Sigma-Aldrich, 47 mm filter diameter) were individually wrapped in aluminum foil and weighed. Then, 40 mL of each fungal culture triplicate was gently filtered onto a combusted and weighed filter, and was dried at 80 °C for 3 days. Thereafter, the sample was weighed again to determine the fungal biomass as dry weight.

2.3. Determination of the Kinetic Parameters of the Extracellular Enzymatic Activity (EEA)

The increase in fluorescence over time (3 h) was transformed into hydrolysis rate [$\mu mol\ L^{-1}\ h^{-1}$] using the equation obtained from the standard calibration lines of MCA and MUF. The resulting hydrolysis rates were fitted directly with the Michaelis–Menten equation using nonlinear least-squares regression analysis with R software [31]. The enzymatic kinetic parameters maximum velocity (V_{max}) and half-saturation constant (K_m) were calculated. To obtain the biomass-specific activity, the V_{max} was normalized to the fungal biomass obtained from the dry weight. Additionally, the Q_{10} value was calculated to identify the dependence of enzyme activity on temperature [32].

2.4. Statistical Analyses

For the kinetic parameters V_{max} and K_m, a one-way analysis of variance (ANOVA) was performed to determine differences among species and temperature. Tukey's Honestly Significant Difference (Tukey's HSD) was used for a multiple and simultaneous comparison between species and to identify significance at the species level. A Student's T-test was used to test the normal distribution of the data. A principal component analysis (PCA) was performed to analyze the fungal enzymatic activity by species and substrate. For this purpose, the prcomp command of the R software was used. To maximize the sum of the variance of the squared loadings, Varimax rotation was performed.

3. Results

Remarkably, all the fungal strains were hydrolyzing all the fluorogenic substrates offered (Figures 1 and 2). The kinetic parameters V_{max} and K_m of the EEA varied, however, among the different fungal strains and substrates (Figures 3 and 4). Additionally, the response of EEA to temperature was species and substrate dependent (Figures 5 and 6).

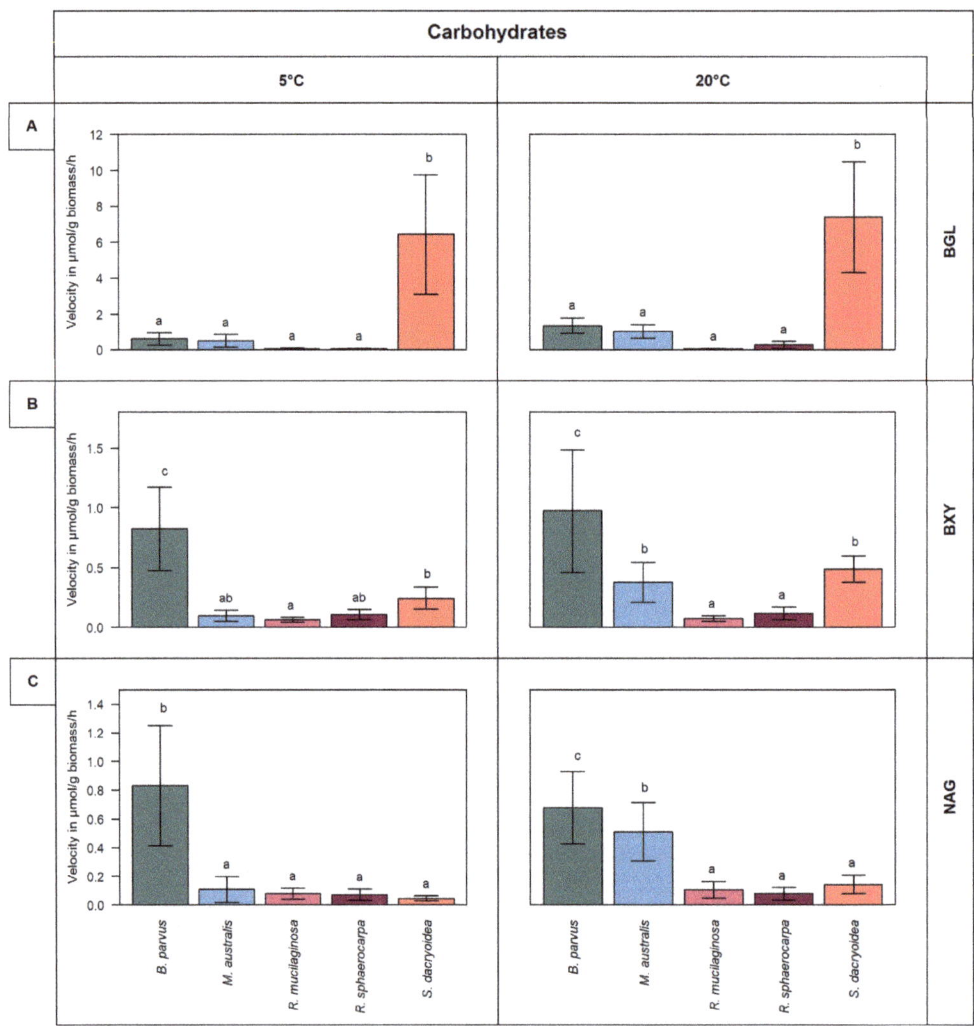

Figure 1. V_{max} in µmol/g biomass/h obtained from the total enzymatic activity for the substrates representing carbohydrates such as (**A**) β-glucosidase (BGL), (**B**) β-xylosidase (BXY), and (**C**) *N*-acetyl-β-D-glucosaminidase (NAG) of five marine fungal isolates *B. parvus, M. australis, R. mucilaginosa, R. sphaerocarpa*, and *S. dacryoidea* at 5 °C and 20 °C. According to Tukey's HSD, bars denoted by a different letter (a, b, and c) are significantly different ($p < 0.05$), whereas bars denoted by a common letter (ab) are not significantly different.

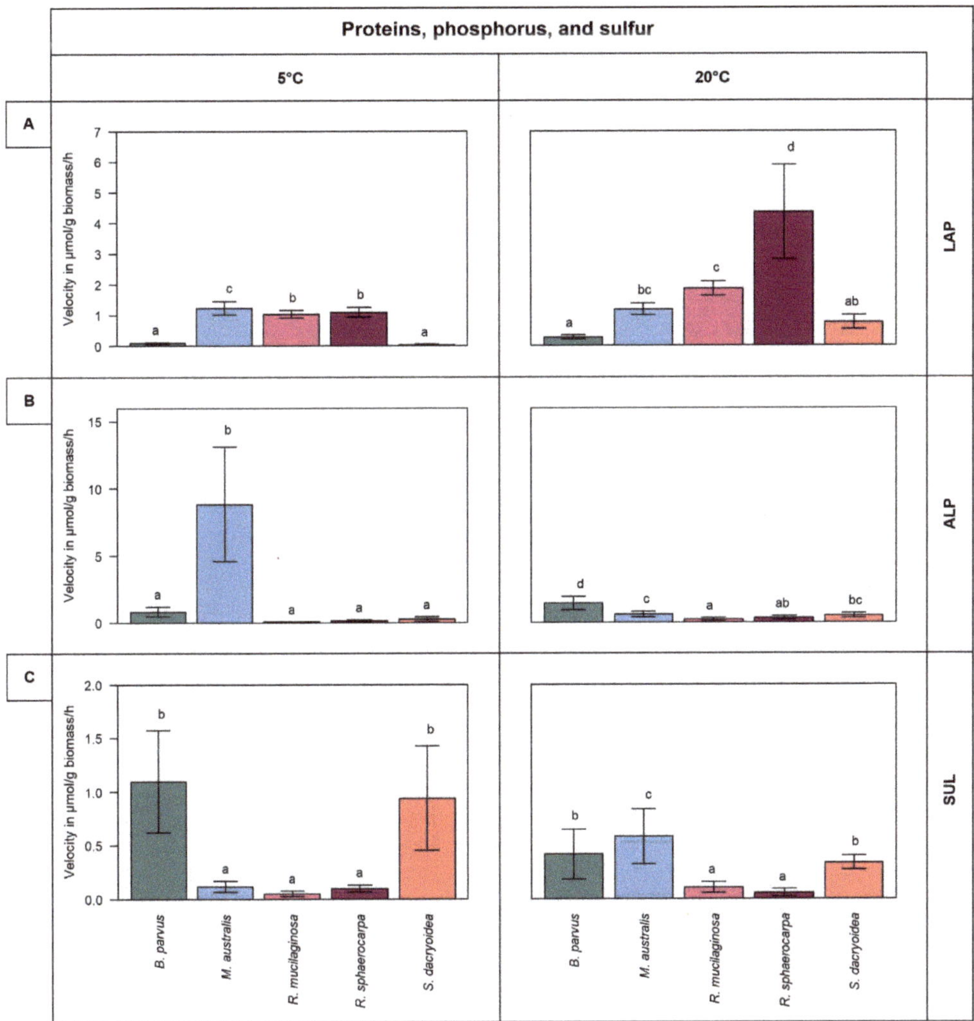

Figure 2. V_{max} in µmol/g biomass/h obtained from the total enzymatic activity for the substrates representing proteins, phosphorus, and sulfur, such as (**A**) leucine aminopeptidase (LAP), (**B**) alkaline phosphatase (ALP), and (**C**) sulfatase (SUL) of the five marine fungal isolates *B. parvus*, *M. australis*, *R. mucilaginosa*, *R. sphaerocarpa*, and *S. dacryoidea* at 5 °C and 20 °C. According to Tukey's HSD, bars denoted by a different letter (a, b, c, and d) are significantly different ($p < 0.05$), whereas bars denoted by a common letter (ab and bc) are not significantly different.

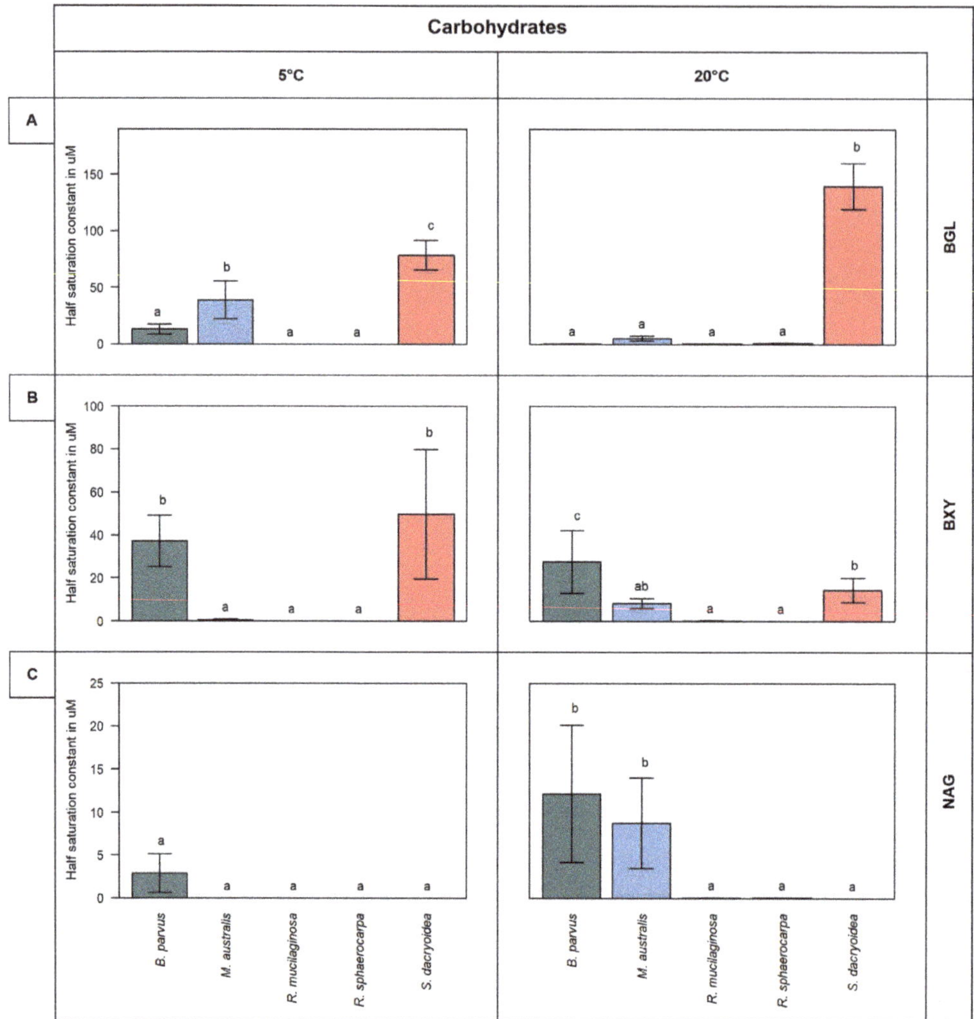

Figure 3. K_m in μM obtained from the total enzymatic activity for the substrates representing carbohydrates, such as (**A**) β-glucosidase (BGL), (**B**) β-xylosidase (BXY), and (**C**) *N*-acetyl-β-D-glucosaminidase (NAG) of the five marine fungal isolates (*B. parvus, M. australis, R. mucilaginosa, R. sphaerocarpa*, and *S. dacryoidea*). Measurements were performed at 5 °C and 20 °C in the exponential phase. According to Tukey's HSD, bars denoted by a different letter (a, b, and c) are significantly different ($p < 0.05$), whereas bars denoted by a common letter (ab) are not significantly different.

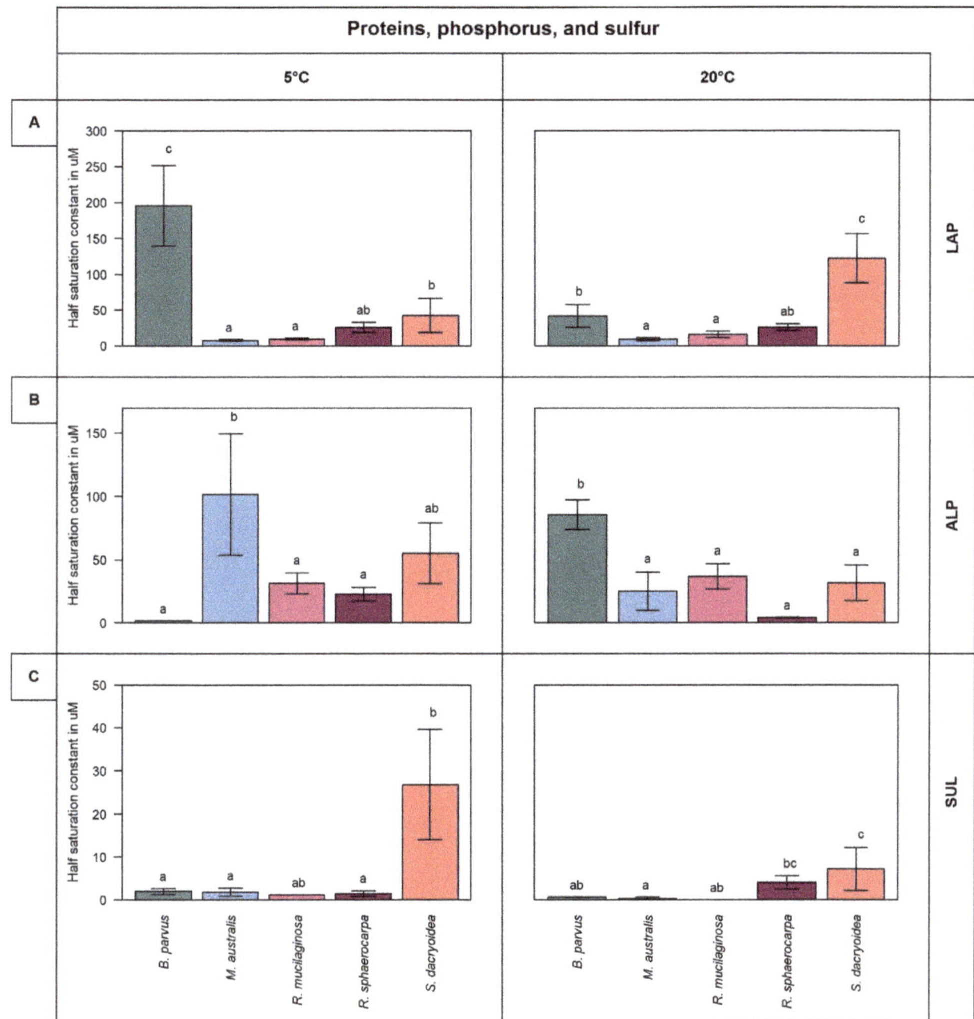

Figure 4. K_m in µM obtained from the total enzymatic activity for the substrates representing proteins, phosphorus, and sulfur, such as (**A**) leucine aminopeptidase (LAP), (**B**) alkaline phosphatase (ALP), and (**C**) sulfatase (SUL) of the five marine fungal isolates (*B. parvus*, *M. australis*, *R. mucilaginosa*, *R. sphaerocarpa*, and *S. dacryoidea*). Measurements were performed at 5 °C and 20 °C in the exponential phase. According to Tukey's HSD, bars denoted by a different letter (a, b, and c) are significantly different ($p < 0.05$), whereas bars denoted by a common letter (ab and bc) are not significantly different.

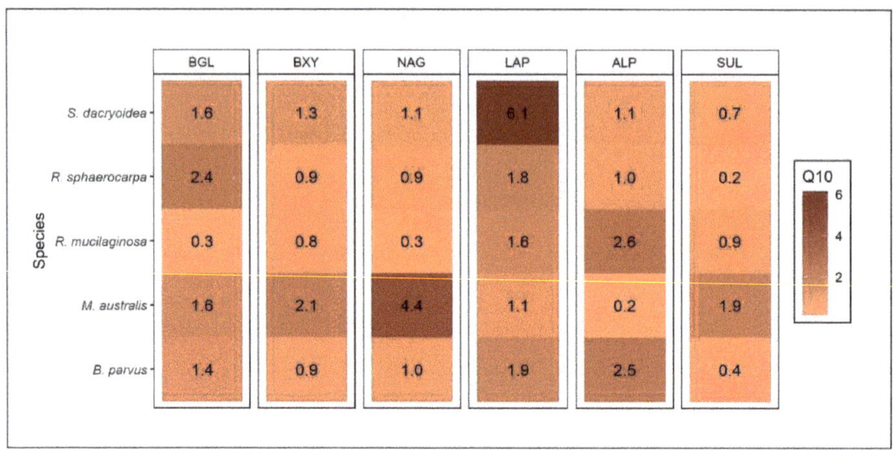

Figure 5. Q_{10} of the normalized total enzymatic activity (V_{max}) with the biomass (dry weight) for the substrates β-glucosidase (BGL), β-xylosidase (BXY), *N*-acetyl-β-D-glucosaminidase (NAG), leucine aminopeptidase (LAP), alkaline phosphatase (ALP), and sulfatase (SUL) and the species *B. parvus*, *M. australis*, *R. mucilaginosa*, *R. sphaerocarpa*, and *S. dacryoidea*.

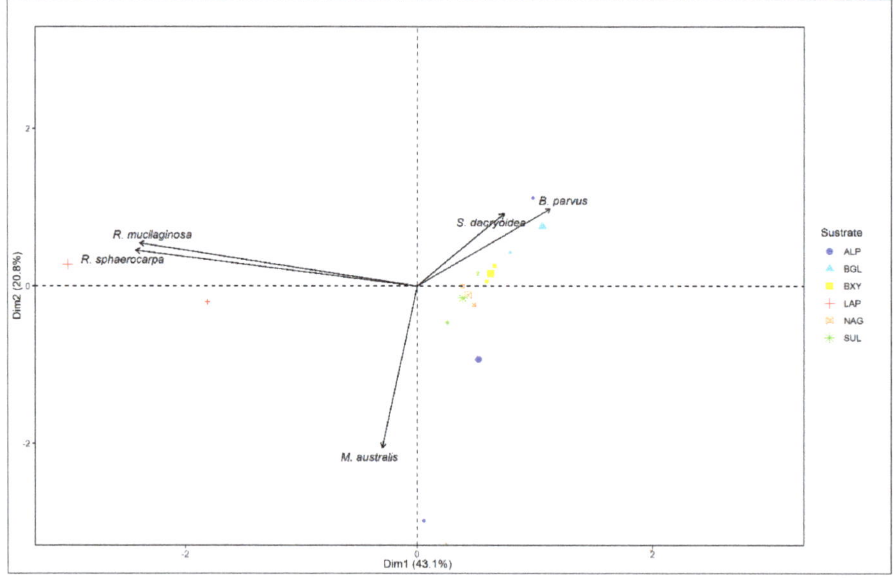

Figure 6. PCA plot from the normalization of the total enzymatic activity (V_{max}) with the biomass (dry weight) at 5 °C and 20 °C for all the species. This was compared for each substrate, which corresponds to alkaline phosphatase (ALP), β-glucosidase (BGL), β-xylosidase (BXY), leucine aminopeptidase (LAP), *N*-acetyl-β-D-glucosaminidase (NAG), and sulfatase (SUL).

3.1. Carbohydrate Active Enzymes

3.1.1. β-Glucosidase (BGL)

S. dacryoidea exhibited a significantly higher V_{max} for BGL than the other fungal strains (*t*-test; $p < 0.001$) at 5 °C (6.4 ± 3.2 µmol/g biomass/h) and 20 °C (7.4 ± 3.0 µmol/g biomass/h) (Figure 1A). The other fungal strains exhibited generally low BGL activity, particularly at 5 °C. At 20 °C, the BGL activity was slightly higher than at 5 °C, except for *R. mucilaginosa*, which maintained a low BGL activity at both temperatures. Consequently, V_{max} was a species- and temperature-specific value, with Q_{10} values varying between 1.4 and 2.4, except for *R. mucilaginosa* (Figure 5).

The K_m was significantly higher in *S. dacryoidea* than in the other fungal strains (*t*-test; $p < 0.001$) and ranged between 139.6 ± 19.1 µM and 78.6 ± 11.9 µM (Figure 3A). Moreover, the K_m of *S. dacryoidea* was significantly higher at 20 °C than at 5 °C (*t*-test; $p < 0.001$). Nonetheless, the other two species of the phylum Basidiomycota, *R. sphaerocarpa* and *R. mucilaginosa*, showed low K_m values, especially the latter. Finally, for the Ascomycota species, the K_m was highest at 5 °C (*t*-test; $p = 0.01$) with 12.9 ± 3.4 µM for *B. parvus* and 38.7 ± 14.6 for *M. australis*.

3.1.2. β-Xylosidase (BXY)

All fungal strains tested were capable to cleave xylose, however, at low rates (Figure 1B). *B. parvus* exhibited a significantly higher V_{max} for BXY than the other strains (*t*-test; $p < 0.001$). In *B. parvus*, the V_{max} was 0.8 ± 0.3 µmol/g biomass/h at 5 °C and 1.0 ± 0.5 µmol/g biomass/h at 20 °C. In the other strains, the hydrolysis rates were higher at 20 °C than at 5 °C, amounting to 0.5 and 0.1 µmol/g biomass/h, respectively. When all the fungal species were compared, the temperature had a greater effect in *M. australis* and *S. dacryoidea* with Q_{10} values of 2.1 and 1.8, respectively, than in the other fungal strains (Figure 5). For *B. parvus* and both species of the genus *Rhodotorula*, Q_{10} values close to 1 suggested that V_{max} was independent of the temperature.

The K_m varied significantly between both temperatures (*t*-test; $p = 0.90$) (Figure 3B). At 5 °C, *S. dacryoidea* showed a high K_m (49.7 ± 27.6 µM) followed by *B. parvus* with a K_m value of 37.2 ± 8.5 µM. In contrast, at 20 °C, the K_m decreased in both *B. parvus* and *S. dacryoidea* but increased in *M. australis*. The other two species, *R. sphaerocarpa* and *R. mucilaginosa*, showed low K_m values corresponding to their low V_{max}.

3.1.3. *N*-acetyl-β-D-glucosaminidase (NAG)

B. parvus exhibited similar NAG hydrolysis rates at 5 °C and 20 °C (*t*-test; $p = 0.39$) (Figure 1C). At 5 °C, the V_{max} was 0.8 ± 0.4 µmol/g biomass/h and at 20 °C 0.7 ± 0.2 µmol/g biomass/h. Although the other strains exhibited low NAG enzymatic activity, it increased from 5 °C to 20 °C, particularly in *M. australis* and *S. dacryoidea*. In *M. australis* the V_{max} increased from 0.1 ± 0.08 µmol/g biomass/h to 0.5 ± 0.2 µmol/g biomass/h and in *S. dacryoidea* from 0.04 ± 0.02 µmol/g biomass/h to 0.1 ± 0.06 µmol/g biomass/h. The Q_{10} values for NAG in *M. australis* and *S. dacryoidea* were 2.1 and 1.1, respectively (Figure 5).

The high NAG enzymatic activity detected in *B. parvus* coincided with a high K_m (Figure 3C). At 5 °C, the K_m was 2.9 ± 1.8 µM, while at 20 °C it was 12.1 ± 5.6 µM. At 20 °C, the K_m of *M. australis* increased to 8.7 ± 4.6 µM. Thus, the K_m values of these two Ascomycota species increased with temperature, but they remained low in the three Basidiomycota species.

3.2. Extracellular Enzymes Targeting Proteins, Phosphorus, and Sulfur Compounds

3.2.1. Leucine Aminopeptidase (LAP)

Generally, the LAP activity was higher at 20 °C than at 5 °C in all the fungal strains (*t*-test; $p < 0.001$), except in *M. australis* (*t*-test; $p = 0.46$) (Figure 2A). The LAP activity in *R. sphaerocarpa* at 20 °C was 4.3 ± 1.5 µmol/g biomass/h; however, it was only 1.1 ± 0.2 µmol/g biomass/h (*t*-test; $p < 0.001$) at 5 °C resulting in a Q_{10} of 6.1 (Figure 5). All the other fungal strains had Q_{10} values ranging from 1.9 to 1.1. Thus, the LAP activity was the only extracellular enzymatic activity tested that showed a temperature dependency in all the fungal strains (Figure 5).

B. parvus and *S. dacryoidea* exhibited significantly higher K_m values than the other fungal strains (*t*-test; $p < 0.001$) (Figure 4A). The K_m values varied between 195.4 µM and 41.6 µM in *B. parvus* and between 122.3 µM and 42.2 µM in *S. dacryoidea*. While in

B. parvus, the K_m decreased with increasing temperature, and in *S. dacryoidea*, the K_m value incremented with increasing temperature.

3.2.2. Alkaline Phosphatase (ALP)

At 5 °C, *M. australis* exhibited a significantly higher V_{max} (8.8 ± 3.7 μmol/g biomass/h) than the other fungal isolates (*t*-test; $p = 0.03$) (Figure 2B). At 20 °C, however, the V_{max} was about 10-fold lower than at 5 °C (0.6 ± 0.2 μmol/g biomass/h). In contrast, in all the other fungal strains, V_{max} slightly increased with temperature (Figure 2B). The Q_{10} values ranging from 2.6 to 1.0 indicated a moderate temperature dependency of ALP in all the fungal strains examined, except in *M. australis* (Figure 5).

The K_m was significantly higher in the Ascomycota than in the Basidiomycota strains (*t*-test; $p = 0.01$) (Figure 4B). *M. australis* exhibited a higher K_m at 5 °C (101.5 ± 43.5 μM) than at 20 °C, whereas the K_m in *B. parvus* was higher at 20 °C (85.7 ± 10.5 μM) than at 5 °C. The other fungal species belonging to the Basidiomycota phylum exhibited K_m values ranging from 55.1 μM to 4.1 μM (Figure 4B).

3.2.3. Sulfatase (SUL)

Sulfatase (SUL) activity was generally low compared to the other extracellular enzymatic activities (Figure 2C). Higher V_{max} values were determined for *B. parvus* and *S. dacryoidea* at 5 °C (1.1 and 0.9 μmol/g biomass/h, respectively) (*t*-test; $p = 0.01$) than at 20 °C (0.4 and 0.3 μmol/g biomass/h, respectively). In contrast, for *M. australis, R. mucilaginosa*, and *R. sphaerocarpa*, the SUL activity increased with temperature. The Q_{10} value in *M. australis* was 1.9 while all the other fungal strains were <1.0 (Figure 5). With the exception of *R. sphaerocarpa*, all the fungal strains exhibited higher K_m values at 5 °C than at 20 °C (*t*-test; $p \leq 0.002$) (Figure 4C).

3.3. Relation between Enzyme Kinetic Parameters, Enzyme Types, and Phylogeny

PCA analysis allowed for the comparison of the five studied marine fungal species and the different extracellular enzymatic activities determined. The explained variance of the dataset was 63.9% (Figure 6). The minor angle between *R. mucilaginosa* and *R. sphaerocarpa* belonging to the same genera (*Rhodotorula*) indicated very similar activity levels and extracellular enzyme characteristics. In contrast, the other fungal strains substantially differed in their extracellular enzymatic activity and enzyme characteristics, independent of their taxonomic affiliation.

4. Discussion

All the studied marine pelagic fungi species, *B. parvus, M. australis, R. mucilaginosa, R. sphaerocarpa*, and *S. dacryoidea*, produced extracellular enzymes to degrade substrate analogues of carbohydrates such as cellulose, chitin, and xylan. Additionally, they all produced enzymes to cleave off amino acids, phosphate, and sulfate esters from organic compounds.

4.1. Extracellular Enzymatic Activities of Pelagic Fungal Isolates

4.1.1. Influence of Taxonomy/Diversity on the Different EEAs

Since each organism has specific enzymatic capabilities and substrate preferences [33], extracellular enzymatic activities (EEA) can be used as functional traits to investigate functional diversity [34]. Hence, the different EEAs detected in the studied marine fungi can be used to infer their influence on marine ecological processes. Interestingly, we found that the two species belonging to the genus *Rhodotorula* (*R. mucilaginosa* and *R. sphaerocarpa*) exhibited similar kinetic parameters (V_{max} and K_m) for the majority of extracellular enzymes (Figures 1–4 and 6). This might indicate some degree of trait conservation among organisms on the genus level, although more species of this genus need to be investigated before a firm conclusion can be drawn. Differences were observed, however, in the EEAs of all the other fungal strains.

Polysaccharides, as the most abundant organic compound class, but also the most complex one, require a wide range of enzymes to degrade them [35,36]. We measured three EEAs responsible for the cleavage of cellulose, xylan, and chitin (Figure 1). In terrestrial environments, the plant cell wall is composed mainly of cellulose and protected by lignin [37]. In marine ecosystems, cellulose is present in the algae cell wall and covered by distinct polymers [38]. Nonetheless, as marine cellulose is more accessible, but less frequent than other substrates, only specific microorganisms are capable of degrading cellulose [39]. Some studies have reported cellulose degradation by marine fungi like *Arthrinium saccharicola* [40] and *Lulworthia floridana* [41]. Other studies have also described cellulose hydrolysis by wider distributed fungal species, for instance, *Aspergillus niger* [42] and *Trichoderma virens* [43]. Vaz, et al. [44] showed that 76% of the studied marine fungi exhibit cellulolytic activity. In this case, even though all the marine fungal species used were able to cleave cellulose, *S. dacryoidea* dominated this EEA (Figure 1A). Hudson [45] stated that each fungi species have different capacities to decompose cellulose due to diverse enzymatic machinery. *S. dacryoidea* exhibited a high K_m indicating a low affinity to the substrate [31]. This also suggests that, even though the overall enzymatic activity is high, the substrate dissociates easily from the enzyme [46].

Xylan is a polysaccharide formed by residual monosaccharides called xylose [29]. Similar to cellulose, in terrestrial environments, xylan can be found in plants [47], whereas in the ocean, xylan can be present in algae [48,49]. Fungi can degrade xylose via the oxidoreductase pathway [50] as it is a primary carbon source [28]. Even though the general EEA was low (Figure 1B), we can deduce that the studied marine fungi are capable of releasing enzymes related to the hydrolysis of xylose. Raghukumar, et al. [51] identified low xylanase activity rates of fungal coastal strains. Duarte, et al. [52] also reported xylanase activity of Antarctic fungal strains but highlighted a higher activity of Basidiomycota over Ascomycota strains. In this study, we could not identify a clear difference between these two phyla. Nonetheless, the low K_m suggests a high affinity of the enzyme to the substrate at low concentrations (Figure 3B). Thus, it seems likely that pelagic marine fungi might use xylose as a carbon source even when present at low concentrations.

Chitin is one of the most abundant naturally occurring polysaccharides, and it is an essential component of the cell wall of fungi, the exoskeleton of arthropods, the radula of mollusks, and the beak of cephalopods [53,54]. Although the overall chitinase activity was low, we can infer that the studied marine fungi are capable of degrading chitin (Figure 1C). The chitin degradation by marine fungi has been reported in species such as *Lecanicillium muscarium* [55] and *Verticillium lecanii* [56]. Interestingly, the number of fungal enzymes involved in the degradation of chitin is related to the fungal chitin content, which varies strongly between fungal species and is dependent on the growth mode [57]. For instance, the hyphae-like fungal cell wall consists of 10 to 20% of chitin [54], whereas yeast-like fungi have a rather low chitin content of 0.5 to 5% [58]. The filamentous fungus *B. parvus* exhibited high chitinase activity (Figure 1C). Moreover, the low K_m obtained for this species suggests that only a low substrate concentration is needed to saturate its chitinase. Thus, pelagic marine fungi likely use chitin as a carbon source even when present at only low concentrations.

Leucine aminopeptidase is a critical biological enzyme due to its key role in the degradation of proteins [59]. In the present study, this enzymatic activity was different for each fungal species. Despite the majority of microbial LAP being intracellular, extracellular enzymes have been reported in filamentous fungi [60]. All the species we tested showed LAP activity with *R. sphaerocarpa* exhibiting the highest V_{max}. The substrate concentration needed to achieve half V_{max} varied among species (Figure 4A). Although *B. parvus* had one of the lowest V_{max}, its K_m was the highest (Figures 2A and 4A). In contrast, *R. sphaerocarpa* exhibited high substrate affinity and a low K_m (Figure 4A) Thus, as the kinetics for LAP varied among the fungal species, the protein hydrolysis in the ocean might be species dependent.

Inorganic phosphate (Pi) is the preferred phosphorus source for microbial uptake. In surface waters, however, Pi frequently limits phytoplankton productivity [61]. To overcome this P-limitation, microorganisms use dissolved organic phosphorus (DOP) [62]. For prokaryotic microorganisms, Baltar et al. [63] reported that irrespective of the phosphate bioavailability, the activation of alkaline phosphatase was related to sporadic pulses of organic matter. Thus, a high K_m might be beneficial to allow for high cleavage rates when organic substrate availability is high. The species *R. sphaerocarpa*, *R. mucilaginosa*, and *S. dacryoidea*, belonging to the phylum Basidiomycota, exhibited a low enzymatic activity but a high K_m (Figures 2B and 4B). In contrast, the species of the phylum Ascomycota (*B. parvus* and *M. australis*) seem to be more suitable to overcome this P-limitation, but a higher amount of substrate, and hence of organic matter, might be needed for this purpose.

In living organisms, sulfur is the sixth most abundant element as it can be found in amino acids, such as cysteine and methionine, but also in polysaccharides and proteoglycans [35]. In contrast to terrestrial polysaccharides, several marine polysaccharides, especially in the cell wall of macroalgae, are highly sulfated [35]. The terrestrial fungus *Fusarium proliferatum* was reported to produce sulfatase for the assimilation of sulfated fucoidans of brown algae [64]. In the present study, the species of the genus *Rhodotorula* maintained a low sulfatase activity, whereas it varied among the other species. The V_{max} and K_m were high in *S. dacryoidea*, indicating fast hydrolysis and low substrate affinity. As all the fungal species showed different sulfatase activity, we can deduce that these marine fungi might be capable to use sulfated amino acids as well as carbohydrates.

In this study, even though we analyzed just a few hydrolysis possibilities, the functional diversity seems to be broad in marine fungi. As suggested by Berlemont [65] and Baltar et al. [10], marine fungi are potentially involved in carbohydrates' degradation. Based on our results, we can infer that marine pelagic fungi are actively utilizing carbohydrates such as cellulose, chitin, and xylose, as well as some carbohydrates with sulfur content, potentially of algal origin [47].

The ocean is a complex and diverse ecosystem composed of microorganisms such as archaea, bacteria, fungi, protists, and viruses. As osmoheterotrophic organisms, fungi might provide intermediate decomposition products needed for other microorganisms. These also might lead to the proliferation or inhibition of other microbes as well as enzymatic activities. As marine fungi can utilize a wide range of organic substrates [66], nutrient availability can impact the magnitude and distribution of extracellular enzymatic activities [33].

4.1.2. Temperature Influence

Temperature is considered one of the most important abiotic factors because it influences essentially all biochemical reactions [67]. Enzymes are sensitive to temperature [46] influencing the kinetics along with the substrate binding property and stability [33,68]. According to the Van't Hoff rule, a temperature increase of 10 °C can double a reaction rate. Q_{10} values lower than 1.0 would indicate a reaction rate completely independent of temperature, whereas values above 1 indicate thermodependency [69].

The results obtained in this study indicate that the majority of enzymatic activities were lower at 5 °C than at 20 °C (Figures 1 and 2). In general terms, at 5 °C the species *R. mucilaginosa* and *R. sphaerocarpa* maintained V_{max} values as low as 0.1 μmol/g biomass/h, whereas the other species exhibited only half of the activity at 5 °C, particularly for BGL and ALP (Figures 1 and 2). This reduced enzymatic synthesis at a low temperature might be due to limited transcriptional and translational activity, limited protein folding, and DNA and RNA secondary structures' stabilization [11]. *B. parvus*, for all the substrates except ALP, had a higher K_m at 5 °C, which suggests that the affinity of this species for some substrates increases with increasing temperature. Taken together, the effect of temperature on the characteristics of extracellular enzymes depends on the fungal species and the type of enzyme.

Psychrophiles microorganisms have evolved a complex range of adaptation strategies, such as production of antifreeze proteins [70] and exopolysaccharides (EPS) [71], high levels of unsaturated fatty acids to maintain the membrane fluidity [72], and certain enzymes adapted to those temperatures [11]. Gerday et al. [73] described a peculiar type of extracellular enzymes known as "cold-adapted enzymes" produced by microorganisms living at low temperatures. For these enzymes, the reaction rate is dependent on the encounter rate of the enzyme and substrate, so it is controlled mainly by diffusion, and it is temperature independent [14,74].

Aghajari et al. [75] suggested that the main structural feature of these "cold-adapted enzymes" is flexibility or plasticity. The structures involved in the catalytic cycle are more flexible, whereas other structures that do not participate in the catalytic cycle might be more rigid [67,76]. For instance, the chitinase of *Glaciozyma antarctica* presented fewer salt bridges and hydrogen bonds, which increased its flexibility [12,77]. Another key structural feature of these enzymes is stability [73,74], with for example, amino acids modifications in key regions of the protein [77–80]. Nonetheless, there is not a single strategy, as each cold-adapted enzyme can perform different ways to enhance its activity at low temperatures [12,74].

Cold-adapted enzymes have been reported from a wide variety of marine fungi [12,44,52,55,56,77,81,82]. *M. australis* and *R. sphaerocarpa* were one of the few species that showed a noticeable enzymatic activity at 5 °C for ALP and SUL, respectively (Figure 2B,C). *M. australis* is an endemic species of Antarctic waters [19,83], whereas *R. sphaerocarpa* was originally isolated close to Marguerite Bay on the west side of the Antarctic Peninsula but has a wider distribution including the Caribbean Sea [84] and the Andaman Sea [85], among others. Low temperatures exert high selective pressure on endemic organisms [86], such as alkalinity phosphatase in *M. australis*. For this species, the substrate-binding affinity was lower at 5 °C, whereas for *R. sphaerocarpa*, K_m was lower at this temperature (Figure 4). This suggested that the enzyme–substrate complex of *M. australis* ALP is more effective at higher substrate concentration typical for Antarctic waters known as a major high-nutrient low-chlorophyll region in the global ocean [87].

Fungal cold-adapted chitinases have been previously reported [77,88,89]. Ramli et al. [77] found that the chitinase sequence of *G. antarctica* had a low sequence identity with other chitinases. Moreover, they found that the enzyme flexibility was due to certain amino acids substitutions in the surface and loop regions. In this study, at 5 °C, we could only identify a higher chitinase activity for *B. parvus*. For the rest of the species, there was a positive enzymatic activity, but higher at 20 °C.

Microorganisms isolated from cold environments can also display kinetic parameters similar to those of their mesophilic counterparts [69,90]. Ito et al. [91] deduced that a high Q_{10} value (>2) is due to a conformational change in proteins and indicates the need for high activation energy. We found that only for LAP, all the examined fungal species expressed Q_{10} values higher than 1. Generally, the temperature where an enzyme can achieve its highest activity does not match the optimal growth temperature of the microorganism that is producing it [78]. Apparently, cold-adapted species, such as *B. parvus*, *M. australis*, *R. sphaerocarpa*, and *S. dacryoidea*, can respond to a temperature rise by increasing enzymatic activity. According to the Arrhenius equation, temperature can influence the activation energy needed to initiate a chemical reaction, and hence, its rate. At a higher temperature, the molecules gain energy to move faster, which also increases the collisions between enzymes and substrates. As a result, elevated extracellular enzymatic activity at increasing temperatures in the surface waters might lead to changes in cleavage and uptake rate of organic matter in oceanic fungi.

5. Conclusions

In the present study, we have shown that different marine fungal strains exhibit varying extracellular enzyme characteristics with V_{max} and K_m values varying over a range of one order of magnitude. Although the fungal species were isolated from coastal Antarctic waters (*B. parvus*, *M. australis*, *R. sphaerocarpa*, and *S. dacryoidea*), and hence are potentially adapted to low temperatures, they exhibited higher extracellular enzymatic activity at

20 °C than at 5 °C, with some exceptions. While in some fungal strains the K_m values for specific extracellular enzymes were higher at low temperatures, for other enzymes they were lower. Additionally, there was considerable species-specific variability in the extracellular enzymatic activity. Taken together, our study indicates that temperature might be one of most important physical factors controlling marine fungal extracellular enzymatic activity. Thus, species composition and temperature determine the role of marine fungi in organic matter cleavage in the global ocean.

Author Contributions: Conceptualization, K.S.A. and F.B.; data curation, K.S.A.; formal analysis, K.S.A.; funding acquisition, G.J.H. and F.B.; investigation, K.S.A.; methodology, K.S.A. and F.B.; project administration, F.B.; software, K.S.A.; supervision, F.B.; visualization, K.S.A.; writing—original draft, K.S.A.; writing—review and editing, K.S.A., G.J.H., and F.B. All authors have read and agreed to the published version of the manuscript.

Funding: F.B. was supported by the Austrian Science Fund (FWF) projects OCEANIDES (P34304-B), ENIGMA (TAI534), and EXEBIO (P35248). G.J.H. was supported by the Austrian Science Fund (FWF) project ARTEMIS (P28781-B21), the projects I486-B09 and P23234-B11 by the European Research Council (ERC) under the European Community's Seventh Framework Programme (FP7/2007–2013)/ERC grant agreement 268595 (MEDEA project). Open Access Funding by the Austrian Science Fund (FWF).

Institutional Review Board Statement: Not applicable.

Informed Consent Statement: Not applicable.

Data Availability Statement: The raw data supporting the conclusions of this article will be made available by the authors, without undue reservation to any qualified researcher.

Acknowledgments: We would like to thank Charlotte Doebke, Marilena Heitger, Marina Montserrat Díez, and Katarína Tamášová for their valuable help in the laboratory work.

Conflicts of Interest: The authors declare no conflict of interest.

References

1. Nevalainen, H.; Kautto, L.; Te'o, J. Methods for isolation and cultivation of filamentous fungi. *Methods Mol. Biol.* **2014**, *1096*, 3–16.
2. Richards, T.A.; Jones, M.D.; Leonard, G.; Bass, D. Marine fungi: Their ecology and molecular diversity. *Annu. Rev. Mar. Sci.* **2012**, *4*, 495–522. [CrossRef] [PubMed]
3. Arrieta, J.M.; Herndl, G.J. Changes in bacterial β-glucosidase diversity during a coastal phytoplankton bloom. *Limnol. Oceanogr.* **2002**, *47*, 594–599. [CrossRef]
4. Baltar, F.; Arístegui, J.; Sintes, E.; Van Aken, H.M.; Gasol, J.M.; Herndl, G.J. Prokaryotic extracellular enzymatic activity in relation to biomass production and respiration in the meso-and bathypelagic waters of the (sub) tropical Atlantic. *Environ. Microbiol.* **2009**, *11*, 1998–2014. [CrossRef]
5. Gutiérrez, M.H.; Pantoja, S.; Tejos, E.; Quiñones, R.A. The role of fungi in processing marine organic matter in the upwelling ecosystem off Chile. *Mar. Biol.* **2011**, *158*, 205–219. [CrossRef]
6. Cunliffe, M.; Hollingsworth, A.; Bain, C.; Sharma, V.; Taylor, J.D. Algal polysaccharide utilisation by saprotrophic planktonic marine fungi. *Fungal Ecol.* **2017**, *30*, 135–138. [CrossRef]
7. Gao, Z.; Johnson, Z.I.; Wang, G. Molecular characterization of the spatial diversity and novel lineages of mycoplankton in Hawaiian coastal waters. *ISME J.* **2010**, *4*, 111–120. [CrossRef]
8. Bochdansky, A.B.; Clouse, M.A.; Herndl, G.J. Eukaryotic microbes, principally fungi and labyrinthulomycetes, dominate biomass on bathypelagic marine snow. *ISME J.* **2017**, *11*, 362–373. [CrossRef]
9. Morales, S.E.; Biswas, A.; Herndl, G.J.; Baltar, F. Global Structuring of Phylogenetic and Functional Diversity of Pelagic Fungi by Depth and Temperature. *Front. Mar. Sci.* **2019**, *6*, 131. [CrossRef]
10. Baltar, F.; Zhao, Z.; Herndl, G.J. Potential and expression of carbohydrate utilization by marine fungi in the global ocean. *Microbiome* **2021**, *9*, 106. [CrossRef]
11. D'Amico, S.; Collins, T.; Marx, J.C.; Feller, G.; Gerday, C.; Gerday, C. Psychrophilic microorganisms: Challenges for life. *EMBO Rep.* **2006**, *7*, 385–389. [CrossRef] [PubMed]
12. Yusof, N.A.; Hashim, N.H.F.; Bharudin, I. Cold adaptation strategies and the potential of psychrophilic enzymes from the antarctic yeast, Glaciozyma antarctica PI12. *J. Fungi* **2021**, *7*, 528. [CrossRef] [PubMed]
13. Morita, R.Y. Psychrophilic bacteria. *Bacteriol. Rev.* **1975**, *39*, 144–167. [CrossRef] [PubMed]
14. Feller, G.; Gerday, C. Psychrophilic enzymes: Hot topics in cold adaptation. *Nat. Rev. Microbiol.* **2003**, *1*, 200–208. [CrossRef]
15. Vonnahme, T.R.; Dietrich, U.; Hassett, B.T. Progress in microbial ecology in ice-covered seas. In *YOUMARES 9—The Oceans: Our Research, Our Future*; Springer: Cham, Switzerland, 2020; p. 261.

16. Amend, A.; Burgaud, G.; Cunliffe, M.; Edgcomb, V.P.; Ettinger, C.L.; Gutiérrez, M.H.; Heitman, J.; Hom, E.F.Y.; Ianiri, G.; Jones, A.C.; et al. Fungi in the Marine Environment: Open Questions and Unsolved Problems. *mBio* **2019**, *10*, e01189-18. [CrossRef]
17. Taylor, J.D.; Cunliffe, M. Multi-year assessment of coastal planktonic fungi reveals environmental drivers of diversity and abundance. *ISME J.* **2016**, *10*, 2118–2128. [CrossRef]
18. Fell, J.; Statzell, A.C. *Sympodiomyces gen.* n., a yeast-like organism from southern marine waters. *Antonie Van Leeuwenhoek* **1971**, *37*, 359–367. [CrossRef]
19. Fell, J.W.; Hunter, I.L. Isolation of heterothallic yeast strains of Metschnikowia Kamienski and their mating reactions with *Chlamydozyma wickerham* spp. *Antonie Van Leeuwenhoek* **1968**, *34*, 365–376. [CrossRef]
20. Fell, J.W.; Hunter, I.L.; Tallman, A.S. Marine basidiomycetous yeasts (*Rhodosporidium* spp. n.) with tetrapolar and multiple allelic bipolar mating systems. *Can. J. Microbiol.* **1973**, *19*, 643–657. [CrossRef]
21. Newell, S.Y.; Fell, J. The perfect form of a marine-occurring yeast of the genus *Rhodotorula*. *Mycologia* **1970**, *62*, 272–281. [CrossRef]
22. Sampaio, J.P. *Rhodotorula* Harrison (1928). In *The Yeasts*; Elsevier: Burlington, VT, USA, 2011; pp. 1873–1927.
23. Wickerham, L.J. A taxonomic study of Monilia albicans with special emphasis on morphology and morphological variation. *J. Tropical Med. Hyg.* **1939**, *42*, 174–179.
24. Wickerham, L.J. *Taxonomy of Yeasts*; US Department of Agriculture: Washington, DC, USA, 1951.
25. Salazar Alekseyeva, K.; Mähnert, B.; Berthiller, F.; Breyer, E.; Herndl, G.J.; Baltar, F. Adapting an Ergosterol Extraction Method with Marine Yeasts for the Quantification of Oceanic Fungal Biomass. *J. Fungi* **2021**, *7*, 690. [CrossRef]
26. Norkrans, B. Degradation of cellulose. *Annu. Rev. Phytopathol.* **1963**, *1*, 325–350. [CrossRef]
27. Pointing, S.B.; Vrijmoed, L.L.P.; Jones, E.B.G. A Qualitative Assessment of Lignocellulose Degrading Enzyme Activity in Marine Fungi. *Bot. Mar.* **1998**, *41*, 293–298. [CrossRef]
28. Collins, T.; Gerday, C.; Feller, G. Xylanases, xylanase families and extremophilic xylanases. *FEMS Microbiol. Rev.* **2005**, *29*, 3–23. [CrossRef] [PubMed]
29. dos Santos, J.A.; Vieira, J.M.F.; Videira, A.; Meirelles, L.A.; Rodrigues, A.; Taniwaki, M.H.; Sette, L.D. Marine-derived fungus *Aspergillus* cf. *tubingensis* LAMAI 31: A new genetic resource for xylanase production. *AMB Express* **2016**, *6*, 25.
30. Hoppe, H.-G. Significance of exoenzymatic activities in the ecology of brackish water: Measurements by means of methylumbelliferyl-substrates. *Mar. Ecol. Prog. Ser.* **1983**, *11*, 299–308. [CrossRef]
31. Huitema, C.; Horsman, G. Analyzing Enzyme Kinetic Data Using the Powerful Statistical Capabilities of R. *bioRxiv* **2018**, *1*, 316588.
32. Sterratt, D.C. *Q10: The Effect of Temperature on Ion Channel Kinetics, in Encyclopedia of Computational Neuroscience*; Jaeger, D., Jung, R., Eds.; Springer: New York, NY, USA, 2015; pp. 2551–2552.
33. Arnosti, C.; Bell, C.; Moorhead, D.L.; Sinsabaugh, R.L.; Steen, A.D.; Stromberger, M.; Wallenstein, M.; Weintraub, M.N. Extracellular enzymes in terrestrial, freshwater, and marine environments: Perspectives on system variability and common research needs. *Biogeochemistry* **2014**, *117*, 5–21. [CrossRef]
34. Cullings, K.; Courty, P.-E. Saprotrophic capabilities as functional traits to study functional diversity and resilience of ectomycorrhizal community. *Oecologia* **2009**, *161*, 661–664. [CrossRef]
35. Helbert, W. Marine Polysaccharide Sulfatases. *Front. Mar. Sci.* **2017**, *4*, 6. [CrossRef]
36. Metreveli, E.; Khardziani, T.; Elisashvili, V. The Carbon Source Controls the Secretion and Yield of Polysaccharide-Hydrolyzing Enzymes of Basidiomycetes. *Biomolecules* **2021**, *11*, 1341. [CrossRef]
37. Houtman, C.J.; Atalla, R.H. Cellulose-Lignin Interactions (A Computational Study). *Plant Physiol.* **1995**, *107*, 977–984. [CrossRef] [PubMed]
38. Domozych, D.; Ciancia, M.; Fangel, J.; Mikkelsen, M.; Ulvskov, P.; Willats, W. The Cell Walls of Green Algae: A Journey through Evolution and Diversity. *Front. Plant Sci.* **2012**, *3*, 82. [CrossRef]
39. Arora, D.K. *Handbook of Applied Mycology: Volume 1: Soil and Plants*; CRC Press: Boca Raton, FL, USA, 1991.
40. Hong, J.-H.; Jang, S.; Heo, Y.M.; Min, M.; Lee, H.; Lee, Y.M.; Lee, H.; Kim, J.-J. Investigation of Marine-Derived Fungal Diversity and Their Exploitable Biological Activities. *Mar. Drugs* **2015**, *13*, 4137–4155. [CrossRef] [PubMed]
41. Meyers, S.; Scott, E. Cellulose degradation by *Lulworthia floridana* and other lignicolous marine fungi. *Mar. Biol.* **1968**, *2*, 41–46. [CrossRef]
42. Baraldo Junior, A.; Borges, D.G.; Tardioli, P.W.; Farinas, C.S. Characterization of β-glucosidase produced by *Aspergillus niger* under solid-state fermentation and partially purified using MANAE-Agarose. *Biotechnol. Res. Int.* **2014**, *2014*, 317092. [CrossRef]
43. Ab Wahab, A.F.F.; Karim, N.A.A.; Ling, J.G.; Hasan, N.S.; Yong, H.Y.; Bharudin, I.; Kamaruddin, S.; Bakar, F.D.A.; Murad, A.M. Functional characterisation of cellobiohydrolase I (Cbh1) from Trichoderma virens UKM1 expressed in *Aspergillus niger*. *Protein Expr. Purif.* **2019**, *154*, 52–61. [CrossRef]
44. Vaz, A.B.M.; Rosa, L.H.; Vieira, M.L.A.; de Garcia, V.; Brandão, L.R.; Teixeira, L.C.R.S.; Moliné, M.; Libkind, D.; van Broock, M.; Rosa, C. The diversity, extracellular enzymatic activities and photoprotective compounds of yeasts isolated in Antarctica. *Braz. J. Microbiol.* **2011**, *42*, 937–947. [CrossRef]
45. Hudson, H.J. *Fungal Saprophytism*; Edward Arnold: London, UK, 1972.
46. Roussel, M.R. *A Life Scientist's Guide to Physical Chemistry*; Cambridge University Press: Cambridge, UK, 2012.
47. Balabanova, L.; Slepchenko, L.; Son, O.; Tekutyeva, L. Biotechnology Potential of Marine Fungi Degrading Plant and Algae Polymeric Substrates. *Front. Microbiol.* **2018**, *9*, 1527. [CrossRef]

48. Goddard-Borger, E.D.; Sakaguchi, K.; Reitinger, S.; Watanabe, N.; Ito, M.; Withers, S.G. Mechanistic insights into the 1,3-xylanases: Useful enzymes for manipulation of algal biomass. *J. Am. Chem. Soc.* **2012**, *134*, 3895–3902. [CrossRef] [PubMed]
49. Synytsya, A.; Čopíková, J.; Kim, W.J.; Park, Y.I. Cell Wall Polysaccharides of Marine Algae. In *Springer Handbook of Marine Biotechnology*; Kim, S.-K., Ed.; Springer: Berlin/Heidelberg, Germany, 2015; pp. 543–590.
50. Domingues, R.; Bondar, M.; Palolo, I.; Queirós, O.; de Almeida, C.D.; Cesário, M.T. Xylose Metabolism in Bacteria—Opportunities and Challenges towards Efficient Lignocellulosic Biomass-Based Biorefineries. *Appl. Sci.* **2021**, *11*, 8112. [CrossRef]
51. Raghukumar, C.; Raghukumar, S.; Chinnaraj, A.; Chandramohan, D.; D'souza, T.; Reddy, C. Laccase and Other Lignocellulose Modifying Enzymes of Marine Fungi Isolated from the Coast of India. *Botanica Marina.* **1994**, *37*, 515–523. [CrossRef]
52. Duarte, A.W.F.; Dayo-Owoyemi, I.; Nobre, F.; Pagnocca, F.C.; Chaud, L.C.S.; Pessoa, A.; Felipe, M.G.A.; Sette, L.D. Taxonomic assessment and enzymes production by yeasts isolated from marine and terrestrial Antarctic samples. *Extremophiles* **2013**, *17*, 1023–1035. [CrossRef] [PubMed]
53. Numata, K.; Kaplan, D.L. 20—Biologically derived scaffolds. In *Advanced Wound Repair Therapies*; Farrar, D., Ed.; Woodhead Publishing: Sawston, UK, 2011; pp. 524–551.
54. Ruiz-Herrera, J. *Fungal Cell Wall: Structure, Synthesis, and Assembly*; CRC Press: Boca Raton, FL, USA, 1991.
55. Fenice, M. The psychrotolerant Antarctic fungus *Lecanicillium muscarium* CCFEE 5003: A powerful producer of cold-tolerant chitinolytic enzymes. *Molecules* **2016**, *21*, 447. [CrossRef]
56. Fenice, M.; Selbmann, L.; Di Giambattista, R.; Federici, F. Chitinolytic activity at low temperature of an Antarctic strain (A3) of *Verticillium lecanii*. *Res. Microbiol.* **1998**, *149*, 289–300. [CrossRef]
57. Hartl, L.; Zach, S.; Seidl-Seiboth, V. Fungal chitinases: Diversity, mechanistic properties and biotechnological potential. *Appl. Microbiol. Biotechnol.* **2012**, *93*, 533–543. [CrossRef]
58. Garcia-Rubio, R.; de Oliveira, H.C.; Rivera, J.; Trevijano-Contador, N. The Fungal Cell Wall: *Candida*, *Cryptococcus*, and *Aspergillus* Species. *Front. Microbiol.* **2020**, *10*, 2993. [CrossRef]
59. Burley, S.K.; David, P.R.; Taylor, A.; Lipscomb, W.N. Molecular structure of leucine aminopeptidase at 2.7—A resolution. *Proc. Natl. Acad. Sci. USA* **1990**, *87*, 6878–6882. [CrossRef]
60. Nampoothiri, K.M.; Nagy, V.; Kovacs, K.; Szakacs, G.; Pandey, A. L-leucine aminopeptidase production by filamentous *Aspergillus* fungi. *Lett. Appl. Microbiol.* **2005**, *41*, 498–504. [CrossRef]
61. Karl, D.M. Phosphorus, the staff of life. *Nature* **2000**, *406*, 31–33. [CrossRef] [PubMed]
62. Srivastava, A.; Saavedra, D.E.M.; Thomson, B.; García, J.A.L.; Zhao, Z.; Patrick, W.M.; Herndl, G.J.; Baltar, F. Enzyme promiscuity in natural environments: Alkaline phosphatase in the ocean. *ISME J.* **2021**, *15*, 3375–3383. [CrossRef] [PubMed]
63. Baltar, F.; Lundin, D.; Palovaara, J.; Lekunberri, I.; Reinthaler, T.; Herndl, G.J.; Pinhassi, J. Prokaryotic Responses to Ammonium and Organic Carbon Reveal Alternative CO2 Fixation Pathways and Importance of Alkaline Phosphatase in the Mesopelagic North Atlantic. *Front. Microbiol.* **2016**, *7*, 1670. [CrossRef] [PubMed]
64. Shvetsova, S.V.; Zhurishkina, E.V.; Bobrov, K.S.; Ronzhina, N.L.; Lapina, I.M.; Ivanen, D.R.; Gagkaeva, T.Y.; Kulminskaya, A.A. The novel strain *Fusarium proliferatum* LE1 (RCAM02409) produces α-l-fucosidase and arylsulfatase during the growth on fucoidan. *J. Basic Microbiol.* **2015**, *55*, 471–479. [CrossRef] [PubMed]
65. Berlemont, R. Distribution and diversity of enzymes for polysaccharide degradation in fungi. *Sci. Rep.* **2017**, *7*, 222. [CrossRef] [PubMed]
66. Loque, C.P.; Medeiros, A.O.; Pellizzari, F.M.; Oliveira, E.C.; Rosa, C.A.; Rosa, L.H. Fungal community associated with marine macroalgae from Antarctica. *Polar Biol.* **2010**, *33*, 641–648. [CrossRef]
67. Georlette, D.; Blaise, V.; Collins, T.; D'Amico, S.; Gratia, E.; Hoyoux, A.; Marx, J.-C.; Sonan, G.; Feller, G.; Gerday, C. Some like it cold: Biocatalysis at low temperatures. *FEMS Microbiol. Rev.* **2004**, *28*, 25–42. [CrossRef]
68. John, A.B.; Diana, E.V.; Paul, J.H. Effects of temperature on growth rate, cell composition and nitrogen metabolism in the marine diatom *Thalassiosira pseudonana* (*Bacillariophyceae*). *Mar. Ecol. Prog. Ser.* **2002**, *225*, 139–146.
69. Collins, T.; D'Amico, S.; Marx, J.C.; Feller, G.; Gerday, C. Cold-Adapted Enzymes. *Physiol. Biochem. Extrem.* **2007**, *1*, 165–179.
70. Kawahara, H.; Iwanaka, Y.; Higa, S.; Muryoi, N.; Sato, M.; Honda, M.; Omura, H.; Obata, H. A novel, intracellular antifreeze protein in an antarctic bacterium, *Flavobacterium xanthum*. *Cryoletters* **2007**, *28*, 39–49.
71. Krembs, C.; Eicken, H.; Junge, K.; Deming, J. High concentrations of exopolymeric substances in Arctic winter sea ice: Implications for the polar ocean carbon cycle and cryoprotection of diatoms. *Deep. Sea Res. Part I Oceanogr. Res. Pap.* **2002**, *49*, 2163–2181. [CrossRef]
72. Los, D.A.; Murata, N. Membrane fluidity and its roles in the perception of environmental signals. *Biochim. Et Biophys. Acta* **2004**, *1666*, 142–157. [CrossRef] [PubMed]
73. Gerday, C.; Aittaleb, M.; Bentahir, M.; Chessa, J.-P.; Claverie, P.; Collins, T.; D'Amico, S.; Dumont, J.; Garsoux, G.; Georlette, D. Cold-adapted enzymes: From fundamentals to biotechnology. *Trends Biotechnol.* **2000**, *18*, 103–107. [CrossRef]
74. Feller, G.; Gerday, C. Psychrophilic enzymes: Molecular basis of cold adaptation. *Cell. Mol. Life Sci. CMLS* **1997**, *53*, 830–841. [CrossRef] [PubMed]
75. Aghajari, N.; Haser, R.; Feller, G.; Gerday, C. Crystallization and preliminary X-ray diffraction studies of α-amylase from the antarctic psychrophile *Alteromonas haloplanctis* A23. *Protein Sci.* **1996**, *5*, 2128–2129. [CrossRef]
76. Truongvan, N.; Jang, S.-H.; Lee, C. Flexibility and stability trade-off in active site of cold-adapted *Pseudomonas mandelii* esterase EstK. *Biochemistry* **2016**, *55*, 3542–3549. [CrossRef]

77. Ramli, A.N.M.; Mahadi, N.M.; Shamsir, M.S.; Rabu, A.; Joyce-Tan, K.H.; Murad, A.M.A.; Illias, R. Structural prediction of a novel chitinase from the psychrophilic *Glaciozyma antarctica* PI12 and an analysis of its structural properties and function. *J. Comput. Aided Mol. Des.* **2012**, *26*, 947–961. [CrossRef]
78. Baeza, M.; Zúñiga, S.; Peragallo, V.; Barahona, S.; Alcaino, J.; Cifuentes, V. Identification of stress-related genes and a comparative analysis of the amino acid compositions of translated coding sequences based on draft genome sequences of Antarctic yeasts. *Front. Microbiol.* **2021**, *12*, 133. [CrossRef]
79. DasSarma, S.; Capes, M.D.; Karan, R.; DasSarma, P. Amino acid substitutions in cold-adapted proteins from Halorubrum lacusprofundi, an extremely halophilic microbe from Antarctica. *PLoS ONE* **2013**, *8*, e58587. [CrossRef]
80. Michetti, D.; Brandsdal, B.O.; Bon, D.; Isaksen, G.V.; Tiberti, M.; Papaleo, E. A comparative study of cold-and warm-adapted Endonucleases A using sequence analyses and molecular dynamics simulations. *PLoS ONE* **2017**, *12*, e0169586. [CrossRef]
81. Duarte, A.W.F.; Barato, M.B.; Nobre, F.S.; Polezel, D.A.; de Oliveira, T.B.; dos Santos, J.A.; Rodrigues, A.; Sette, L.D. Production of cold-adapted enzymes by filamentous fungi from King George Island, Antarctica. *Polar Biol.* **2018**, *41*, 2511–2521. [CrossRef]
82. Baharudin, I.; Zaki, N.; Bakar, F.A.; Mahadi, N.; NAJMUDIN, N.; Illias, R.; Murad, A. Comparison of RNA extraction methods for transcript analysis from the psychrophilic yeast, Glaciozyma antarctica. *Malays. Appl. Biol.* **2014**, *43*, 71–79.
83. Fell, J.W.; Pitt, J.I. Taxonomy of the yeast genus Metschnikowia: A correction and a new variety. *J. Bacteriol.* **1969**, *98*, 853–854. [CrossRef] [PubMed]
84. Van Uden, N.; Fell, J. Marine yeasts. *Adv. Microbiol. Sea.* **1968**, *1*, 167–201.
85. Hoondee, P.; Wattanagonniyom, T.; Weeraphan, T.; Tanasupawat, S.; Savarajara, A. Occurrence of oleaginous yeast from mangrove forest in Thailand. *World J. Microbiol. Biotechnol.* **2019**, *35*, 1–17. [CrossRef]
86. Gerday, C.; Aittaleb, M.; Arpigny, J.L.; Baise, E.; Chessa, J.-P.; Garsoux, G.; Petrescu, I.; Feller, G. Psychrophilic enzymes: A thermodynamic challenge. *Biochim. Et Biophys. Acta* **1997**, *1342*, 119–131. [CrossRef]
87. Dugdale, R.C.; Wilkerson, F.P. Low specific nitrate uptake rate: A common feature of high-nutrient, low-chlorophyll marine ecosystems. *Limnol. Oceanogr.* **1991**, *36*, 1678–1688. [CrossRef]
88. Plíhal, O.; Sklenář, J.; Hofbauerová, K.; Novák, P.; Man, P.; Pompach, P.; Kavan, D.; Ryšlavá, H.; Weignerová, L.; Charvátová-Pišvejcová, A. Large propeptides of fungal β-N-acetylhexosaminidases are novel enzyme regulators that must be intracellularly processed to control activity, dimerization, and secretion into the extracellular environment. *Biochemistry* **2007**, *46*, 2719–2734. [CrossRef]
89. Reyes, F.; Calatayud, J.; Vazquez, C.; Martínez, M.J. β-N-Acetylglucosaminidase from *Aspergillus nidulans* which degrades chitin oligomers during autolysis. *FEMS Microbiol. Lett.* **1989**, *65*, 83–87. [CrossRef]
90. Santiago, M.; Ramírez-Sarmiento, C.A.; Zamora, R.A.; Parra, L.P. Discovery, molecular mechanisms, and industrial applications of cold-active enzymes. *Front. Microbiol.* **2016**, *7*, 1408. [CrossRef]
91. Ito, E.; Ikemoto, Y.; Yoshioka, T. Thermodynamic implications of high Q 10 of thermo-TRP channels in living cells. *Biophysics* **2015**, *11*, 33–38. [CrossRef] [PubMed]

Journal of *Fungi*

Article

Diversity and N_2O Production Potential of Fungi in an Oceanic Oxygen Minimum Zone

Xuefeng Peng [1,2,*] and David L. Valentine [1,3]

1 Marine Science Institute, University of California, Santa Barbara, CA 93106, USA; valentine@ucsb.edu
2 School of Earth, Ocean and Environment, University of South Carolina, Columbia, SC 29208, USA
3 Department of Earth Science, University of California, Santa Barbara, CA 93106, USA
* Correspondence: xpeng@seoe.sc.edu

Abstract: Fungi in terrestrial environments are known to play a key role in carbon and nitrogen biogeochemistry and exhibit high diversity. In contrast, the diversity and function of fungi in the ocean has remained underexplored and largely neglected. In the eastern tropical North Pacific oxygen minimum zone, we examined the fungal diversity by sequencing the internal transcribed spacer region 2 (ITS2) and mining a metagenome dataset collected from the same region. Additionally, we coupled ^{15}N-tracer experiments with a selective inhibition method to determine the potential contribution of marine fungi to nitrous oxide (N_2O) production. Fungal communities evaluated by ITS2 sequencing were dominated by the phyla *Basidiomycota* and *Ascomycota* at most depths. However, the metagenome dataset showed that about one third of the fungal community belong to early-diverging phyla. Fungal N_2O production rates peaked at the oxic–anoxic interface of the water column, and when integrated from the oxycline to the top of the anoxic depths, fungi accounted for 18–22% of total N_2O production. Our findings highlight the limitation of ITS-based methods typically used to investigate terrestrial fungal diversity and indicate that fungi may play an active role in marine nitrogen cycling.

Keywords: marine fungi; oxygen minimum zone; nitrous oxide; diversity; ^{15}N tracer; size-fractioned; eastern tropical North Pacific; metagenome

Citation: Peng, X.; Valentine, D.L. Diversity and N_2O Production Potential of Fungi in an Oceanic Oxygen Minimum Zone. *J. Fungi* **2021**, *7*, 218. https://doi.org/10.3390/jof7030218

Academic Editor: Federico Baltar

Received: 2 February 2021
Accepted: 15 March 2021
Published: 17 March 2021

1. Introduction

Oceanic oxygen minimum zones (OMZs) are characterized by a sharp oxycline and redox gradient in the water column [1]. As a result, OMZs support diverse microbial communities that directly impact the global biogeochemical cycling of nitrogen, carbon, sulfur, and trace metals [2–6]. As in many other types of marine environments, bacteria and archaea have been the focus of microbial ecology research in OMZs, whereas microbial eukaryotes, in particular fungi, have received much less attention [7,8].

An early cultivation-based survey of marine fungi in the Indian Ocean that included the Arabian Sea OMZ found *Rhodotorula rubra* and *Candida atmosphaerica* to be cosmopolitan, and the yeast population densities ranged from 0–513 cells per liter of seawater [9]. In the eastern tropical South Pacific (ETSP) OMZ off the coast of Chile, high summertime fungal biomass in the water column have been reported [10], including diatom parasites from the phylum Chytridiomycota [11]. In the third and the largest open ocean OMZ, the eastern tropical North Pacific (ETNP), protists diversity has been investigated by sequencing the V4 region of 18S small subunit rRNA genes [7], but little is known about the diversity and function of fungi in this environment.

Fungi in the water column are generally thought to contribute to organic matter recycling, particularly in particle-associated environments [12–14]. High hydrolytic activity on proteinaceous substrates in large size fractions (>25 μm and >90 μm) have been reported in the water column of the ETSP and attributed to fungi given low bacterial biomass in those size fractions [15]. However, it remains unclear if fungi in the water column of OMZs play a role in nitrogen cycling. Discovered in the early 1990s, fungal denitrification is known as a process that reduces nitrate or nitrite with nitrous oxide (N_2O) as the end-product [16,17] This adds to the multiple other pathways and processes (ammonia oxidation, bacterial denitrification, and chemodenitrification) that can produce N_2O [18], a potent greenhouse gas and ozone-depleting agent [19]. Many fungal strains have been found to have the ability to produce N_2O [20], including an *Aspergillus terreus* strain isolated from the Arabian Sea OMZ [21]. In marine environments, fungal denitrification with N_2O as the end-product has been reported from coastal marine sediment in India and Germany [22,23], but its potential contribution in the water column remains unclear.

We investigated the fungal community composition in the eastern tropical North Pacific oxygen minimum zone by sequencing the internal transcribed spacer region 2 (ITS2) and classifying shotgun metagenome reads. To estimate the fungal contribution to N_2O production in the water column, we used a selective inhibition method combined with ^{15}N-labeled tracer incubation experiments. Fungal communities evaluated by ITS2 sequencing were dominated by the phyla *Basidiomycota* and *Ascomycota* at most depths. The metagenome dataset showed that early-diverging fungi accounted for about one third of the fungal community, and the subsurface peaks of fungal abundance coincided with both cyanobacterial abundance and eukaryotic algal abundance. Incubation experiments suggest a possible role of fungi in N_2O production in the water column of the ETNP OMZ.

2. Materials and Methods

2.1. Site Description and Seawater Filtration

In March 2018, aboard the R/V Sally Ride in the eastern tropical North Pacific oxygen minimum zone, two stations were visited to study fungal diversity and the potential fungal contribution to N_2O production (Figure 1a). Dissolved oxygen concentration was determined using the SBE 43 dissolved oxygen sensor attached to the conductivity, temperature, and depth (CTD) rosette. Seawater was collected at multiple depths spanning from the oxycline to the anoxic depths (Figure 1b) using 30 L Niskin bottles.

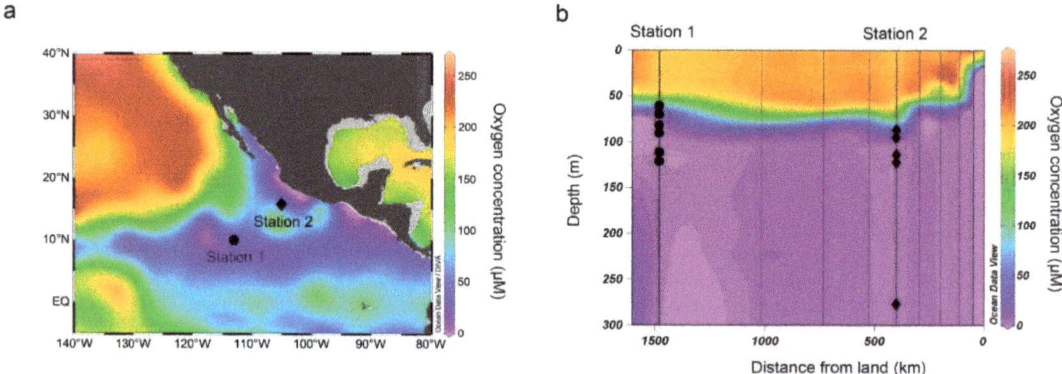

Figure 1. (**a**) Sampling locations in the eastern tropical North Pacific (ETNP) oxygen minimum zone. Color contour shows oxygen concentrations at 100 m depth from World Ocean Atlas 2013 (March average from 1955–2012) [24]. (**b**) Depths sampled for nitrous oxide (N_2O) production experiments are marked by filled symbols. Color contour shows oxygen concentrations measured during this cruise using a Seabird SBE 43 dissolved oxygen sensor.

To collect particulate material at different size fractions, seawater was sequentially filtered through a 47 mm Whatman Grade 541 acid-hardened cellulose filter paper (22 µm nominal particle retention rating, GE Healthcare 1541–047, Marlborough, MA, USA), a 47 mm polycarbonate filter (2.0 µm nominal pore size, Millipore Isopore TTTP-04700, Burlington, MA, USA), and a Sterivex filter (0.22 µm nominal pore size, Millipore SVGP01050, Burlington, MA, USA), using a peristaltic pump filtration at a flow rate < 50 mL/min. For each sample set, 23 to 55 L of seawater was filtered (Table S1). Each 47 mm filter was stored in a 47 mm petri dish and flash-frozen in liquid nitrogen before storage at −80 °C.

2.2. DNA Extraction

In the laboratory, DNA was extracted using the DNeasy Plant Mini Kit (QIAGEN Cat No. 69104, Germantown, MD, USA) following the DNeasy Plant Handbook [25], except that the cell disruption step was customized for our samples. Filter paper from the Sterivex filters were extracted from the plastic case using a snap-blade knife. All filters were first cut into 2 by 2 mm pieces using sterilized scissors and transferred into 2 mL screw cap tubes containing 1 mL of 0.5 mm zirconia/silica beads (Biospec products #11079105z, Bartlesville, OK, USA), 600 mL of buffer AP1, and 6 μL of RNase A. Bead beating of the samples was performed for 90 s using a Biospec Mini-BeadBeater-16 and was followed by incubation at 65 °C for 10 min. After centrifugation at 20,000× *g* for 5 min, the supernatant in each sample tube was transferred to a fresh 2 mL microcentrifuge tube and neutralized with 195 μL of Buffer P3. The remaining DNA extraction steps followed the DNeasy Plant Handbook without modifications. An extraction blank was included for each batch of extraction procedure. DNA yield was quantified using a Qubit fluorometer (ThermoFisher Scientific, Waltham, MA, USA) following the manufacturer's instructions. All extracted DNA samples were stored at −80 °C until amplicon library construction.

2.3. Sequencing and Analysis of the ITS2 Region

The second region of the internal transcribed spacer (ITS2) with flanking regions in the 5.8S and 28S ribosomal RNA was targeted for amplicon library preparation following the Illumina 16S Metagenomic Sequencing Library Preparation [26] with the following modifications. The amplicon PCR was performed using primers ITS3tagmix [27] and ITS4tag001 [28] (Table S2) with Phusion® high-fidelity DNA polymerase (New England BioLabs, M0530, Ipswich, MA, USA). The thermal cycle started with 30 s at 98 °C, followed by 30 cycles of 15 s at 98 °C, 30 s at 55 °C, and 45 s at 72 °C. The final elongation at 72 °C was 10 min long. The quantity and quality of final PCR products were determined using a Qubit (ThermoFisher Scientific, Waltham, MA, USA) and TapeStation 2200 (Agilent, Santa Clara, CA, USA), respectively. Identical quantities of each sample were pooled, and the products were sequenced on an Illumina MiSeq 2 × 250 PE platform in the Biological Nanostructures Lab at UCSB.

Raw sequence reads were first trimmed using ITSxpress to contain only the ITS2 region [29]. Trimmed reads were merged, quality filtered, dereplicated, and denoised following the USEARCH pipeline [30] to generate amplicon sequence variants (ASVs). Taxonomic assignment was performed using a combination of Naïve Bayes classifier implemented in QIIME2 [31] and BLASTn [32] against full UNITE + INSD dataset v02.02.2019 [33] and curated manually. To determine putative fungal denitrifiers in our samples, we performed a closed reference OTU picking [34] of our ASVs and ITS2 sequences from fungi tested for N_2O production [20] using UCLUST [35] implemented in QIIME v1.9.1 [36] against the UNITE database [33]. Raw reads generated in this study are available at the National Center for Biotechnology Information (NCBI) under BioProject PRJNA623945.

2.4. Analysis of Fungal Diversity and Function from Metagenomes

As an independent approach to evaluate the fungal diversity in the eastern tropical North Pacific oxygen minimum zone, we investigated a metagenome dataset sampled at a nearby station in March 2012 when the hydrographic conditions were highly similar (Figure S1) [37]. Raw reads were first filtered using the tool BBDuk Version 38.73 [38] with the options "ktrim=r ordered minlen=51 minlenfraction=0.33 mink=11 tbo tpe rcomp=f k=23 ftm=5". Adapters were trimmed from the BBDuk-filtered reads using the tool Trimmomatic Version 0.39 [39] with the options "ILLUMINACLIP:$adapters:2:30:10 LEADING:3 TRAILING:3 SLIDINGWINDOW:4:15 MINLEN:100". Reads that passed both quality filtering and adapter trimming were queried against the NCBI nr database using DIA-

MOND [40] with an e-value threshold of 1×10^{-5} and the options "–sensitive –min-orf 20". The resultant NCBI taxonomy ID was used to assign taxonomy to each read. In search for the presence of the gene diagnostic for fungal N_2O production, the cytochrome P450 nitric oxide reductase (*P450nor*) [41], we searched the metagenome assemblies under the same BioProject (PRJNA350692) for putative *P450nor* using HMMER v3.2.1 [42] against a *P450nor* profile, which includes both eukaryotic and prokaryotic cytochrome P450 genes [43]. All putative *P450nor* hits were checked against the NCBI nr database using blastp [32] to determine taxonomy.

2.5. Measurements of Potential N_2O Production Rates

To measure potential rates of N_2O production, parallel incubation experiments were performed by adding 0.1 mL of 5 mM 99% pure ^{15}N-labeled potassium nitrate or 0.1 mL of 0.8 mM 99% pure ^{15}N-labeled ammonium chloride (Cambridge Isotopes, Cambridge, MA, USA) to 120 mL glass serum bottles containing freshly collected seawater. To minimize the introduction of atmospheric oxygen, each bottle was overflown three times its volume with water directly from Niskin bottles before filling and crimp-sealing it with grey butyl rubber stopper and aluminum caps. Headspace was created in each bottle by replacing 2 mL of seawater with ultra-high pure helium (Airgas HE UHP300, Radnor, PA, USA). End points were taken by adding 1 mL of 50% (w/v) zinc chloride to separate parallel incubations approximately 0 and 24 h after incubations began at Station 1 and 0, 12, and 24 h after incubations began at Station 2. The concentration of ammonium in seawater was measured onboard according to the fluorometric method of Holmes et al. [44], with a detection limit of 15 nmol L^{-1}. Nitrate concentrations were assayed using a Lachat Flow Injection Analyzer at the Analytical lab at the Marine Science Institute, University of California, Santa Barbara following standard analytical methods [45]. The detection limit of nitrate was 0.2 μmol L^{-1}.

The potential contribution of fungal N_2O production to total N_2O production was determined by incubations with ^{15}N-labeled nitrate ($^{15}NO_3^-$) and chloramphenicol (87.7 mg L^{-1} final concentration), which was applied to each incubation one hour before the addition of ^{15}N tracers to inhibit all prokaryotic activities. All solutions added to the incubation bottles were purged by ultra-high pure helium for one hour at a flow rate of 40 mL min^{-1}. The difference between fungal N_2O production and total N_2O production from incubations with $^{15}NO_3^-$ is attributed to bacterial denitrification (Figure S2). N_2O production rates measured in incubations with ^{15}N-labeled ammonium ($^{15}NH_4^+$) are attributed to archaea and/or bacterial nitrification.

The quantity and isotopic composition of dissolved N_2O was determined using a Delta XP isotope ratio mass spectrometer coupled to a purge-and-trap front end. The detection limit was 1.0 nmol N, and the precision for $\delta^{15}N$ was 2.0‰ ($n \geq 3$). The rate of N_2O production (R_{N2O}) was calculated from the equation [46]:

$$R_{N_2O} = \frac{d^{15}N_2O/dt}{f^{15} \times V}$$

where $d^{15}N_2O/dt$ is the rate of $^{15}N_2O$ production determined from linear regression of the amount of $^{15}N_2O$ against time, f^{15} is the fraction of ^{15}N labeled substrate, and V is the volume of the incubation. The amount of $^{15}N_2O$ at each time point is calculated from the equation:

$$^{15}N_2O = N_2O \times \frac{\left(\frac{\delta^{15}N\_N_2O}{1000} + 1\right) \times R_{ref}}{1 + \left(\frac{\delta^{15}N\_N_2O}{1000} + 1\right) \times R_{ref}}$$

where N_2O is the amount of nitrous oxide determined from in-house N_2O concentration standards (Figure S3), $\delta^{15}N\_N_2O$ is the bulk isotopic composition of sample N_2O, and R_{ref} is isotopic composition of reference gas. The linearity effect for the range of N_2O measured was negligible compared to the enriched $\delta^{15}N\_N_2O$ measured from our incubation samples (Figure S4).

3. Results

3.1. Fungal Diversity Assessed by Sequencing the ITS2 Region

Assessment of the fungal community in the eastern tropical North Pacific using the internal transcribed spacer region 2 (ITS2) revealed that taxa from the phyla *Basidiomycota* and *Ascomycota* dominated at most depths at both stations (Figure 2). The relative abundance of *Basidiomycota* was higher than *Ascomycota* at nearly all depths (Table S3). The most prevalent and abundant taxon is the basidiomycetous yeast family Sporidiobolaceae, primarily consisting of the genera *Rhodotorula*, *Rhodosporidiobolus*, and *Sporobolomyces* (Table S4). Sporidiobolaceae tend to have a higher relative abundance in the 0.2–2 µm size fraction. On the other hand, *Aureobasidium* (Ascomycota) and Exobasidiomycetes (Basidiomycota, primarily *Meira*) were enriched in the larger size fractions (2–22 and >22 µm). In contrast, the basidiomycetous yeast *Malassezia*, when detected, were enriched only in the 2–22 µm size fraction. At Station 2, most of the fungal community cannot be classified even at the phylum level based on ITS2 sequences, indicating the presence of novel fungal lineages in the oxycline of ETNP oxygen minimum zone.

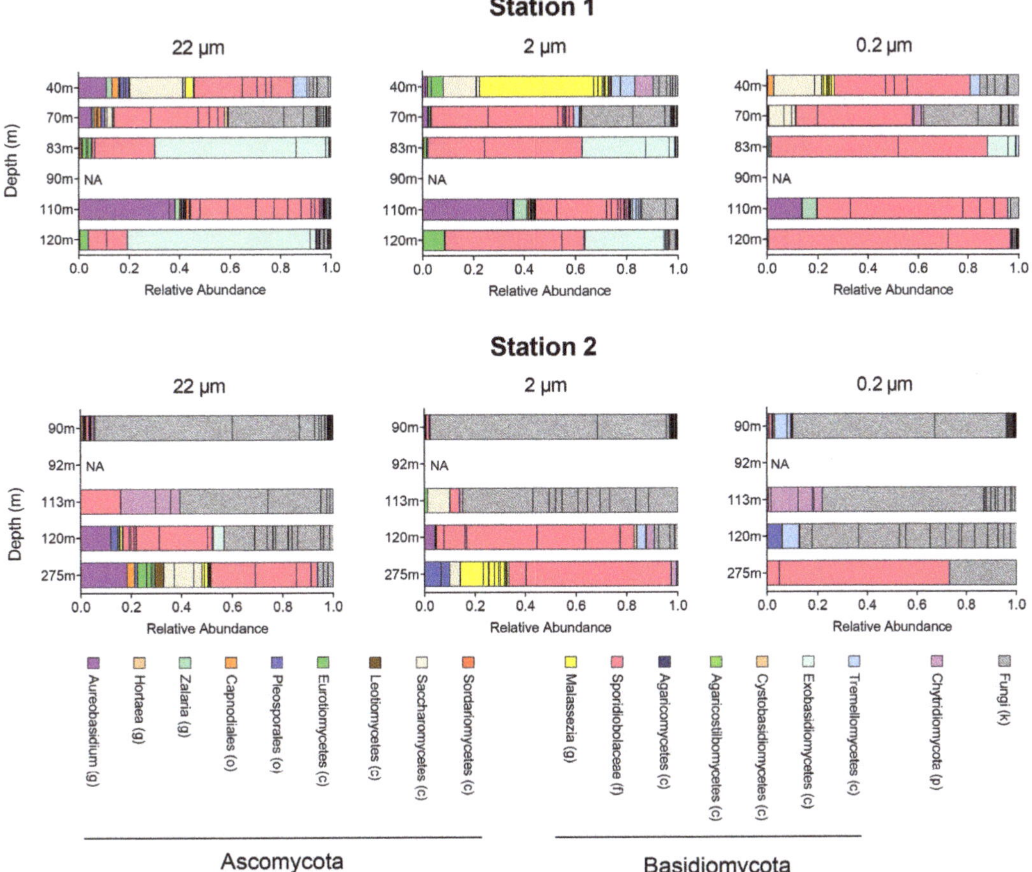

Figure 2. Fungal community composition for three different size fractions including >22 µm (left), 2–22 µm (middle), and 0.2–2 µm (right) at Stations 1 and 2. Each bar represents an amplicon sequence variant (ASV) of the internal transcribed spacer region 2 (ITS2). NA: not available. The level of each taxonomic assignment is indicated in the parenthesis behind the names: g for genus, f for family, o for order, c for class, p for phylum, and k for kingdom.

3.2. Fungal Diversity Assessed from Metagenomes

Fungal community composition assessed by metagenomic reads also showed the dominance of Dikarya fungi, but in contrast to the results from ITS2 sequencing, the relative abundance of Ascomycota was consistently higher than that of Basidiomycota (Figure 3). Surprisingly, over one third of the fungal community belong to early-diverging phyla including Mucoromycota, Zoopagomycota, Chytridiomycota, Blastocladiomycota, Cryptomycota, and Microsporidia. The fungal community composition from 60 to 300 m was uniform, while there was a trend of increasing relative abundance for Ascomycota with depth.

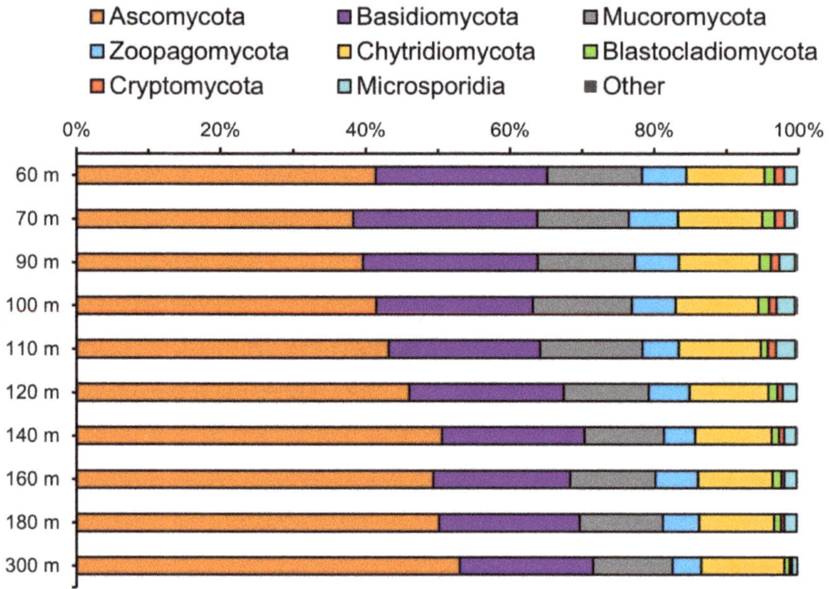

Figure 3. The relative abundance of each fungal phylum identified in the metagenomes [37] collected from a location near Station 2 in this study.

3.3. Relative Abundance of Fungi Compared to Other Taxa

Mining the metagenomes allowed us to estimate the relative abundance of fungi and other taxa as part of the overall microbial community. In each metagenome, 23–57% of the reads had a positive hit against the NCBI nr database with an e-value threshold of 1×10^{-5} (Figure S5a). Fungi accounted for 0.02–0.22% of all classifiable reads, with a subsurface peak at 70 m (Figure 4a). The relative abundance of fungal reads decreased with depth below 70 m but showed a small increase at 140 m, which was the top of the oxygen deficient layer of the water column. Ciliophora, Oomycetes, and Dinophyceae were similar to fungi in both abundance and distribution. The abundance of reads classified as eukaryotic algae also showed a subsurface peak at 70 m (Figure 4b). The abundance of cyanobacteria decreased with depth overall, but there was an increase at 110 m, coinciding with the deep chlorophyll maximum situated immediately above the anoxic depths.

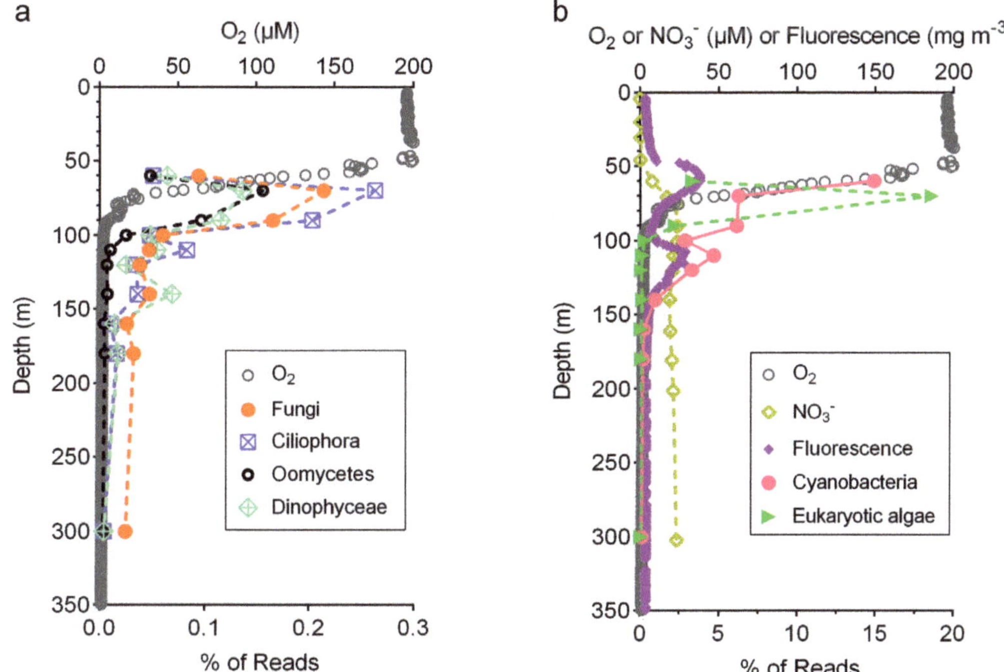

Figure 4. (**a**) Dissolved oxygen concentration (µM) and the relative abundance of reads classified as Fungi, Ciliophora, Oomycetes, and Dinophyceae from the metagenomes [37] collected from a location near Station 2 in this study; (**b**) dissolved oxygen concentration (µM), nitrate (NO_3^-) concentration (µM), fluorescence (mg m^{-3}) [47], and the relative abundance of reads classified as Cyanobacteria and eukaryotic algae. Eukaryotic algae include Haptophyta, Bacillariophyta, Rhodophyta, Cryptophyta, and Viridiplantae.

3.4. Potential Contribution of Fungal N_2O Production

N_2O production rates from incubation experiments with either $^{15}NO_3^-$ or $^{15}NH_4^+$ were detected at multiple oxycline depths of the ETNP oxygen minimum zone (OMZ), and the maximum rate was found at the oxic–anoxic interface (Figure 5). The rates of N_2O production from incubations with $^{15}NO_3^-$ and chloramphenicol were used as an approximation of fungal N_2O production, and they ranged from 0% of total N_2O production from $^{15}NO_3^-$ at 275 m at Station 2 to 56% of total N_2O production from $^{15}NO_3^-$ at 90 m at Station 1. N_2O production from incubations with $^{15}NO_3^-$, both with and without chloramphenicol, was lower at elevated in situ oxygen (O_2) concentration. When integrated from the oxycline (60 m at Station 1 and 90 m at Station 2) to the oxic–anoxic interface, fungal N_2O production approximated by incubations with $^{15}NO_3^-$, and chloramphenicol accounted for 18–22% of total N_2O production (Table S5).

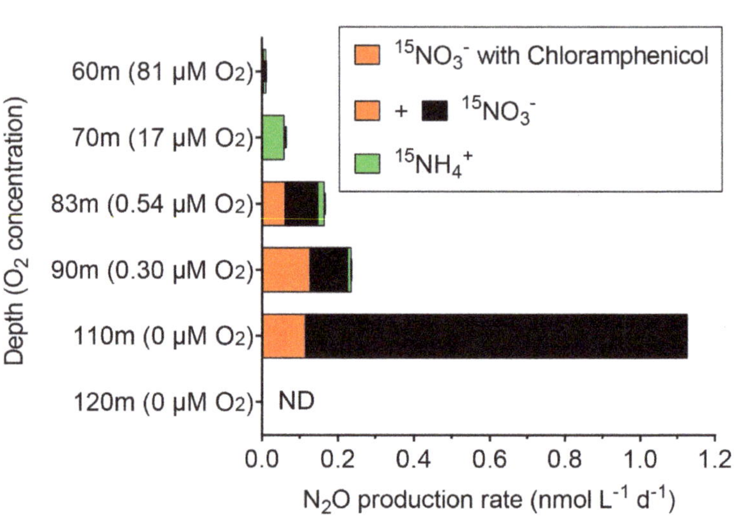

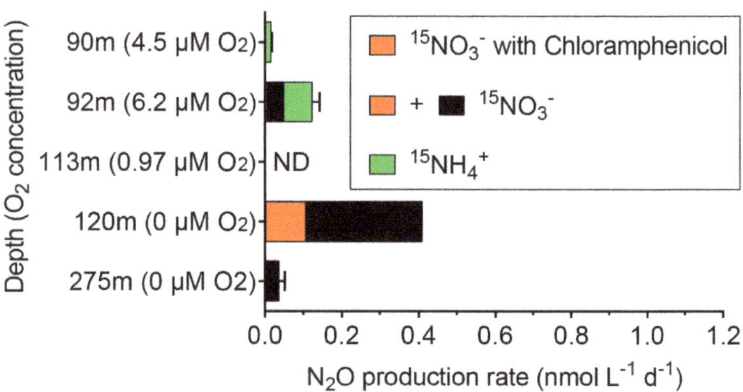

Figure 5. N_2O production rates measured from incubation experiments with $^{15}NO_3^-$ only (orange and black bars combined), with $^{15}NO_3^-$ and chloramphenicol (orange bars), and with $^{15}NH_4^+$ (green bars). Chloramphenicol was intended to inhibit bacterial and archaeal activity.

3.5. Search for Fungal Denitrifiers and Functional Genes

To search for putative fungal denitrifiers in the water column of the eastern tropical North Pacific OMZ, we first analyzed the ITS2 amplicon dataset generated in this study. Of all 237 ASVs, only one (ASV211) was clustered at 97% similarity level with the ITS2 sequences from fungal strains previously shown to produce N_2O (*Chaetomium* sp.) [20]; this ASV was present only in the 0.2–2 μm size fraction at 83 m from Station 1 and at a relative abundance of 0.2% (Table S6). However, one ASV (ASV8) was 96.95% similar to the ITS2 region of the N_2O-producing *Penicillium melinii* [20], which was isolated from seawater [48]. The relative abundance of ASV8 ranged from 0.1 to 8.5% at hypoxic and anoxic depths (Table S6), and it belongs to the same family (Aspergillaceae) as the N_2O-producing fungal strain isolated from the Arabian Sea OMZ [21]. While this putative denitrifier could potentially produce N_2O in the water column of the OMZ, its potential contribution to the total N_2O production is estimated to be less than 0.3% given the low N_2O yield by *P. melinii* in the laboratory (Appendix A). Continuing the search for putative fungal denitrifiers, we analyzed a previously published March 2012 metagenome dataset from a nearby station [37] for the occurrence of *P450nor*. All *P450nor* hits identified by a hidden Markov model (HMM) profile search [43] were prokaryotic genes, most of which from the phylum Actinobacteria (Table S7).

4. Discussion

4.1. Fungal Diversity in Different Size Fractions

In the eastern Tropical North Pacific (ETNP) oxygen minimum zone, the most prevalent and abundant fungal taxa in our ITS2-based survey was the basidiomycetous yeast from the family Sporidiobolaceae, which are usually characterized by the production of carotenoids that color their cells red, pink, or orange and, hence, have the name "red yeasts". The high relative abundance of red yeasts in the 0.2–2 μm size fraction (Figure 2) is consistent with their typically small cell size [49]. A previous study on yeasts in the Indian Ocean, which includes the Arabian Sea oxygen minimum zone in the northern part, has also found red yeasts to be the predominant taxa [9]. On the other hand, *Aureobasidium* (Ascomycota) and Exobasidiomycetes (Basidiomycota, primarily *Meira*) were enriched in the larger size fractions (2–22 μm and >22 μm), consistent with a larger size [50,51], and indicating a likely association with particles. In contrast, the basidiomycetous yeast *Malassezia*, when detected, were enriched only in the 2–22 μm size fraction, suggesting that they were not associated with particles larger than 22 μm.

There is increasing recognition of marine fungi as a key component of the marine microbiome and biogeochemical cycles [12,13,15,52,53], but our knowledge about their diversity and function is far less compared to other microbial eukaryotes [54]. This is particularly the case in the water column of the open ocean, where metabarcoding surveys of fungal diversity are often unable to classify most of the fungal community [55,56]. In one of the two stations sampled in this study, we also could not resolve the taxonomy for most of the fungal community (Figure 2), highlighting the limitation of an ITS-based approach to study fungal diversity. Because the public databases for ITS sequences are primarily based on studies of terrestrial environments or fungal strains, this suggests that the open ocean water column harbors previously undiscovered lineages of fungi.

4.2. Ecology of Marine Fungi in the Oxygen Minimum Zone

By classifying metagenome reads, we showed that about a third of the fungal community were from early-diverging lineages [57], including Mucoromycota, Zoopagomycota, Chytridiomycota, Blastocladiomycota, Cryptomycota, and Microsporidia (Figure 3). Such diversity was previously undiscovered due to both the difficulty in cultivation and the consequent lack of representation in ITS databases. It is unclear what ecological roles these early-diverging fungi play, except for Chytridiomycota, which are typically associated with phytoplankton such as diatoms [58]. Nevertheless, the depth profile of the relative abundance of fungi in relation to other microbial taxa provides a hint (Figure 4). The cooccurrence of subsurface peaks (at 70 m) of the relative abundance of fungi and eukaryotic algae suggests that fungi are directly associated with eukaryotic algae, perhaps as parasites. A secondary peak of the relative abundance of fungi (at 140 m) was observed below that of the deep chlorophyll maximum (at 110 m) typically found at the oxic–anoxic interface of oxygen deficient waters [59]. Hence, the fungal communities at the secondary peak (at 140 m) are likely predominantly saprotrophic, feeding on the particles formed at the base of the mixed layer.

Surprisingly, we did not observe a pronounced effect of oxygen concentration on the fungal community composition, evaluated by either ITS2 amplicon sequencing or metagenomes. In contrast, the protist community in the anoxic depth of the ETNP and eastern tropical South Pacific oxygen minimum zone was enriched in Syndiniales, euglenozoan flagellates, and acantharean radiolarians [7,8]. This may be a result of the versatile respiratory/fermentative metabolisms fungi possess, but it could also be due to the inability of the methods used in this study to detect novel fungal lineages from the oxygen deficient waters. The metagenome-based approach avoids the typical biases associated with amplicon sequencing such as primer bias, but it is limited by the sequences available in the chosen database (NCBI nt in this study).

4.3. Fungal N_2O Production in the Oxygen Minimum Zone

In order to distinguish the N_2O production by fungi from bacteria and archaea, we combined ^{15}N tracer incubation experiments with selective inhibition using the antibiotic chloramphenicol (Figure S2). Chloramphenicol is a broad-spectrum antibiotic that has been used to isolate anaerobic gut fungi from the rumen microbiome [60], and its final concentration used in our incubations was scaled by the typical bacterial cell density in the rumen vs. seawater. Nonetheless, we did not make direct measurements of the specificity and effectiveness of chloramphenicol on inhibiting bacterial and archaeal activities in seawater. Therefore, it is possible that the N_2O production rates we measured from incubations with $^{15}NO_3^-$ and chloramphenicol include N_2O from partially inhibited bacterial denitrification. Consequently, we conservatively interpret those rates as an upper limit of potential fungal N_2O production.

When integrated from the oxycline to the oxic–anoxic interface, fungal N_2O production accounted for 18–22% of total N_2O production (Table S5). While we interpret this as the maximum possible contribution of fungi to N_2O production, we suggest that fungal denitrification could be an important pathway for N_2O production in oceanic oxygen minimum zones. Denitrification, the sequential reduction in NO_3^- to N_2, is known to be inhibited by trace amounts of O_2 [61]. In OMZs, it was demonstrated that 297 nM of O_2 repressed 50% of total N_2O production at the oxic–anoxic interface [62]. In this study, N_2O produced in incubation with $^{15}NO_3^-$ (primarily via denitrification) was inhibited by increasing levels of in situ O_2 concentration. However, the inhibitory effect of O_2 appeared to be less pronounced on fungal denitrification than on bacterial denitrification at low levels, revealing a potential niche ($0.0 < O_2 < 0.93$ µM) for fungi capable of denitrification. Since this potential niche is shallower than the oxygen deficient waters, the potential fungal N_2O production can have a higher chance of reaching the ocean–atmosphere interface than N_2O produced at deeper depths. Alternatively, N_2O produced by fungi can be taken up by bacteria with the atypical ("Clade II") nitrous oxide reductase [63] (primarily Flavobacteria and Chloroflexi), which was shown to have a peak in relative abundance at depths immediately above the oxic–anoxic interface [37].

4.4. Molecular Evidence for Fungal N_2O Production

The search for putative fungal denitrifiers using ITS2 sequence identity suggests an extremely low abundance of known N_2O-producing fungi in the water column of oxygen minimum zones. However, it should be noted that the collection of N_2O-producing fungi used to identify these putative denitrifiers consists of soil fungi exclusively [20], so fungal lineages capable of N_2O production from the open ocean were likely excluded. Therefore, there may be other N_2O-producing fungi from the ETNP OMZ unidentified by this approach.

The absence of fungal *P450nor* in the metagenomic dataset we queried may be attributed to insufficient sequencing depth, given the low percentage (0.02–0.22%) of fungal reads classified by DIAMOND against the NCBI nr database [40] (Figure 4a). Even under the most simplistic assumption that there was only one fungal species present and it possessed *P450nor* in its genome, the sequencing depth in most samples was insufficient to recover just one copy of fungal *P450nor* (Table S8). Additionally, most DNA extraction protocols applied in published metagenome studies (including [37]) are not customized for disrupting chitinous cell walls. This likely resulted in under-sampling DNA from fungi, of which the biomass is low in the ocean water column (0.01–0.12 μg of carbon L^{-1}) [10], especially compared to bacteria (estimated average of 10.5 μg of carbon L^{-1}) [64]. Finally, it should be noted that the detection of fungal *P450nor* genes in metagenomes does not necessarily imply fungal denitrification, as *P450nor* in certain fungal genomes appear to be involved in secondary metabolisms instead [43].

5. Conclusions

Our findings highlight the previously unrecognized fungal diversity in the eastern tropical North Pacific oxygen minimum zone, particularly from the early-diverging taxa as revealed by analysis of shotgun metagenomes. The depth distribution pattern of fungi in relation to cyanobacteria and eukaryotic algae suggest direct association in the mixed layer of the water column and indirect feeding below the deep chlorophyll maximum. Given the limitations of the selective inhibition method using chloramphenicol, we estimate that fungi contribute no more than 18–22% to total N_2O production from the oxycline to the oxic–anoxic interface. It remains challenging for omics-based approaches to provide molecular evidence for fungal denitrification, partially because current databases and recent studies have primarily focused on terrestrial fungi. Overcoming the shortfalls of existing methods necessitates approaches that can target fungal diversity and function, such as the use of RNA-seq combined with eukaryotic messenger RNA enrichment or the combination of fluorescence-activated cell sorting and (meta)genome sequencing. These new methods can increase the likelihood of capturing genetic evidence for fungal activities and functional diversity.

Supplementary Materials: The following are available online at https://www.mdpi.com/2309-608X/7/3/218/s1, Table S1. Volume of seawater filtered for particulate matter collection. Table S2. Primers designed for PCR. The parts in bold font are standard Illumina adapters. The five ITS3tag primers were combined in equimolar as the forward primer. Table S3. Fungal community composition at the phylum level assessed by the internal transcribed spacer region 2 (ITS2) sequencing. Table S4. (Separate spreadsheet) Taxonomy and read count of each amplicon sequence variant (ASV) in each sample. Table S5. N_2O production rates integrated from the oxycline to the oxic–anoxic interface depths and the percentage of potential fungal contribution. Table S6. Relative abundance of putative denitrifiers as part of the total fungal community assessed by the internal transcribed region 2 (ITS2). Non-zero values are highlighted in bold. Table S7. (Separate spreadsheet) Scientific names and kingdoms of the top blastp result against NCBI nr database for genes identified as putative *P450nor* by a hidden Markov model search. All genes were from previously published metagenome assemblies sampled at a station near Station 2 in this study (see Figure S1). Table S8. Comparison of actual number of reads and the estimated minimum number of reads required to recover one copy of *P450nor* in each sample from the Fuchsman et al. (2017) study. The estimated minimum number of reads required were calculated assuming an average fungal genome size of 40,973,539 bp and the length of the *P450nor* gene as 1056 bp. Figure S1: (a) Sampling location from March 2012 ("Fuchsman17") where the metagenome samples investigated in this study were collected, and sampling locations in this study ("Station 1" and "Station 2"). Color contour shows oxygen concentrations at 100 m depth from World Ocean Atlas 2013 (March average from 1955–2012). (b) Comparison of potential density (σ_θ), oxygen concentration (O_2), and fluorescence measured by Seabird profiling sensors from this study (2018) and from the "Fuchsman17" study (2012). Figure S2. Incubation scheme using a selective inhibition method combined with ^{15}N tracer technique to determine the fungal contribution to N_2O production. In each incubation, the final concentration of $^{15}NO_3^-$ was 3 μM and the final concentration of $^{15}NH_4^+$ was 0.5 μM. Chloramphenicol was added to a final concentration of 87.7 mg L^{-1}. Figure S3. Calibration curve used to calculate the amount of N_2O present in each sample based on the total areas under m/z of 44, 45, and 46 ("Area All"). Figure S4. The amount (nmol) and bulk $\delta^{15}N$ (‰) of N_2O measured from incubation samples in which N_2O production rates were non-zero (black empty circles) and of N_2O concentration standards (blue filled diamonds). Error bars represent standard deviations ($n > 5$). Figure S5. (a) The fraction of metagenome reads with a DIAMOND hit. (b) The percentage of metagenome reads that were classified as archaea, bacteria, viruses, and algae. The samples were collected during the same month in 2012 at a station close to the station 2 in this study (Figure S1). Read classification was performed using DIAMOND against the NCBI nr database with an e-value threshold of 1×10^{-5}.

Author Contributions: Conceptualization, X.P. and D.L.V.; methodology, X.P.; formal analysis, X.P.; resources, X.P. and D.L.V.; data curation, X.P.; writing—original draft preparation, X.P.; writing—review and editing, X.P. and D.L.V.; visualization, X.P.; funding acquisition, X.P. and D.L.V. All authors have read and agreed to the published version of the manuscript.

Funding: This research was funded by the Simons Foundation Postdoctoral Fellowship in Marine Microbial Ecology (No. 547606), NSF Grants OCE-1635562, OCE-1756947, OCE-1657663, and the C-BRIDGES program. Supercomputing resources were provided by the Center for Scientific Computing at UCSB, which is supported by NSF Grant CNS-0960316.

Institutional Review Board Statement: Not applicable.

Informed Consent Statement: Not applicable.

Data Availability Statement: Raw reads generated in this study are available at NCBI under BioProject PRJNA623945.

Acknowledgments: We are indebted to members of Bess Ward's and Karen Casciotti's Labs, the crew of R/V Sally Ride, and Frank Kinnaman for general assistance. We thank Alyson Santoro's lab and the Analytical Lab at the Marine Science Institute in UCSB for advice on mass spectrometry method development.

Conflicts of Interest: The authors declare no conflict of interest.

Appendix A

Extrapolation of N_2O Production by a Putative Fungal Denitrifier

ASV8 was 96.95% similar to the ITS2 region of the N_2O-producing *Penicillium melinii* [20], which was isolated from seawater [48]. The relative abundance of ASV8 ranges from 0.1 to 8.5% at hypoxic and anoxic depths (Table S4), and it belongs to the same family (Aspergillaceae) as the N_2O-producing fungal strain isolated from the Arabian Sea OMZ [21]. We estimated the maximum potential contribution of ASV8 to N_2O production by scaling its mass-dependent N_2O production rate (assumed to be identical to *Penicillium melinii* [20]) by its estimated mass-based abundance, assuming the maximum observed (ITS-based) fractional abundance of 8.5% and fungal carbon content from a similar location. The following equation was used:

$$R_{N_2O} = \frac{P_{pm} \times m_f \times f_{ASV8}}{T}$$

where P_{pm} is the measured N_2O production rate for *Penicillium melinii* [20], m_f is the estimated fungal carbon mass (0.12 $\mu g\ L^{-1}$) as measured for the eastern tropical South Pacific oxygen minimum zone [10], f_{ASV8} is the largest fractional contribution of ASV8 to our data set (8.5%), and R_{N_2O} is the estimated maximum rate of N_2O production by ASV8. To make R_{N_2O} an upper bound, we used the largest value of fungal carbon content and an N_2O production rate by *P. melinii* (2 mg N_2O-N g^{-1} $week^{-1}$) reported previously [20]. By this approach, the R_{N_2O} extrapolated for this putative fungal denitrifier is 2.1×10^{-4} nmol L^{-1} d^{-1}. This is accounts for at most 0.3% of any fungal N_2O production rate measured in this study.

References

1. Paulmier, A.; Ruiz-Pino, D. Oxygen minimum zones (OMZs) in the modern ocean. *Prog. Oceanogr.* **2009**, *80*, 113–128. [CrossRef]
2. Thamdrup, B.; Steinsdóttir, H.G.R.; Bertagnolli, A.D.; Padilla, C.C.; Patin, N.V.; Garcia-Robledo, E.; Bristow, L.A.; Stewart, F.J. Anaerobic methane oxidation is an important sink for methane in the ocean's largest oxygen minimum zone. *Limnol. Oceanogr.* **2019**, *64*, 2569–2585. [CrossRef]
3. Callbeck, C.M.; Lavik, G.; Ferdelman, T.G.; Fuchs, B.; Gruber-Vodicka, H.R.; Hach, P.F.; Littmann, S.; Schoffelen, N.J.; Kalvelage, T.; Thomsen, S.; et al. Oxygen minimum zone cryptic sulfur cycling sustained by offshore transport of key sulfur oxidizing bacteria. *Nat. Commun.* **2018**, *9*, 1729. [CrossRef] [PubMed]

4. Ulloa, O.; Canfield, D.E.; DeLong, E.F.; Letelier, R.M.; Stewart, F.J. Microbial oceanography of anoxic oxygen minimum zones. *Proc. Natl. Acad. Sci. USA* **2012**, *109*, 15996–16003. [CrossRef]
5. Beman, J.M.; Vargas, S.M.; Vazquez, S.; Wilson, J.M.; Yu, A.; Cairo, A.; Perez-Coronel, E. Biogeochemistry and hydrography shape microbial community assembly and activity in the eastern tropical North Pacific Ocean oxygen minimum zone. *Environ. Microbiol.* **2020**. [CrossRef]
6. Glass, J.B.; Kretz, C.B.; Ganesh, S.; Ranjan, P.; Seston, S.L.; Buck, K.N.; Landing, W.M.; Morton, P.L.; Moffett, J.W.; Giovannoni, S.J.; et al. Meta-omic signatures of microbial metal and nitrogen cycling in marine oxygen minimum zones. *Front. Microbiol.* **2015**, *6*, 998. [CrossRef]
7. Duret, M.T.; Pachiadaki, M.G.; Stewart, F.J.; Sarode, N.; Christaki, U.; Monchy, S.; Srivasava, A.; Edgcomb, V.P. Size-fractionated diversity of eukaryotic microbial communities in the Eastern Tropical North Pacific oxygen minimum zone. *FEMS Microbiol. Ecol.* **2015**, *91*, fiv037. [CrossRef]
8. Parris, D.J.; Ganesh, S.; Edgcomb, V.P.; DeLong, E.F.; Stewart, F.J. Microbial eukaryote diversity in the marine oxygen minimum zone off northern Chile. *Front. Microbiol.* **2014**, *5*, 543. [CrossRef]
9. Fell, J.W. Distribution of Yeasts in the Indian Ocean. *Bull. Mar. Sci.* **1967**, *17*, 454–470.
10. Gutiérrez, M.H.; Pantoja, S.; Quiñones, R.A.; González, R.R. First record of filamentous fungi in the coastal upwelling ecosystem off central Chile. *Gayana* **2010**, *74*, 66–73.
11. Gutiérrez, M.H.; Jara, A.M.; Pantoja, S. Fungal parasites infect marine diatoms in the upwelling ecosystem of the Humboldt current system off central Chile. *Environ. Microbiol.* **2016**, *18*, 1646–1653. [CrossRef] [PubMed]
12. Richards, T.A.; Jones, M.D.M.; Leonard, G.; Bass, D. Marine Fungi: Their Ecology and Molecular Diversity. *Annu. Rev. Mar. Science* **2012**, *4*, 495–522. [CrossRef] [PubMed]
13. Grossart, H.-P.; Van den Wyngaert, S.; Kagami, M.; Wurzbacher, C.; Cunliffe, M.; Rojas-Jimenez, K. Fungi in aquatic ecosystems. *Nat. Rev. Microbiol.* **2019**, *17*, 339–354. [CrossRef] [PubMed]
14. Amend, A.; Burgaud, G.; Cunliffe, M.; Edgcomb, V.P.; Ettinger, C.L.; Gutiérrez, M.H.; Heitman, J.; Hom, E.F.Y.; Ianiri, G.; Jones, A.C.; et al. Fungi in the Marine Environment: Open Questions and Unsolved Problems. *mBio* **2019**, *10*, e01189-18. [CrossRef] [PubMed]
15. Gutiérrez, M.H.; Pantoja, S.; Tejos, E.; Quiñones, R.A. The role of fungi in processing marine organic matter in the upwelling ecosystem off Chile. *Mar. Biol.* **2011**, *158*, 205–219. [CrossRef]
16. Shoun, H.; Tanimoto, T. Denitrification by the fungus Fusarium oxysporum and involvement of cytochrome P-450 in the respiratory nitrite reduction. *J. Biol. Chem.* **1991**, *266*, 11078–11082. [CrossRef]
17. Shoun, H.; Kim, D.-H.; Uchiyama, H.; Sugiyama, J. Denitrification by fungi. *FEMS Microbiol. Lett.* **1992**, *94*, 277–281. [CrossRef]
18. Stein, L.Y.; Yung, Y.L. Production, Isotopic Composition, and Atmospheric Fate of Biologically Produced Nitrous Oxide. *Annu. Rev. Earth Planet. Sci.* **2003**, *31*, 329–356. [CrossRef]
19. Ravishankara, A.R.; Daniel, J.S.; Portmann, R.W. Nitrous Oxide (N_2O): The Dominant Ozone-Depleting Substance Emitted in the 21st Century. *Science* **2009**, *326*, 123–125. [CrossRef]
20. Maeda, K.; Spor, A.; Edel-Hermann, V.; Heraud, C.; Breuil, M.-C.; Bizouard, F.; Toyoda, S.; Yoshida, N.; Steinberg, C.; Philippot, L. N_2O production, a widespread trait in fungi. *Sci. Rep.* **2015**, *5*, srep09697. [CrossRef]
21. Stief, P.; Fuchs-Ocklenburg, S.; Kamp, A.; Manohar, C.-S.; Houbraken, J.; Boekhout, T.; de Beer, D.; Stoeck, T. Dissimilatory nitrate reduction by Aspergillus terreus isolated from the seasonal oxygen minimum zone in the Arabian Sea. *BMC Microbiol.* **2014**, *14*, 35. [CrossRef]
22. Wankel, S.D.; Ziebis, W.; Buchwald, C.; Charoenpong, C.; de Beer, D.; Dentinger, J.; Xu, Z.; Zengler, K. Evidence for fungal and chemodenitrification based N_2O flux from nitrogen impacted coastal sediments. *Nat. Commun.* **2017**, *8*, 15595. [CrossRef]
23. Cathrine, S.J.; Raghukumar, C. Anaerobic denitrification in fungi from the coastal marine sediments off Goa, India. *Mycol. Res.* **2009**, *113*, 100–109. [CrossRef] [PubMed]
24. Garcia, H.E.; Locarnini, R.A.; Boyer, T.P.; Antonov, J.I.; Baranova, O.K.; Zweng, M.M.; Reagan, J.R.; Johnson, D.R. World Ocean Atlas 2013, Volume 4: Dissolved Inorganic Nutrients (phosphate, nitrate, silicate). In *NOAA Atlas NESDIS*; Levitus, S., Ed.; U.S. Department of Commerce: Silver Spring, MD, USA, 2013; pp. 1–25.
25. QIAGEN. DNeasy Plant Handbook. 2015. Available online: https://www.qiagen.com/us/products/discovery-and-translational-research/dna-rna-purification/dna-purification/genomic-dna/dneasy-plant-mini-kit/#resources (accessed on 8 March 2019).
26. Illumina. 16S Metagenomic Sequencing Library Preparation. Available online: https://support.illumina.com/downloads/16s_metagenomic_sequencing_library_preparation.html (accessed on 8 March 2019).
27. Tedersoo, L.; Anslan, S.; Bahram, M.; Põlme, S.; Riit, T.; Liiv Kõljalg, U.; Kisand, V.; Nilsson, H.; Hildebrand, F.; Bork, P.; et al. Shotgun metagenomes and multiple primer pair-barcode combinations of amplicons reveal biases in metabarcoding analyses of fungi. *MycoKeys* **2015**, *10*, 1–43. [CrossRef]
28. Tedersoo, L.; Bahram, M.; Põlme, S.; Kõljalg, U.; Yorou, N.S.; Wijesundera, R.; Ruiz, L.V.; Vasco-Palacios, A.M.; Thu, P.Q.; Suiji, A.; et al. Global diversity and geography of soil fungi. *Science* **2014**, *346*, 1256688. [CrossRef] [PubMed]
29. Rivers, A.R.; Weber, K.C.; Gardner, T.G.; Liu, S.; Armstrong, S.D. ITSxpress: Software to rapidly trim internally transcribed spacer sequences with quality scores for marker gene analysis. *F1000Research* **2018**, *7*, 1418. [CrossRef] [PubMed]
30. Edgar, R.C. UNOISE2: Improved error-correction for Illumina 16S and ITS amplicon sequencing. *bioRxiv* **2016**, 081257. [CrossRef]

31. Bolyen, E.; Rideout, J.R.; Dillon, M.R.; Bokulich, N.A.; Abnet, C.C.; Al-Ghalith, G.A.; Alexander, H.; Alm, E.J.; Arumugam, M.; Asnicar, F.; et al. Reproducible, interactive, scalable and extensible microbiome data science using QIIME 2. *Nat. Biotechnol.* **2019**, *37*, 852–857. [CrossRef]
32. Camacho, C.; Coulouris, G.; Avagyan, V.; Ma, N.; Papadopoulos, J.; Bealer, K.; Madden, T.L. BLAST+: Architecture and applications. *BMC Bioinform.* **2009**, *10*, 421. [CrossRef]
33. UNITE Community. UNITE QIIME Release for Fungi. Version 02.02.2019. Available online: https://unite.ut.ee/repository.php (accessed on 8 March 2019).
34. Kearns, P.J.; Bulseco-McKim, A.N.; Hoyt, H.; Angell, J.H.; Bowen, J.L. Nutrient Enrichment Alters Salt Marsh Fungal Communities and Promotes Putative Fungal Denitrifiers. *Microb. Ecol.* **2018**, *77*, 358–369. [CrossRef]
35. Edgar, R.C. Search and clustering orders of magnitude faster than BLAST. *Bioinformatics* **2010**, *26*, 2460–2461. [CrossRef] [PubMed]
36. Caporaso, J.G.; Kuczynski, J.; Stombaugh, J.; Bittinger, K.; Bushman, F.D.; Costello, E.K.; Fierer, N.; Peña, A.G.; Goodrich, J.K.; Gordon, J.I.; et al. QIIME allows analysis of high-throughput community sequencing data. *Nat. Methods* **2010**, *7*, 335–336. [CrossRef] [PubMed]
37. Fuchsman, C.A.; Devol, A.H.; Saunders, J.K.; McKay, C.; Rocap, G. Niche Partitioning of the N Cycling Microbial Community of an Offshore Oxygen Deficient Zone. *Front. Microbiol* **2017**, *8*, 2384. [CrossRef] [PubMed]
38. Bushnell, B. BBMap Short-Read Aligner, and Other Bioinformatics Tools. Available online: https://sourceforge.net/projects/bbmap/ (accessed on 8 March 2019).
39. Bolger, A.M.; Lohse, M.; Usadel, B. Trimmomatic: A flexible trimmer for Illumina sequence data. *Bioinformatics* **2014**, *30*, 2114–2120. [CrossRef]
40. Buchfink, B.; Xie, C.; Huson, D.H. Fast and sensitive protein alignment using DIAMOND. *Nat. Methods* **2015**, *12*, 59–60. [CrossRef]
41. Takaya, N.; Shoun, H. Nitric oxide reduction, the last step in denitrification by Fusarium oxysporum, is obligatorily mediated by cytochrome *P450nor*. *Mol. Gen. Genet.* **2000**, *263*, 342–348. [CrossRef]
42. Eddy, S.R. Accelerated Profile HMM Searches. *PLoS Comput. Biol.* **2011**, *7*, e1002195. [CrossRef]
43. Higgins, S.A.; Schadt, C.W.; Matheny, P.B.; Löffler, F.E. Phylogenomics Reveal the Dynamic Evolution of Fungal Nitric Oxide Reductases and Their Relationship to Secondary Metabolism. *Genome Biol. Evol.* **2018**, *10*, 2474–2489. [CrossRef]
44. Holmes, R.M.; Aminot, A.; Kérouel, R.; Hooker, B.A.; Peterson, B.J. A simple and precise method for measuring ammonium in marine and freshwater ecosystems. *Can. J. Fish. Aquat. Sci* **1999**, *56*, 1801–1808. [CrossRef]
45. Hales, B.; van Geen, A.; Takahashi, T. High-frequency measurement of seawater chemistry: Flow-injection analysis of macronutrients. *Limnol. Oceanogr. Methods* **2004**, *2*, 91–101. [CrossRef]
46. Ji, Q.; Grundle, D.S. An automated, laser-based measurement system for nitrous oxide isotope and isotopomer ratios at nanomolar levels. *Rapid Commun. Mass Spectrom.* **2019**, *33*, 1553–1564. [CrossRef]
47. Peng, X.; Fuchsman, C.A.; Jayakumar, A.; Oleynik, S.; Martens-Habbena, W.; Devol, A.H.; Ward, B.B. Ammonia and nitrite oxidation in the Eastern Tropical North Pacific. *Glob. Biogeochem. Cycles* **2015**, *29*, 2034–2049. [CrossRef]
48. Balabanova, L.A.; Gafurov, Y.M.; Pivkin, M.V.; Terentyeva, N.A.; Likhatskaya, G.N.; Rasskazov, V.A. An Extracellular S1-Type Nuclease of Marine Fungus Penicillium melinii. *Mar. Biotechnol.* **2012**, *14*, 87–95. [CrossRef]
49. Kurtzman, C.P.; Fell, J.W.; Boekhout, T. *The Yeasts: A Taxonomic Study*, 5th ed.; Elsevier: Amsterdam, The Netherlands, 2011; pp. 678–692.
50. Boekhout, T.; Theelen, B.; Houbraken, J.; Robert, V.; Scorzetti, G.; Gafni, A.; Gerson, U.; Sztejnberg, A. Novel anamorphic mite-associated fungi belonging to the Ustilaginomycetes: *Meira geulakonigii* gen. nov., sp. nov., *Meira argovae* sp. nov. and *Acaromyces ingoldii* gen. nov., sp. nov. *Int. J. Syst. Evol. Microbiol.* **2003**, *53*, 1655–1664. [CrossRef]
51. Zalar, P.; Gostinčar, C.; de Hoog, G.S.; Uršič, V.; Sudhadham, M.; Gunde-Cimerman, N. Redefinition of *Aureobasidium pullulans* and its varieties. *Stud. Mycol.* **2008**, *61*, 21–38. [CrossRef]
52. Hyde, K.D.; Jones, E.B.G.; Leaño, E.; Pointing, S.B.; Poonyth, A.D.; Vrijmoed, L.L.P. Role of fungi in marine ecosystems. *Biodivers. Conserv.* **1998**, *7*, 1147–1161. [CrossRef]
53. Hassett, B.T.; Vonnahme, T.R.; Peng, X.; Jones, E.B.G.; Heuzé, C. Global diversity and geography of planktonic marine fungi. *Bot. Mar.* **2020**, *63*, 121–139. [CrossRef]
54. Worden, A.Z.; Follows, M.J.; Giovannoni, S.J.; Wilken, S.; Zimmerman, A.E.; Keeling, P.J. Rethinking the marine carbon cycle: Factoring in the multifarious lifestyles of microbes. *Science* **2015**, *347*, 1257594. [CrossRef]
55. Jeffries, T.C.; Curlevski, N.J.; Brown, M.V.; Harrison, D.P.; Doblin, M.A.; Petrou, K.; Ralph, P.J.; Seymour, J.R. Partitioning of fungal assemblages across different marine habitats. *Environ. Microbiol. Rep.* **2016**, *8*, 235–238. [CrossRef] [PubMed]
56. Wang, X.; Singh, P.; Gao, Z.; Zhang, X.; Johnson, Z.I.; Wang, G. Distribution and Diversity of Planktonic Fungi in the West Pacific Warm Pool. *PLoS ONE* **2014**, *9*, e101523.
57. Berbee, M.L.; James, T.Y.; Strullu-Derrien, C. Early Diverging Fungi: Diversity and Impact at the Dawn of Terrestrial Life. *Annu. Rev. Microbiol.* **2017**, *71*, 41–60. [CrossRef] [PubMed]
58. Hassett, B.T.; Gradinger, R. Chytrids dominate arctic marine fungal communities. *Environ. Microbiol.* **2016**, *18*, 2001–2009. [CrossRef] [PubMed]
59. Garcia-Robledo, E.; Padilla, C.C.; Aldunate, M.; Stewart, F.J.; Ulloa, O.; Paulmier, A.; Gregori, G.; Revsbech, N.P. Cryptic oxygen cycling in anoxic marine zones. *Proc. Natl. Acad. Sci. USA* **2017**, *114*, 8319–8324. [CrossRef] [PubMed]

60. Bauchop, T.; Mountfort, D.O. Cellulose Fermentation by a Rumen Anaerobic Fungus in Both the Absence and the Presence of Rumen Methanogens. *Appl. Environ. Microbiol.* **1981**, *42*, 1103–1110. [CrossRef] [PubMed]
61. Zumft, W.G. Cell biology and molecular basis of denitrification. *Microbiol. Mol. Biol. Rev.* **1997**, *61*, 533–616. [CrossRef]
62. Dalsgaard, T.; Stewart, F.J.; Thamdrup, B.; Brabandere, L.D.; Revsbech, N.P.; Ulloa, O.; Candield, D.E.; DeLong, E.F. Oxygen at Nanomolar Levels Reversibly Suppresses Process Rates and Gene Expression in Anammox and Denitrification in the Oxygen Minimum Zone off Northern Chile. *mBio* **2014**, *5*, e01966. [CrossRef]
63. Hallin, S.; Philippot, L.; Löffler, F.E.; Sanford, R.A.; Jones, C.M. Genomics and Ecology of Novel N2O-Reducing Microorganisms. *Trends Microbiol.* **2018**, *26*, 43–55. [CrossRef]
64. Watson, S.W.; Novitsky, T.J.; Quinby, H.L.; Valois, F.W. Determination of bacterial number and biomass in the marine environment. *Appl. Environ. Microbiol.* **1977**, *33*, 940–946. [CrossRef]

Journal of
Fungi

MDPI

Review

Diversity, Ecological Role and Biotechnological Potential of Antarctic Marine Fungi

Stefano Varrella [1,*], Giulio Barone [2], Michael Tangherlini [3], Eugenio Rastelli [4], Antonio Dell'Anno [5] and Cinzia Corinaldesi [1,*]

1 Department of Materials, Environmental Sciences and Urban Planning, Polytechnic University of Marche, Via Brecce Bianche, 60131 Ancona, Italy
2 Institute for Biological Resources and Marine Biotechnologies, National Research Council (IRBIM-CNR), Largo Fiera della Pesca, 60125 Ancona, Italy; giulio.barone@irbim.cnr.it
3 Department of Research Infrastructures for Marine Biological Resources, Stazione Zoologica "Anton Dohrn", Fano Marine Centre, Viale Adriatico 1-N, 61032 Fano, Italy; michael.tangherlini@szn.it
4 Department of Marine Biotechnology, Stazione Zoologica "Anton Dohrn", Fano Marine Centre, Viale Adriatico 1-N, 61032 Fano, Italy; eugenio.rastelli@szn.it
5 Department of Life and Environmental Sciences, Polytechnic University of Marche, Via Brecce Bianche, 60131 Ancona, Italy; a.dellanno@univpm.it
* Correspondence: s.varrella@univpm.it (S.V.); c.corinaldesi@univpm.it (C.C.)

Abstract: The Antarctic Ocean is one of the most remote and inaccessible environments on our planet and hosts potentially high biodiversity, being largely unexplored and undescribed. Fungi have key functions and unique physiological and morphological adaptations even in extreme conditions, from shallow habitats to deep-sea sediments. Here, we summarized information on diversity, the ecological role, and biotechnological potential of marine fungi in the coldest biome on Earth. This review also discloses the importance of boosting research on Antarctic fungi as hidden treasures of biodiversity and bioactive molecules to better understand their role in marine ecosystem functioning and their applications in different biotechnological fields.

Keywords: marine fungi; mycology; fungal diversity; Antarctica; bioprospecting; psychrophiles; cold-adapted enzymes; industrial applications; blue biotechnologies

Citation: Varrella, S.; Barone, G.; Tangherlini, M.; Rastelli, E.; Dell'Anno, A.; Corinaldesi, C. Diversity, Ecological Role and Biotechnological Potential of Antarctic Marine Fungi. *J. Fungi* **2021**, *7*, 391. https://doi.org/10.3390/jof7050391

Academic Editor: Federico Baltar

Received: 20 April 2021
Accepted: 13 May 2021
Published: 17 May 2021

1. Introduction

The Antarctic ecosystem is one of the most hostile environments on Earth [1]. Despite the harsh environmental conditions (extremely low temperatures, prolonged periods of darkness, and high levels of ultraviolet radiations), Antarctica hosts a variety of unique organisms, from penguins and other endemic birds to whales, seals, fish, and invertebrates inhabiting both the land and the Southern Ocean [2]. Although the number of species inhabiting the Antarctic mainland is low compared to other terrestrial environments [3], the marine ecosystems host an unexpected high biodiversity and an ever-increasing number of species is being reported every year [4–7]. For example, the number of invertebrate marine Antarctic species has been estimated to range from 17,000 to 20,000, only 8000 of which have been described to date [6,8]. However, these estimates are likely underrating the overall Antarctic marine diversity due to the low sampling effort and limited spatial coverage of the studies conducted [5,7,9]. Moreover, molecular techniques are now enabling the identification of cryptic and previously unknown species, thus boosting our current ability in assessing Antarctic biodiversity [10].

Many factors can promote the high biodiversity in the Southern Ocean, including high environmental heterogeneity, isolation, and low human impact [4]. Indeed, coastal habitats in Antarctica are characterized by a wide spatial heterogeneity caused by high variability in nutrient dynamics, light availability, and extensive seascape variations due to ice formation and melting which determine major changes in thermohaline conditions, biological productivity, and sedimentation processes. Moreover, Antarctica's geographic and oceanographic isolation has allowed many new species to evolve in the absence of competition from lower latitudes' vicariants [4]. Overall, these factors have contributed to shape Antarctic biodiversity in a unique way [11].

The diversity of large organisms inhabiting Antarctic ecosystems has received a larger attention than the diversity of microbial assemblages, although the microbial component represents an important fraction of the whole biomass and plays pivotal roles in biogeochemical cycles and marine food web functioning [12–16]. There is also evidence that microbial diversity represents a major reservoir of novel taxa, biochemical pathways, genes and compounds with biotechnological applications [17–20].

Despite the harsh conditions, fungi are ubiquitously present in Antarctic ecosystems [21]. This success is largely due to the vast array of fungal enzymes, secondary metabolites, and bioactive molecules, which allowed fungal colonization and diversification in almost every habitat on Earth [22]. Their peculiar biological adaptions to low temperatures lead to the production of structurally novel enzymes and bioactive metabolites, which provide fungi competitive advantages over other microorganisms through chemically-mediated interactions, defense, and virulence factors for plants and animals [23–26]. In the last decade, several studies, exploring the fungal diversity in Antarctic marine environments, have revealed promising properties of fungi relevant for biotechnological applications (e.g., pharmaceutical, food, and cosmeceutical industries [27–29]). Indeed, increasing evidence indicates that cold-adapted fungi are a relevant target to the present and future scientific research for their possible biotechnological applications, including the development of new drugs and exploitation in several industrial processes [30,31].

In this review we collected the current literature about fungi in Antarctic marine ecosystems, focusing on their taxonomic diversity and ecological functions, as well as their potential for developing new blue biotechnologies.

2. Fungal Diversity and Ecology in Antarctic Marine Environments

Fungi are widely distributed in marine Antarctic ecosystem, and their occurrence has been recorded in seawater and sediments as well as associated with macroalgae and invertebrates [27]. Marine fungi are supposed to contribute to population dynamics, C and nutrient cycles in the oceans [32], and yet only a limited number of studies have investigated fungal diversity and ecology of Antarctic marine fungi. These studies mainly focused on the identification of fungi and yeasts isolated from waters, sediments, animals, and/or macroalgae (Figure 1). So far, most of the information has been acquired through culture-dependent approaches on samples collected from the Shetland Islands, leaving most coastal and offshore habitats unexplored.

2.1. Fungal Diversity

2.1.1. Fungi in Antarctic Marine Environment

In the last years, culture-dependent and molecular techniques have allowed us to describe a large number of fungal taxa in seawater [33,34]. Likewise, fungi have been identified in several polar environments [35–37]. To date, only a few studies have successfully identified filamentous fungi and yeasts in both coastal and offshore Antarctic waters [38,39]. These studies, employing culture-dependent methods, have been able to isolate a number of fungal genera belonging to the orders *Eurotiales, Hypocreales, Chaetothyriales,* and *Kriegeriales.* In particular, Antarctic waters seem to exclusively host the genera *Exophiala, Graphium, Simplicillium, Purpureocillium,* and *Akanthomyces.* These genera are known to include parasites, pathogens, and likely saprotrophs, which may be involved in complex interactions within the water column [40–42]. However, other genera found in Antarctic waters such as *Penicillium, Metschnikowia, Rhodotorula,* and *Glaciozyma* have also been found elsewhere in Antarctica (Figure 2; Table S2). Within the water column, fungi can have significant effects on primary production dynamics and carbon fluxes within the marine food webs, by acting as saprotrophs and interacting with marine phytoplankton [37,43,44].

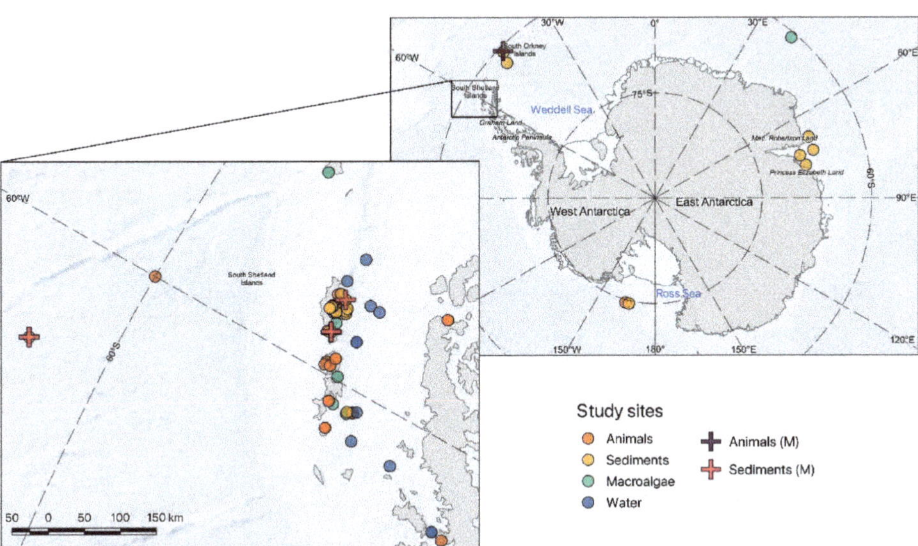

Figure 1. Locations of the fungi identified from different Antarctic marine substrates: animals (orange circle), sediments (yellow circle), macroalgae (green circle), and water (blue circle) based on culture dependent approaches or identified through metagenomic analysis: animals (purple cross) and sediments (pink cross) (for detailed elucidation on the samples where fungal taxa were isolated, and coordinates see Table S1).

Despite the increasing number of investigations carried out worldwide, to date, fungal diversity in Antarctic sediments has been explored in a relatively limited number of studies, which mainly focused on the diversity of culturable fungal species [45–61]. Recently, for the first time, molecular tools have been employed to investigate the fungal diversity in Antarctic marine sediments and allowed the identification of a large number of fungal taxa although much of the fungal diversity in Antarctic marine sediments still remains unknown [49].

Antarctic marine sediments have been shown to host a plethora of fungal taxa. For example, the genera *Pseudocercosporella, Toxicocladosporium, Trichoderma, Humicola, Paraconiothyrium, Phaeoacremonium*, and *Phenoliferia* have been documented exclusively in marine sediments and to be potentially involved in the degradation of organic matter. Nonetheless, fungal diversity in marine sediments also include many other genera found in other Antarctic habitats, such as *Penicillium, Metschnikowia, Rhodotorula, Cladosporium* and *Glaciozyma* (Figure 2). The genus *Pseudogymnoascus* genus found in Antarctic sediments has been also recorded in other cold environments including polar regions and glaciers [27,49,62,63], and the genus *Phaeosphaeria*, whose members include plant pathogens, has been also found in association with the Antarctic macroalgae *Adenocystis utricularis* [64,65].

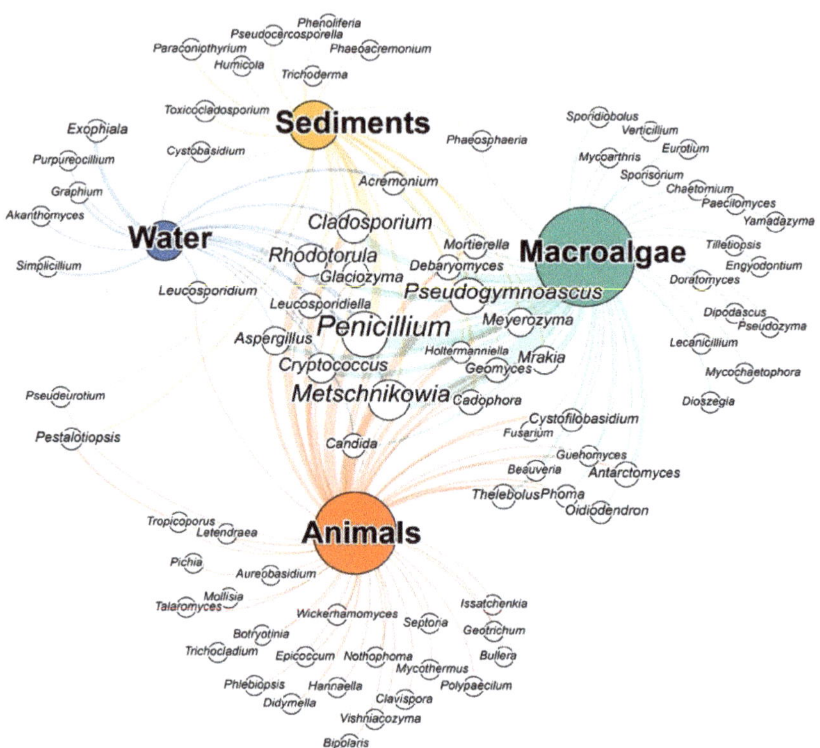

Figure 2. Network diagram displaying the records of Antarctic fungal taxa belonging to different genera identified through culture dependent approaches or metagenomic analyses in the four most commonly marine matrices (water, sediments, animals, macroalgae). The size of the white nodes is proportional to the number of records in the studies in which the genus has been found, while the size of coloured nodes is proportional to the overall number of genera retrieved.

2.1.2. Fungi Associated with Antarctic Macroalgae and Animals

Most of the research on Antarctic fungi focused on the assemblages associated with benthic macroalgae of the South Shetland Islands (Antarctic Peninsula) [54,64,66–70]. In particular, the macroalgae *Adenocystis utricularis, Desmarestia anceps* and *Palmaria decipiens* host a rich fungal diversity, including genera like *Pseudogymnoascus*, *Antarctomyces*, *Oidiodendron*, *Penicillium*, *Phaeosphaeria*, *Cryptococcus*, *Leucosporidium*, *Metschnikowia*, and *Rhodotorula*. Some of the species belonging to these genera appear to be endemic of Antarctica (i.e., *Antarctomyces pellizariae*, *Antarctomyces psychrotrophicus*, *Cryptococcus victoriae*, *Cryptococcus adeliensis*, *Metschnikowia australis*, *Pseudogymnoascus pannorum*, *Mortierella antarctica*), while others appear rather ubiquitously distributed (i.e., *Penicillium* sp., *Aureobasidium* sp. and *Rhodotorula* sp.) [64,66,68,69]. Despite this, fungi associated with macroalgae are apparently unique since several genera have never been found in any other marine sample (Figure 2). This rich and diverse fungal assemblages can have an important ecological role since they may produce enzymes with the potential to degrade algal detritus and may be involved in organic matter cycling and energy transfer within the marine food web [69].

Analogously to what has been reported for macroalgae, fungi have been isolated also from marine animals, with which they can interact as pathogens, parasites, or symbionts [71–78]. Nevertheless, the nature and the strength of the relationships between the host and the associated fungi have yet to be fully understood [78]. Fungal diversity has been investigated in a variety of Antarctic organisms such as sponges, annelids, crustaceans, molluscs, and echinoderms collected in the South Shetland Islands (Figure 1), mainly through culture-dependent approaches [54,79–83]. Although culture-based studies allowed isolating and investigating a variety of fungal taxa, molecular tools have allowed the identification of a larger fraction of the fungal diversity provided. In particular, such molecular methods provided new information about the fungal diversity associated with the Antarctic Krill, *Euphausia superba* [84], and the marine sponges *Leucetta antarctica* and *Myxilla* sp. [85]. Several other fungal taxa, including members of the orders Saccharomycetales and Eurotiales, and of the families *Saccharomycetaceae* and *Aspergillaceae,* have been identified. In particular, the most represented taxa included ubiquitous genera such as *Rhodotorula, Penicillium, Metschnikowia, Aspergillus,* as well as 23 different exclusive genera such as *Wickerhamomyces, Geotrichum, Letendraea,* and *Bullera* (Figure 2). Among them, *Wickerhamomyces* spp. has particular characteristics: It can in fact grow under extreme environmental stressful conditions, such as low and high pH, high osmotic pressure, absence of oxygen, and also shows antimycotic activity [86]. Despite further studies being needed to better understand the ecological significance of fungal–host interactions, there is evidence that these fungi can be involved in defensive mechanisms by producing cytotoxic, antimycotic, and antibiotic molecules, which could increase the animal's wellbeing [83]. For instance, fungi associated with the Antarctic krill can be involved in the defense mechanisms of the host against pathogenic bacteria [83]. Nevertheless, a number of fungal pathogens (i.e., *Rhodotorula, Debaryomyces*) have been isolated from marine organisms which have the ability to compromise host health and fitness, but their possible detrimental effects on marine fauna have not yet been estimated.

2.2. Contribution of Fungi to Ecological Processes in Antarctic Marine Ecosystems

While on terrestrial ecosystems the role of fungi is largely recognized, their ecological role in Antarctic marine ecosystems has yet to be understood. In the last decade, several studies highlighted that fungi are widely distributed in marine environments, from coastal waters [87] to the deep-sea surface and subsurface sediments [88–94], extreme habitats such as hypersaline anoxic basins [95–97], cold seeps [98–100], hydrothermal vents [101,102], and oil reservoirs [103]. Polar environments are also characterized by a considerably high number of fungal taxa [36,55,104–106], which can be involved in organic matter degradation and nutrient cycling, as well as in intimate relationships with a variety of organisms [27,32,49,107].

In Antarctic marine waters, owing to their unique enzymatic capabilities and metabolic versatility [29,108,109], fungi can be important components for the biogeochemical processes and the functioning of the food webs [110]. As saprotrophs, some fungi may utilize phytoplankton-derived detritus contributing to organic matter fluxes [111], while others can interact with planktonic microorganisms influencing ecological dynamics and food webs [112]. The same Authors have provided a theoretical framework to describe how aquatic fungi interact with their environment, introducing the concept "mycoflux" to refer to the interactions between pelagic fungi and other microbes and their effects on the carbon pump [112]. Previously, Kagami et al. (2014) introduced the concept "mycoloop" to indicate the parasitic interaction between fungi and other planktonic components, suggesting that parasitic fungi can facilitate energy transfer from phytoplankton to zooplankton [113]. This concept is also supported by recent metabarcoding and metagenomic studies carried out in different aquatic environments [37,114,115], which revealed a high diversity of parasitic (or facultative parasitic) zoosporic fungi associated with phytoplankton and zooplankton [43,113,116–120]. Overall, these findings suggest that fungal parasites can be important in influencing the aquatic food webs as other planktonic parasites [121].

Even in Antarctic benthic ecosystems, fungi can have an important role in C cycling and nutrient regeneration processes. Benthic fungi, acting as decomposers of organic matter can be involved in the degradation of recalcitrant organic compounds, which otherwise accumulate in marine sediments [93,122,123], and may mediate C and energy transfer to higher trophic levels [112,124–126]. The association of fungi with microalgal [43] and macroalgal detritus can improve the nutritional value of organic matter by lowering the carbon:nitrogen:phosphorus ratio [127,128]. Moreover, fungi in Antarctic benthic ecosystems are likely involved in ecological interactions with other eukaryotes, similarly to what observed in other extreme marine environments acting as pathogens and parasites [129,130]

In Antarctic benthic coastal ecosystems, an important relationship between fungi and macroalgae has been reported [64]. Macroalgae are ecosystem engineers that contribute to primary production in cold and temperate coastal marine environments [131]. Macroalgae represent the second biggest reservoir of fungal diversity after sponges [132], and the relationship between the host and its fungal assemblage encompasses mutualism, parasitism, and saprophytism [133–135]. In this relationship, fungi can provide important advantages for the growth, defence, development, and nutrition of the algal host [136–138]. However, several fungi can also act as pathogens, compromising the host's health and its ecological functions [139].

Fungi have also been documented in association with Antarctic benthic marine animals [27,54], but so far, the nature and the mechanisms of these relationships remain mostly unknown. To date, only one study based on a functional analysis of fungi associated with sponges highlighted that fungi can have an important role in the degradation of the organic matter, contributing to nutrient cycling and in turn influencing the carbon fixation pathways of prokaryotes and other micro-eukaryotes within the microbial assemblages [85].

Paradoxically, the biotechnological focus of the Antarctic marine fungi has contributed to accumulate more information on their potential industrial applications than on their quantitative relevance and role in biodiversity and functioning of Antarctic marine ecosystems. Only one study, indeed, has so far addressed the ecological role of fungi in Antarctic marine habitats [85]. This knowledge gap highlights the importance to increase studies based on molecular and biochemical tools to better comprehend fungal diversity and ecology and to elucidate the nature and strength of the relationships between fungi and their hosts especially in extreme environments of the Earth, such as the Antarctic ecosystems.

3. Biotechnological Potential of Fungi Inhabiting Marine Antarctic Environments

The extreme conditions of Antarctic marine environments have forced microorganisms to evolve peculiar metabolic pathways as well as adaptive mechanisms, which allow them to thrive in cold ecosystems [140]. Psychrophilic and psychrotrophic fungi hold outstanding biological features arising from the harsh environmental conditions with which they must cope [141,142]. For these reasons, they are considered treasure of unique enzymes and bioactive molecules with an exceptional application potential [29,109,143–146]. Therefore, Antarctic fungi can greatly contribute to the discovery of new compounds of marine origin to be exploited in the industrial "white" and bio-pharmaceutical "red" biotechnology [28,147].

3.1. Antarctic Marine Fungi: Promising Candidates for Bioprospecting

With the increasing demand for novel antimicrobial and chemotherapeutic drugs, the discovery of biologically active molecules to improve human health is one of the most important challenges for mankind [148,149]. There is a general consensus that natural products offer extraordinary advantages over chemical molecules, and this makes marine microorganisms an astonishing potential source of new drugs [150–154]. Marine extremophilic microorganisms, including fungi, can also represent a huge reservoir of bioactive molecules that have recently triggered interest in bioprospecting research because of their promising therapeutical properties [21,22,29,109,152,155–159]. In this regard, marine fungi isolated from polar environments reported their ability to synthesize metabolites with unique structures and a wide range of biological activities, compared to mesophilic fungi, highlighting that psychrophilic fungi can be a new resource for several applications in biotechnology [28,30,160–162]. However, the search for natural bioactive products has been focused so far on a very small number of fungi isolated from Antarctic marine sediments, seawater and few organisms such as sponges and macroalgae [27].

Crude extracts of fungal strains have been isolated from fresh thalli of Antarctic macroalgal species and tested for their antibiotic, antifungal, antiviral, and antiparasitic activity [66]. Among these, extracts of two *Penicillium* species associated to the endemic macroalgae *Palmaria decipiens* and *Monostroma hariotii* contained compounds with high and selective antifungal and trypanocidal activities [68]. In addition, extracts of *Pseudogymnoascus* species, *Guehomyces pullulans*, and *Metschnikowia australis* showed high antifungal activity against *Candida albicans*, *Candida krusei*, and *Cladosporium sphaerospermum*, whereas the extract of *Penicillium steckii* greatly inhibited BHK-21 cell line expressing the yellow fever virus [68]. Moreover, *Geomyces* species associated with Antarctic marine sponges, have been suggested as a source of several promising antimicrobial and antitumoral compounds [82].

Interestingly, extracts of *Pseudogymnoascus* 5A-1C315IIII, isolated from marine sediments of Admiralty Bay (South Shetland Islands, Antarctica), can inhibit different phytopathogenic *Xanthomonas* species [57,58]. Another survey conducted on marine sediments collected at Deception Island (South Shetland Islands, Antarctica) led to the isolation of *Pseudogymnoascus* specie and *Simplicillium lamellicola*, which showed high and selective antifungal activity against *Paracoccidioides brasiliensis* [104].

Despite the considerable number of crude fungal extracts, only a few bioactive compounds have been actually tested so far (Table 1). Most of the research on the bioactive compounds carried out in Antarctic environments mainly focus on a few *Penicillium* strains, leaving most of the actual fungal biodiversity largely unexplored. The bioprospection of psychrophilic and psychrotolerant polar *Penicillium* strains have resulted in the collection of many promising bioactive molecules with a complex and peculiar structure and a broad range of biological activities, highlighting their outstanding potential [162]. For instance, eremophilane-type sesquiterpene isolated from *Penicillium* sp. PR19N-1, showed potent inhibitory activity against A549 tumor cells with IC50 value of 5.2 μM [51,163]. In a recent study, neuchromenin, extracted from *Penicillium glabrum* SF-7123, has shown great anti-inflammatory effects inactivating the NF-κB and p38 MAPK pathways in BV2 and RAW264.7 cells [50].

Table 1. Bioactive molecules isolated from Antarctic marine fungi. Available chemical structures of the bioactive compounds have been downloaded from [164] and used to produce Figures S1–S3 for ease of visualization.

Fungal Taxa	Product	Bioactivity	Source	Ref.
Penicillium citrinum OUCMDZ4136	2,4-Dihydroxy-3,5,6-trimethylbenzoic acid; Citreorosein; Pinselin; Citrinin; Dihydrocitrinone; Pennicitrinone A; Quinolactacin A1	Cytotoxic activities against MCF-7, A549, K562 cell lines	Antarctic krill *Euphasia superba*	[83]
Penicillium citreonigrum SP-6	Diketopiperazine, phenols	Inhibitory activity against HCT116 cancer cell line	Marine sediment, Great Wall Station	[165]
Penicillium crustosum HDN153086	Diketopiperazine	Cytotoxic activities against K562 cell line	Marine sediment, Pridz Bay	[61]
Penicillium crustosum PRB-2	Penilactone A	NF-KB inhibitory activities of HCT-8, Bel-7402, BGC-823, A549 and A2780 tumor cell lines	Deep-sea sediment, Prydz Bay	[46]
Penicillium glabrum SF-7123	Citromycetin derivative, neuchromenin; myxotrichin C, deoxyfunicone;	Anti-inflammatory; tyrosine phosphatase 1B inhibition	Marine sediment, Ross Sea	[50]
Penicillium granulatum MCCC 3A00475	Spirograterpene A	Antiallergic effect on immunoglobulin E (IgE)-mediated rat mast RBL-2H3 cells	Deep-sea sediment, Prydz Bay	[52]
Penicillium sp. PR19N-1	Chlorinated eremophilane sesquiterpenes, eremofortine C, eremophilane-type sesquiterpenes, eremophilane-type lactam	Cytotoxic activity against HL-60 and A549 cancer cell lines	Deep-sea sediment, Prydz Bay	[51,163]
Penicillium sp. S-1–18	Butanolide A, guignarderemophilane F, xylarenone A	Butanolide: inhibitory activity against tyrosine phosphatase 1B; xylarenone A: antitumor activity against HeLa and HepG2 cells and growth-inhibitory effects against pathogenic microbes	Sea-bed sediment	[60,166]
Penicillium sp. UFMGCB 6034 and UFMGCB 6120	Aromatic compounds	Antifungal and trypanocidal activities	Macroalgae: *Palmaria decipiens* and *Monostroma hariotii*	[66]
Pseudogymnoascus sp.	Pseudogymnoascin A, B, C, 3-nitroasterric acid; Geomycins B, C	Antibacterial and antifungal activities	Sponge genus *Hymeniacidon*	[167,168]
Trichoderma asperellum	Asperelines A-F, peptaibols	Not assayed	Marine sediment, Penguin Island	[56,169]

The fungus *Penicillium crustosum* HDN153086, isolated from Antarctic sediment from Prydz Bay, can synthesize a new diketopiperazine with moderate cytotoxic activity against K652 cell lines [61]. Interestingly, seven bioactive compounds extracted from *Penicillium citrinum* OUCMDZ4136, associated with the Antarctic krill *Euphausia superba,* showed moderate to strong cytotoxicity against A549, K562, and MCF-7 cell lines [83]. Conversely, *Pseudogymnoascus, Trichoderma, Aspergillus,* and other fungal genera isolated from macroorganisms living in Antarctic environments, have not received as much attention as the *Penicillium* genus. In fact, to date, only a few studies investigated the properties of bioactive molecules extracted from these isolated genera and tested their bioactivity [56,167–169]. For instance, different compounds have been extracted from *Pseudogymnoascus;* however, only two bioactive molecules (Geomycins B, C) from a sediment sample showed antifungal and antimicrobial activities [168].

3.2. Antarctic Fungi as Novel Source of Cold-Active Enzymes

Over the last 20 years, the interest in cold-adapted microorganisms as a source of new enzymes for industrial processes has intensely grown [170–172]. Nowadays, a considerable number of industrial processes and products take advantage of microbial enzymes [173,174]. Nevertheless, studies prospecting and characterizing enzymes have mainly focused on prokaryotes [175,176], while there is little information on enzymes from psychrophilic or psychrotolerant eukaryotes inhabiting Antarctic systems [162,177,178].

Despite the increasing demand for new biocatalysts, very few enzymes are currently isolated from extreme environments, the so-called "extremozymes" [179]. Cold extremozymes display a greater versatility and adaptability with respect to their non-extreme counterparts that can be advantageous for modern industries [180]. These enzymes are an optimal alternative to their mesophilic equivalents thanks to their higher stability under various physicochemical conditions (i.e., pH, temperature, salinity) and their advantage to reduce costs and energy consumption [181–185]. Indeed, the use of these enzymes represents an eco-friendly method compared to the chemical procedures employed in many industrial processes, allowing to avoid the use of organic solvents and other hazardous compounds that can be seriously harmful to the environment [186]. Nevertheless, only some of these potentially useful enzymes have been successfully introduced in the market so far, and their use is spreading rather slowly [187]. This is mainly due to several steps including chemical characterization, condition optimization, and process validation that need to be passed for commercializing novel biocatalysts [188]

Antarctic environments are a relatively new frontier for the isolation of cold-active enzymes. These molecules have distinct features that meet the need of green industry applications [162,177,189,190]. In this view, psychrophilic and psychrotolerant fungi are specialized in producing extracellular and intracellular cold active-enzymes [20,189,191], which they use to live in harsh conditions to degrade molecules and for the uptake of nutrients [141,170,192–194]. Recently, these cold-adapted enzymes have attracted growing interest because of their potential benefits in several industrial fields [175,185,195]. The main characteristics of fungal extremozymes are the high activity at low temperatures and thermolability, which have been manly gained the attention for being applied as detergent additives (e.g., lipases) for eco-friendly cold-water washing and for food, biofuel, and textile processing [184,185]. Among them, cold-adapted hydrolases (EC 3.x, proteases, lipases, cellulases, glycosidases) can be employed and useful in a variety of biotechnological processes (i.e., food, beverage, cleaning agents, textiles, biofuels, and pulp and paper; see Table 2). This class of enzymes is extremely important since it covers over 90% of the total industrial enzymes market [196,197].

Table 2. Examples of extremophilic fungi as a source of cold-adapted enzymes utilized in industrial applications. The fungal taxa reported are isolated from Antarctic marine environments: seawater, marine sediments, and organisms.

Enzyme	Reaction	Fungi	Source of (Isolate) Sample	Applications/Potential Uses	Ref.
Carragenase (EC 3.2.1.83)	Hydrolysis of 1,4-β-linkages between galactose 4-sulfate and 3,6-anhydro-galactose to produce kappa-carrageenans	*Pseudogymnoascus* sp. UFMGCB 10054	Macroalga: *Iridaea cordata*,	Biomedical field, textile industry, bioethanol production, and detergent additive	[69]
Cellulase (EC 3.2.1.4)	Cellulose hydrolysis into glucose	*Cystofilobasidium infirmominiatum* 071209-E8-C1-liblev; *Metschnikowia australis*, *Rhodotorula glacialis*; *Candida spencermartinsiae*, *Leucosporidiella creatinivora*, *Leucosporidium scottii*	Marine sponge: *Tedania*; marine sediments; seawater	Food industry, animal feed, beer and wine, textile and laundry, pulp and paper industry, agriculture, biofuel, pharmaceutical industries, and waste management	[45,80]
Chitinase (EC 3.2.1.14)	Cleavage of glycosidic linkages in chitin and chitodextrins generating chitooligosaccharides	*Lecanicillium muscarium* CCFEE-5003; *Glaciozyma antarctica* PI12	Shrimp wastes; seawater	Cosmetic, pharmaceutic fields, fermentation research, and biomedicine	[45,198–202]
Endo-β-1,3(4)-glucanase (EC 3.2.1.6)	Endohydrolysis of (1→3)- or (1→4)-linkages in β-D-glucans	*Glaciozyma antarctica* PI12	Seawater	Brewing and animal, feed-stuff industry, biofuel production, and pharmaceuticals	[202–204]
Esterase (EC 3.1.1.1)	Hydrolyis of short acyl-chain soluble esters	*Cryptococcus victoriae*, *Metschnikowia australis*, *Rhodotorula glacialis*, *Leucosporidium scottii*, *Leucosporidiella creatinivora*; *Glaciozyma antarctica*	Marine sediments; seawater, sea ice	Paper bleaching, bioremediation, degradation, and removal of xenobiotics and toxic compounds	[45,205]
Invertase (EC 3.2.1.26)	Hydrolysis of the terminal non-reducing β-fructofuranoside residue in sucrose, raffinose and related β-D-fructofuranosides	*Glaciozyma antarctica* 17 (formerly *Leucosporidium antarcticum*)	Seawater	Beverage, confectionary, bakery, invert sugar, high fructose syrup, artificial honey, calf feed, food for honeybees	[38]
Laccase (EC 1.10.3.2)	Oxidation of phenolic compound like lignin	*Cadophora malorum* A2B, *Cadophora malorum* AS2A, *Cadophora luteo-olivacea* P1	Marine sediments	Biosensors, microfuel and bioelectrocatalysis, food, pharmaceutic, cosmetic, pulp and paper, textile industries, and bioremediation	[206]
Lignin peroxidase (EC 1.11.1.14)	Oxidative breakdown of lignin	*Cadophora malorum* M7, *Cadophora* sp. OB-4B	Marine sediments	Pulp and paper, cosmetics (treatment of hyperpigmentation, and skin-lightening through melanin oxidation), textile, bioremediation (degradation of azo, heterocyclic, reactive, and polymeric dyes, xenobiotic, and pesticides), and bioethanol production	[206]

Table 2. *Cont.*

Enzyme	Reaction	Fungi	Source of (Isolate) Sample	Applications/Potential Uses	Ref.
Lipase (EC 3.1.1.3)	Hydrolysis of long-chain triacylglycerol substances with the formation of an alcohol and a carboxylic acid	*Leucosporidium scottii* L117, *Metschnikowia* sp. CRM1589; *Mrakia blollopis* SK-4; *Cystofilobasidium infirmominiatum* 071209-E8-C1-IIa-lev and isolate 131209-E2A-C1-II-lev; *Metschnikowia australis* 131209-E3-C1-(GPY)-lev and isolate 131209-E2A-C4-II-lev; *Rhodotorula pinicola* 071209-E4-C9-lev; *Candida zeylanoides*, *Cryptococcus victoriae*, *Leucosporidiella creatinivora*, *Leucosporidium scottii*, *Candida sake*, *Candida spencermartinsiae*	Marine sediments; Algal mat in sediment; marine sponges: *Tedania*, *Hymeniacidon*, *Dendrilla*; Seawater	Food, beverage, detergent, biofuel production, animal feed, textiles, leather, paper processing, and cosmetic industry	[45,47,48,207–209]
L-asparaginase (EC 3.5.1.1)	Degradation of asparagine into ammonia and aspartate	*Cosmospora* sp 0B4B, *Cosmospora* sp 0B1B, *Cosmospora* sp 0B2, *Geomyces* sp. S2B	Marine sediments	Food industry and medical applications as anti-cancer, antimicrobial, infectious diseases, autoimmune diseases	[206]
Pectinase (EC 3.2.1.15)	Hydrolisis of polysaccharides to produce pectate and other galacturonans	*Geomyces* sp. strain F09-T3-2, *Pseudogymnoascus* sp., *Cladosporium* sp. *F09-T12-1*, *Cryptococcus victoriae*, *Leucosporidiella muscorum*, *Metschnikowia australis*, *Rhodotorula glacialis*; *Leucosporidiella creatinivora*, *Leucosporidium scottii*	Marine sponges; marine sediments; Seawater	Food and textile industry, coffee and tea fermentation, wine processing, oil extraction, vegetable and fruit processing industry for juice clarification, color, and yield enhancer. Applications in paper and pulp making, recycling of wastepaper, pretreatment of pectic wastewaters, and retting of plant fibers	[45,82,210]
Phytase (EC 3.1.3.26)	Hydrolysis of phytate to produce phosphorylated myo-inositol derivatives	*Rhodotorula mucilaginosa* JMUY14	Deep-sea sediments	Food and feed industry, pharmaceutical use as neuro protective agents, anti-inflammatory, antioxidant and anti-cancer agents	[211]
Protease (EC 3.4)	Cleavage of peptide bonds	*Rhodotorula mucilaginosa* L7; *Pseudogymnoascus* sp. CRM1533, *Leucosporidiella muscorum*; *Leucosporidiella* sp. 131209-E2A-C3-II-lev, *Leucosporidiella creatinivora* 071209-E8-C4-II-lev; *Rhodotorula glacialis*; *Leucosporidiella creatinivora*, *Leucosporidium scottii*	Marine macroalgae; marine sediments; marine sponges: *Tedania*, *Hymeniacidon*; Seawater	Food, feed, pharmacology (anticancer and antihemolytic activity) cosmetic (keratin-based preparation) industries, cleaning processes (e. g. detergent additive), waste management	[45,47,48,212,213]

Table 2. *Cont.*

Enzyme	Reaction	Fungi	Source of (Isolate) Sample	Applications/Potential Uses	Ref.
Protease (Subtilase) (EC 3.4.21)	Cleavage of peptide bonds	*Glaciozyma antarctica* 17 (formerly *Leucosporidium antarcticum*)	Sub-glacial waters (depth of 200 m)	Food and beverage industries	[214, 215]
Transglutami-nase (EC 2.3.2.13)	Acyl transfer reaction between gamma-carboxyamide groups of glutamine residues in proteins and various primary amines	*Penicillium chrysogenum*	Marine macroalga *Gigartinas kosttbergii*	Food, pharmaceutical, leather, textile, biotechnology industry, biomedical research	[216]
Xylanase (EC 3.2.1.8)	Hydrolysis of the main chain of xylan to oligosaccharides, which in turn are degraded to xylose	*Cladosporium* sp.; *Penicillium* sp. E2B *Penicillium* sp. N5, *Penicillium* sp. E2-1	Marine sponge; marine sediments	Food (bread making), feed, paper and pulp industries, and also used to increase the sugar recovery from agricultural residues for biofuel production	[206, 217, 218]
α-amylase (EC 3.2.1.1)	Cleavage of α-1,4-glycosidic linkages within starch molecules, which generate smaller polymers of glucose units	*Glaciozyma antarctica* PI12 (formerly *Leucosporidium antarcticum*); *Cystofilobasidium infirmominiatum* 071209-E8-C1-IIa-lev, 131209-E2A-C1-II-lev, 131209-E2A-C5-II-lev and isolate 071209-E8-C1-IIb-lev; *Metschnikowia* australis 071209-E8-C3-II-lev and isolate 071209-E8-C1-II-lev; *Leucosporidiella* sp. 131209-E2A-C3-II-lev	Seawater; marine sponges: *Tedania*, *Hymeniacidon*	Pharmaceutical and chemical industry; employed as additives in processed food, in detergents for cold washing, in waste-water treatment, in bioremediation in cold climates, and in molecular biology protocols	[80,202, 219]
β-agarase (EC 3.2.1.81)	Hydrolysis of beta-(1–>4) linkages of agarose to produce oligosaccharides	*Penicillium* sp., *Cladosporium* sp. 2, *Penicillium* sp., *Pseudogymnoascus* sp. UFMGCB 10054, *Doratomyces* sp.	Macroalgae: *Ascoseira mirabilis*, *Georgiella confluens*, *Iridaea cordata*, *Palmaria decipiens*	Food, cosmetic, medical industries, and as a tool enzyme for biological, physiological, and cytological studies	[69]
β-galactosidase (EC 3.2.1.23)	Hydrolysis of lactose into its constituent monosaccharides	*Tausonia pullulans* 17-1 (formerly *Guehomyces, pullulans*)	Marine sediments	Food, biofuel, and agricultural industries; surfactant production	[220]

Tausonia pullulans 17-1 isolated from Antarctic marine sediments can produce cold-active β-galactosidases, which can be a tool for hydrolyzing the lactose present in milk and milk derivatives at low temperatures in the milk-processing industry, allowing intolerant people to consume lactose-free foods and beverages [221,222].

Psychrophilic fungi isolated from Antarctic marine organisms are promising sources of cold-active xylanases, which have interesting applications in the food industry for bread-making as well as in agricultural industry and biofuel production [206,218,223–225]. For instance, the fungus *Cladosporium* sp. isolated from a marine sponge displayed high xylanase activity at a lower temperature than the mesophilic fungus *Penicillium purpurogenum* MYA-30, used as a control [218]. In addition, the *Penicillium* species isolated from different Antarctic marine organisms (i.e., sea stars, molluscs, macroalgae) were able to produce more than 10 U mL^{-1} of xylanase molecules after seven days of cultivation at 20.0 °C [206].

Microbial lipases are important enzymes employed in a variety of applications in the dairy, bakery, oil, meat and fish processing, and beverage industries, for enhancing the food quality, as well as for the detergent and cosmetic industry [226,227]. It is forecasted to reach a market size of 590.2 USD Million by 2023, with an annual growth rate of 6.8% from 2018 [228]. For example, Lipoclean® marketed by Novozymes is a cold-active lipase that is suitable as a detergent additive for its activity at $\simeq$20 °C, high stability in the presence of other enzymes, at alkaline pH and also resistance to oxidizing and chelating agents [229,230]. An interesting lipase activity was reported in some yeasts belonging to *Cryptococcus*, *Leucosporidium*, and *Metschnikowia* genera, which were isolated from Antarctic marine samples [47]. In particular, the highest activity (0.88 U mL^{-1}) was observed in *Metschnikowia* sp. CRM 1589 isolated from marine sediments and *Salpa* sp. when cultured at 15 °C [47]. At the same temperature *Cryptococcus laurentii* L59, *Cryptococcus adeliensis* L121, and *Leucosporidium scottii* L117 showed an enzyme production between 0.1 and 0.23 U mL^{-1} after six days of incubation [54]. Generally, yeasts belonging to genera *Candida*, *Yarrowia*, and *Saccharomyces* can produce lipases [231]. In fact, *Candida antarctica* isolated from sediments of Lake Vanda in Antarctica (a lake permanently covered by ice) can produce two forms of distinct lipases (Lipase A and B), whose production has been patented in 2005 with different industrial and environmental applications [232–234].

Assays using Tween 80 as substrate for testing esterase activity testing identified *Cryptococcus*, *Metschnikowia*, *Rhodotorula*, *Leucosporidium*, and *Leucosporidiella* marine genera as cold-active esterase producers [45,208,235]. Hashim et al. (2018) discovered a new cold-active esterase-like protein with putative dienelactone hydrolase (GaDlh) activity produced by the psychrophilic yeast *Glaciozyma antarctica* isolated from sea ice near the Casey Research Station [205]. This pioneering study on the bioprospection of cold-active enzymes performing the isolation, heterologous expression, and biochemical characterization of recombinant GaDlh highlighted interesting cold-adapted features in the predicted protein structure at a temperature of 10 °C and pH 8.0 [205]. Overall, esterase enzymes are exploited in fine chemicals production and pharmaceutical industries for improving the production of optically pure compounds, such as ibuprofen, ketoprofen, and naproxen [236].

Since the beginning of the new millennium, very few studies have addressed the potential of Antarctic marine fungi as protease producers [237]. Proteases account for 60% of the total enzyme market and it is amongst the most precious commercial enzymes for the wide uses in different kind of industries (i.e., food, pharmacology, detergent) [237–239]. One of the first studies on microbial Antarctic proteases was carried out on *Leucosporidium antarcticum* 171, which can produce a novel extracellular serine protease, lap2 with an optimal temperature as low as 25 °C, high catalytic efficiency in the range 0–25 °C [214]. Afterward, halotolerant extracellular protease produced by *Rhodotorula mucilaginosa* L7 was characterized with optimal catalytic activity at 50 °C and pH 5.0, after a selection of protease positive strains isolated from marine organisms [54,212]. Recently, *Pseudogymnoascus* sp. CRM1533, isolated from Antarctic marine sediments, showed a protease activity of 6.21 U mL^{-1}, even though further studies are needed to characterize the functional potential of this enzyme [47]. The genera *Metschnikowia*, *Cystofilobasidium*, and *Leucosporidiella*, associated with Antarctic marine sponges, displayed extracellular amylase between 4 and 20 °C, whereas only *Leucosporidiella* also showed protease activity [80].

Finally, different fungal genera such as *Penicillium*, *Cladosporium*, *Geomyces*, and *Pseudogymnoascus* isolated from Antarctic macroalgae and sponges showed carrageenolytic and agarolytic activities which can be useful in processes involving the extraction of the algal biomass for the production of bioethanol [69,210]. Among the tested strains, *Geomyces* sp. strain F09-T3-2 displayed also high activity pectinase: 121 U/mg after 5 days at 30 °C [210].

Overall, these findings indicate that the Antarctic marine ecosystems host promising fungal assemblages that display a wide array of unique and novel enzymes. These enzymes offer new horizons for a broad range of biotechnological applications and have great potential to reduce resource and energy consumption, thus promoting eco-sustainability. Obtaining further genetic and functional information on extremophilic fungi inhabiting Antarctic marine ecosystems, coupled with the development of specific bioinformatic pipelines for bioprospecting, are of fundamental importance for the identification of new fungal enzymes and molecules useful for enhancing the growth and competitiveness of the blue biotechnologies.

3.3. Emerging Bioprospecting Methods: Pitfalls and Future Perspectives

Isolation techniques usually employed for characterizing extremophilic fungi typically foster isolation of selected fungal taxa (e.g., faster-growing generalists, mesophilic strains), thus hampering our ability of bioprospection of natural molecules produced by the currently unculturable fungi [32,240,241]. Indeed, there is evidence that cultivability is a significant bottle-neck for the discovery of natural products from extremophilic marine fungi [242,243]. The implementation and development of methodologies aimed at the isolation of extremophilic fungi are urgently needed to fill this gap [244]. Novel cultivation methods as well as culture-independent approaches, can help to overcome current limitations in our understanding of the fungal biodiversity in extreme environments and in the discovery of new enzymes and molecules with biotechnological potential [245]. Indeed, culture-independent techniques coupled with genomics-based approaches are becoming valuable and fast tools to analyse the functional potential of fungal secondary metabolites useful for biotechnological applications [85,246,247]. Metagenomics, metatranscriptomics, and metaproteomics, as well as single-cell genomics, followed by heterologous expression of selected genes of potential interest, represent promising tools to shed new light on the possible biotechnological exploitation of still-uncultured Antarctic fungi [248–251]. Finally, the bioinformatics mining of the still poorly described but rich genetic biodiversity of Antarctic fungi will certainly enhance the rate of discovery of bioactive molecules potentially useful for biotechnological purposes [30,252].

Supplementary Materials: The following are available online at https://www.mdpi.com/article/10.3390/jof7050391/s1, Figures S1–S3: Chemical structures of available bioactive compounds; Table S1: Coordinates of different Antarctic samples collected for fungal identification; Table S2: Fungal taxa isolated with culture-dependent methods and identified through metagenomic analyses.

Author Contributions: C.C. and S.V. conceived the study. S.V. and G.B., M.T., E.R., A.D. conducted literature analysis. G.B. performed analyses on the available data. S.V. and G.B. wrote the first draft of the manuscript, and all authors participated in the writing and revision of the manuscript. All authors have read and agreed to the published version of the manuscript.

Funding: This study has been conducted in the framework of the project PNRA16_00173 "Diversity and Evolution of Marine Microbial Communities associated with Antarctic Benthic Invertebrates (DEMBAI)".

Institutional Review Board Statement: Not applicable.

Informed Consent Statement: Not applicable.

Conflicts of Interest: The authors declare that they have no conflict of interest.

References

1. Rogers, A.D.; Johnston, N.M.; Murphy, E.J.; Clarke, A. *Antarctic Ecosystems*; Rogers, A.D., Johnston, N.M., Murphy, E.J., Clarke, A., Eds.; John Wiley & Sons, Ltd.: Chichester, UK, 2012; ISBN 9781444347241.
2. Chown, S.L.; Brooks, C.M. The State and Future of Antarctic Environments in a Global Context. *Annu. Rev. Environ. Resour.* **2019**, *44*, 1–30. [CrossRef]
3. Wall, D.H. Biodiversity and ecosystem functioning in terrestrial habitats of Antarctica. *Antarct. Sci.* **2005**, *17*, 523–531. [CrossRef]
4. Peck, L.S. Antarctic marine biodiversity: Adaptations, environments and responses to change. *Oceanogr. Mar. Biol.* **2018**, *56*, 105–236.
5. Clarke, A.; Johnston, N. Antarctic Marine Benthic Diversity. In *Oceanography and Marine Biology, An Annual Review*; CRC Press: Boca Raton, FL, USA, 2003; Volume 41, pp. 55–57.
6. De Broyer, C.; Koubbi, P.; Griffiths, H.J.; Raymond, B.; d'Acoz, C.U.; Van de Putte, A.P.; Danis, B.; David, B.; Grant, S.; Gutt, J.; et al. *Biogeographic Atlas of the Southern Ocean*; Scientific Committee on Antarctic Research: Cambridge, UK, 2014; ISBN 9780948277283.
7. Griffiths, H.J. Antarctic marine biodiversity—What do we know about the distribution of life in the southern ocean? *PLoS ONE* **2010**, *5*, e11683. [CrossRef]
8. Gutt, J.; Sirenko, B.I.; Smirnov, I.S.; Arntz, W.E. How many macrozoobenthic species might inhabit the Antarctic shelf? *Antarct. Sci.* **2004**, *16*, 11–16. [CrossRef]
9. Gutt, J.; Constable, A.; Cummings, V.; Hosie, G.; McIntyre, T.; Mintenbeck, K.; Murray, A.; Peck, L.; Ropert-Coudert, Y.; Saba, G.K. Vulnerability of Southern Ocean biota to climate change. *Antarct. Environ. Portal* **2016**. [CrossRef]
10. Convey, P.; Peck, L.S. Antarctic environmental change and biological responses. *Sci. Adv.* **2019**, *5*, eaaz0888. [CrossRef]
11. Clarke, A.; Crame, J.A. Evolutionary dynamics at high latitudes: Speciation and extinction in polar marine faunas. *Philos. Trans. R. Soc. B Biol. Sci.* **2010**, *365*, 3655–3666. [CrossRef]
12. Lo Giudice, A.; Azzaro, M. Diversity and Ecological Roles of Prokaryotes in the Changing Antarctic Marine Environment. In *The Ecological Role of Micro-Organisms in the Antarctic Environment*; Springer Polar Sciences: Cham, Switzerland; Berlin/Heidelberg, Germany, 2019; pp. 109–131.
13. Cavicchioli, R. Microbial ecology of Antarctic aquatic systems. *Nat. Rev. Microbiol.* **2015**, *13*, 691–706. [CrossRef]
14. Worden, A.Z.; Follows, M.J.; Giovannoni, S.J.; Wilken, S.; Zimmerman, A.E.; Keeling, P.J. Rethinking the marine carbon cycle: Factoring in the multifarious lifestyles of microbes. *Science* **2015**, *347*, 1257594. [CrossRef]
15. Wilkins, D.; Yau, S.; Williams, T.J.; Allen, M.A.; Brown, M.V.; Demaere, M.Z.; Lauro, F.M.; Cavicchioli, R. Key microbial drivers in Antarctic aquatic environments. *FEMS Microbiol. Rev.* **2013**, *37*, 303–335. [CrossRef]
16. Falkowski, P.G.; Fenchel, T.; Delong, E.F. The microbial engines that drive Earth's biogeochemical cycles. *Science* **2008**, *320*, 1034–1039. [CrossRef]
17. Margesin, R.; Feller, G. Biotechnological applications of psychrophiles. *Environ. Technol.* **2010**, *31*, 835–844. [CrossRef]
18. Rizzo, C.; Lo Giudice, A. The variety and inscrutability of polar environments as a resource of biotechnologically relevant molecules. *Microorganisms* **2020**, *8*, 1422. [CrossRef]
19. Correa, T.; Abreu, F. Antarctic microorganisms as sources of biotechnological products. In *Physiological and Biotechnological Aspects of Extremophiles*; Salwan, R., Sharma, V., Eds.; Elsevier: Amsterdam, The Netherlands, 2020; pp. 269–284. ISBN 978-0-12-818322-9.
20. Bruno, S.; Coppola, D.; di Prisco, G.; Giordano, D.; Verde, C. Enzymes from Marine Polar Regions and Their Biotechnological Applications. *Mar. Drugs* **2019**, *17*, 544. [CrossRef]
21. Tiquia-Arashiro, S.M.; Grube, M. *Fungi in Extreme Environments: Ecological Role and Biotechnological Significance*; Tiquia-Arashiro, S.M., Grube, M., Eds.; Springer International Publishing: Cham, Switzerland; Berlin/Heidelberg, Germany, 2019; ISBN 978-3-030-19030-9.
22. Sayed, A.M.M.; Hassan, M.H.A.H.A.; Alhadrami, H.A.A.; Hassan, H.M.M.; Goodfellow, M.; Rateb, M.E.E. Extreme environments: Microbiology leading to specialized metabolites. *J. Appl. Microbiol.* **2020**, *128*, 630–657. [CrossRef]
23. Macheleidt, J.; Mattern, D.J.; Fischer, J.; Netzker, T.; Weber, J.; Schroeckh, V.; Valiante, V.; Brakhage, A.A. Regulation and role of fungal secondary metabolites. *Annu. Rev. Genet.* **2016**, *50*, 371–392. [CrossRef]
24. Spiteller, P. Chemical ecology of fungi. *Nat. Prod. Rep.* **2015**, *32*, 971–993. [CrossRef]
25. Brakhage, A.A. Regulation of fungal secondary metabolism. *Nat. Rev. Microbiol.* **2013**, *11*, 21–32. [CrossRef]
26. Yu, J.-H.; Keller, N. Regulation of secondary metabolism in filamentous fungi. *Annu. Rev. Phytopathol.* **2005**, *43*, 437–458. [CrossRef]
27. Rosa, L.H. *Fungi of Antarctica: Diversity, Ecology and Biotechnological Applications*; Springer International Publishing: Cham, Switzerland; Berlin/Heidelberg, Germany, 2019; ISBN 978-3-030-18367-7.
28. Yarzábal, L.A. Antarctic Psychrophilic Microorganisms and Biotechnology: History, Current Trends, Applications, and Challenges. In *Microbial Models: From Environmental to Industrial Sustainability*; Springer: Singapore, 2016; pp. 83–118.
29. Corinaldesi, C.; Barone, G.; Marcellini, F.; Dell'Anno, A.; Danovaro, R. Marine microbial-derived molecules and their potential use in cosmeceutical and cosmetic products. *Mar. Drugs* **2017**, *15*, 118. [CrossRef]
30. Chávez, R.; Fierro, F.; García-Rico, R.O.; Vaca, I. Filamentous fungi from extreme environments as a promising source of novel bioactive secondary metabolites. *Front. Microbiol.* **2015**, *6*, 903. [CrossRef] [PubMed]
31. Cavicchioli, R.; Amils, R.; Wagner, D.; Mcgenity, T. Life and applications of extremophiles. *Environ. Microbiol.* **2011**, *13*, 1903–1907. [CrossRef] [PubMed]

32. Amend, A.; Burgaud, G.; Cunliffe, M.; Edgcomb, V.P.; Ettinger, C.L.; Gutiérrez, M.H.; Heitman, J.; Hom, E.F.Y.; Ianiri, G.; Jones, A.C.; et al. Fungi in the marine environment: Open questions and unsolved problems. *MBio* **2019**, *10*, 2021. [CrossRef] [PubMed]
33. Wang, Y.; Sen, B.; He, Y.; Xie, N.; Wang, G. Spatiotemporal Distribution and Assemblages of Planktonic Fungi in the Coastal Waters of the Bohai Sea. *Front. Microbiol.* **2018**, *9*, 584. [CrossRef] [PubMed]
34. Hassett, B.T.; Vonnahme, T.R.; Peng, X.; Jones, E.B.G.; Heuzé, C. Global diversity and geography of planktonic marine fungi. *Bot. Mar.* **2020**, *63*, 121–139. [CrossRef]
35. López-García, P.; Rodríguez-Valera, F.; Pedrós-Alió, C.; Moreira, D. Unexpected diversity of small eukaryotes in deep-sea Antarctic plankton. *Nature* **2001**, *409*, 603–607. [CrossRef]
36. Zhang, T.; Wang, N.F.; Zhang, Y.Q.; Liu, H.Y.; Yu, L.Y. Diversity and distribution of fungal communities in the marine sediments of Kongsfjorden, Svalbard (High Arctic). *Sci. Rep.* **2015**, *5*, 14524. [CrossRef]
37. Hassett, B.T.; Ducluzeau, A.L.L.; Collins, R.E.; Gradinger, R. Spatial distribution of aquatic marine fungi across the western Arctic and sub-arctic. *Environ. Microbiol.* **2017**, *19*, 475–484. [CrossRef]
38. Turkiewicz, M.; Pazgier, M.; Donachie, S.P.; Kalinowska, H. Invertase and α-glucosidase production by the endemic Antarctic marine yeast *Leucosporidium antarcticum*. *Polish Polar Res.* **2005**, *26*, 125–136.
39. Gonçalves, V.N.; Vitoreli, G.A.; de Menezes, G.C.A.; Mendes, C.R.B.; Secchi, E.R.; Rosa, C.A.; Rosa, L.H. Taxonomy, phylogeny and ecology of cultivable fungi present in seawater gradients across the Northern Antarctica Peninsula. *Extremophiles* **2017**, *21*, 1005–1015. [CrossRef]
40. Ward, N.A.; Robertson, C.L.; Chanda, A.K.; Schneider, R.W. Effects of *Simplicillium lanosoniveum* on phakopsora pachyrhizi, the soybean rust pathogen, and its use as a biological control agent. *Phytopathology* **2012**, *102*, 749–760. [CrossRef]
41. Guerra, R.S.; do Nascimento, M.M.F.; Miesch, S.; Najafzadeh, M.J.; Ribeiro, R.O.; Ostrensky, A.; de Hoog, G.S.; Vicente, V.A.; Boeger, W.A. Black Yeast Biota in the Mangrove, in Search of the Origin of the Lethargic Crab Disease (LCD). *Mycopathologia* **2013**, *175*, 421–430. [CrossRef]
42. Dayarathne, M.; Jones, E.; Maharachchikumbura, S.; Devadatha, B.; Sarma, V.; Khongphinitbunjong, K.; Chomnunti, P.; Hyde, K. Morpho-molecular characterization of microfungi associated with marine based habitats. *Mycosphere* **2020**, *11*, 1–188. [CrossRef]
43. Hassett, B.T.; Gradinger, R. Chytrids dominate arctic marine fungal communities. *Environ. Microbiol.* **2016**, *18*, 2001–2009. [CrossRef]
44. Hassett, B.T.; Borrego, E.J.; Vonnahme, T.R.; Rämä, T.; Kolomiets, M.V.; Gradinger, R. Arctic marine fungi: Biomass, functional genes, and putative ecological roles. *ISME J.* **2019**, *13*, 1484–1496. [CrossRef]
45. Vaz, A.B.M.; Rosa, L.H.; Vieira, M.L.A.; de Garcia, V.; Brandão, L.R.; Teixeira, L.C.R.S.; Moliné, M.; Libkind, D.; van Broock, M.; Rosa, C.A. The diversity, extracellular enzymatic activities and photoprotective compounds of yeasts isolated in Antarctica. *Braz. J. Microbiol.* **2011**, *42*, 937–947. [CrossRef]
46. Wu, G.; Ma, H.; Zhu, T.; Li, J.; Gu, Q.; Li, D. Penilactones A and B, two novel polyketides from Antarctic deep-sea derived fungus *Penicillium crustosum* PRB-2. *Tetrahedron* **2012**, *68*, 9745–9749. [CrossRef]
47. Wentzel, L.C.P.; Inforsato, F.J.; Montoya, Q.V.; Rossin, B.G.; Nascimento, N.R.; Rodrigues, A.; Sette, L.D. Fungi from Admiralty Bay (King George Island, Antarctica) Soils and Marine Sediments. *Microb. Ecol.* **2019**, *77*, 12–24. [CrossRef]
48. Ogaki, M.B.; Teixeira, D.R.; Vieira, R.; Lírio, J.M.; Felizardo, J.P.S.; Abuchacra, R.C.; Cardoso, R.P.; Zani, C.L.; Alves, T.M.A.; Junior, P.A.S.; et al. Diversity and bioprospecting of cultivable fungal assemblages in sediments of lakes in the Antarctic Peninsula. *Extremophiles* **2020**, *124*, 601–611. [CrossRef]
49. Ogaki, M.B.; Pinto, O.H.B.; Vieira, R.; Neto, A.A.; Convey, P.; Carvalho-Silva, M.; Rosa, C.A.; Câmara, P.E.A.S.; Rosa, L.H. Fungi Present in Antarctic Deep-Sea Sediments Assessed Using DNA Metabarcoding. *Microb. Ecol.* **2021**, 1–8. [CrossRef]
50. Ha, T.M.; Kim, D.-C.C.; Sohn, J.H.; Yim, J.H.; Oh, H. Anti-Inflammatory and Protein Tyrosine Phosphatase 1B Inhibitory Metabolites from the Antarctic Marine-Derived Fungal Strain *Penicillium glabrum* SF-7123. *Mar. Drugs* **2020**, *18*, 247. [CrossRef] [PubMed]
51. Wu, G.; Lin, A.; Gu, Q.; Zhu, T.; Li, D. Four new chloro-eremophilane sesquiterpenes from an antarctic deep-sea derived fungus, *Penicillium* sp. PR19N-1. *Mar. Drugs* **2013**, *11*, 1399–1408. [CrossRef] [PubMed]
52. Niu, S.; Fan, Z.-W.; Xie, C.-L.; Liu, Q.; Luo, Z.-H.; Liu, G.; Yang, X.-W. Spirograterpene A, a Tetracyclic Spiro-Diterpene with a Fused 5/5/5/5 Ring System from the Deep-Sea-Derived Fungus *Penicillium granulatum* MCCC 3A00475. *J. Nat. Prod.* **2017**, *80*, 2174–2177. [CrossRef]
53. Laich, F.; Vaca, I.; Chávez, R. *Rhodotorula portillonensis* sp. nov., a basidiomycetous yeast isolated from Antarctic shallow-water marine sediment. *Int. J. Syst. Evol. Microbiol.* **2013**, *63*, 3884–3891. [CrossRef]
54. Duarte, A.W.F.; Dayo-Owoyemi, I.; Nobre, F.S.; Pagnocca, F.C.; Chaud, L.C.S.; Pessoa, A.; Felipe, M.G.A.; Sette, L.D. Taxonomic assessment and enzymes production by yeasts isolated from marine and terrestrial Antarctic samples. *Extremophiles* **2013**, *17*, 1023–1035. [CrossRef]
55. Gonçalves, V.N.; Campos, L.S.; Melo, I.S.; Pellizari, V.H.; Rosa, C.A.; Rosa, L.H. *Penicillium solitum*: A mesophilic, psychrotolerant fungus present in marine sediments from Antarctica. *Polar Biol.* **2013**, *36*, 1823–1831. [CrossRef]
56. Ren, J.; Yang, Y.; Liu, D.; Chen, W.; Proksch, P.; Shao, B.; Lin, W. Sequential determination of new peptaibols asperelines G-Z12 produced by marine-derived fungus *Trichoderma asperellum* using ultrahigh pressure liquid chromatography combined with electrospray-ionization tandem mass spectrometry. *J. Chromatogr. A* **2013**, *1309*, 90–95. [CrossRef]

57. Purić, J.; Vieira, G.; Cavalca, L.B.B.; Sette, L.D.D.; Ferreira, H.; Vieira, M.L.C.L.C.; Sass, D.C.C. Activity of Antarctic fungi extracts against phytopathogenic bacteria. *Lett. Appl. Microbiol.* **2018**, *66*, 530–536. [CrossRef]
58. Vieira, G.; Purić, J.; Morão, L.G.G.; dos Santos, J.A.A.; Inforsato, F.J.J.; Sette, L.D.D.; Ferreira, H.; Sass, D.C.C. Terrestrial and marine Antarctic fungi extracts active against *Xanthomonas citri* subsp. citri. *Lett. Appl. Microbiol.* **2018**, *67*, 64–71. [CrossRef]
59. Li, W.; Luo, D.; Huang, J.; Wang, L.; Zhang, F.; Xi, T.; Liao, J.; Lu, Y. Antibacterial constituents from Antarctic fungus, *Aspergillus sydowii* SP-1. *Nat. Prod. Res.* **2018**, *32*, 662–667. [CrossRef]
60. Zhou, Y.; Li, Y.H.; Yu, H.B.; Liu, X.Y.; Lu, X.L.; Jiao, B.H. Furanone derivative and sesquiterpene from Antarctic marine-derived fungus *Penicillium* sp. S-1-18. *J. Asian Nat. Prod. Res.* **2018**, *20*, 1108–1115. [CrossRef]
61. Liu, C.-C.; Zhang, Z.-Z.; Feng, Y.-Y.; Gu, Q.-Q.; Li, D.-H.; Zhu, T.-J. Secondary metabolites from Antarctic marine-derived fungus *Penicillium crustosum* HDN153086. *Nat. Prod. Res.* **2019**, *33*, 414–419. [CrossRef]
62. Mercantini, R.; Marsella, R.; Moretto, D.; Finotti, E. Keratinophilic fungi in the antarctic environment. *Mycopathologia* **1993**, *122*, 169–175. [CrossRef]
63. Dos Santos, J.A.; Meyer, E.; Sette, L.D. Fungal community in antarctic soil along the retreating collins glacier (Fildes peninsula, King George Island). *Microorganisms* **2020**, *8*, 1145. [CrossRef]
64. Loque, C.P.; Medeiros, A.O.; Pellizzari, F.M.; Oliveira, E.C.; Rosa, C.A.; Rosa, L.H. Fungal community associated with marine macroalgae from Antarctica. *Polar Biol.* **2010**, *33*, 641–648. [CrossRef]
65. Putzke, J.; Pereira, A.B.; Pereira, A.B. *Phaeosphaeria deschampsii* (Ascomycota): A new parasite species of *Deschampsia antarctica* (Poaceae) described to Antarctica. *An. Acad. Bras. Cienc.* **2016**, *88*, 1967–1969. [CrossRef]
66. Godinho, V.M.; Furbino, L.E.; Santiago, I.F.; Pellizzari, F.M.; Yokoya, N.S.; Pupo, D.; Alves, T.M.A.; Junior, P.A.S.; Romanha, A.J.; Zani, C.L.; et al. Diversity and bioprospecting of fungal communities associated with endemic and cold-adapted macroalgae in Antarctica. *ISME J.* **2013**, *7*, 1434–1451. [CrossRef]
67. Fernandes Duarte, A.W.; Zambrano Passarini, M.R.; Delforno, T.P.; Pellizzari, F.M.; Zecchin Cipro, C.V.; Montone, R.C.; Petry, M.V.; Putzke, J.; Rosa, L.H.; Sette, L.D. Yeasts from macroalgae and lichens that inhabit the South Shetland Islands, Antarctica. *Environ. Microbiol. Rep.* **2016**, *8*, 874–885. [CrossRef]
68. Furbino, L.E.; Godinho, V.M.; Santiago, I.F.; Pellizari, F.M.; Alves, T.M.A.A.; Zani, C.L.; Junior, P.A.S.S.; Romanha, A.J.; Carvalho, A.G.O.O.; Gil, L.H.V.G.V.G.; et al. Diversity Patterns, Ecology and Biological Activities of Fungal Communities Associated with the Endemic Macroalgae Across the Antarctic Peninsula. *Microb. Ecol.* **2014**, *67*, 775–787. [CrossRef]
69. Furbino, L.E.; Pellizzari, F.M.; Neto, P.C.; Rosa, C.A.; Rosa, L.H. Isolation of fungi associated with macroalgae from maritime Antarctica and their production of agarolytic and carrageenolytic activities. *Polar Biol.* **2018**, *41*, 527–535. [CrossRef]
70. Martorell, M.M.; Lannert, M.; Matula, C.V.; Quartino, M.L.; de Figueroa, L.I.C.; Mac Cormack, W.P.; Ruberto, L.A.M. Studies toward the comprehension of fungal-macroalgae interaction in cold marine regions from a biotechnological perspective. *Fungal Biol.* **2020**, *125*, 218–230. [CrossRef]
71. Alker, A.P.; Smith, G.W.; Kim, K. Characterization of *Aspergillus sydowii* (Thom et Church), a fungal pathogen of Caribbean sea fan corals. *Hydrobiologia* **2001**, *460*, 105–111. [CrossRef]
72. Van Dover, C.L.; Ward, M.E.; Scott, J.L.; Underdown, J.; Anderson, B.; Gustafson, C.; Whalen, M.; Carnegie, R.B. A fungal epizootic in mussels at a deep-sea hydrothermal vent. *Mar. Ecol.* **2007**, *28*, 54–62. [CrossRef]
73. Cawthorn, R.J. Diseases of American lobsters (*Homarus americanus*): A review. *J. Invertebr. Pathol.* **2011**, *106*, 71–78. [CrossRef]
74. Amend, A. From Dandruff to Deep-Sea Vents: Malassezia-like Fungi Are Ecologically Hyper-diverse. *PLoS Pathog.* **2014**, *10*, e1004277. [CrossRef]
75. Nagahama, T.; Hamamoto, M.; Nakase, T.; Horikoshi, K. *Rhodotorula lamellibrachii* sp. nov., a new yeast species from a tubeworm collected at the deep-sea floor in Sagami Bay and its phylogenetic analysis. *Antonie Leeuwenhoek* **2001**, *80*, 317–323. [CrossRef]
76. Yu, Z.; Zhang, B.; Sun, W.; Zhang, F.; Li, Z. Phylogenetically diverse endozoic fungi in the South China Sea sponges and their potential in synthesizing bioactive natural products suggested by PKS gene and cytotoxic activity analysis. *Fungal Divers.* **2013**, *58*, 127–141. [CrossRef]
77. Nguyen, M.T.H.D.; Thomas, T. Diversity, host-specificity and stability of sponge-associated fungal communities of co-occurring sponges. *PeerJ* **2018**, *2018*, e4965. [CrossRef] [PubMed]
78. Raghukumar, S. Animals in Coastal Benthic Ecosystem and Aquaculture Systems. In *Fungi in Coastal and Oceanic Marine Ecosystems*; Raghukumar, S., Ed.; Springer International Publishing: Cham, Switzerland; Berlin/Heidelberg, Germany, 2017; pp. 163–183. ISBN 978-3-319-54304-8.
79. Laich, F.; Chávez, R.; Vaca, I. *Leucosporidium escuderoi* f.a., sp. nov., a basidiomycetous yeast associated with an Antarctic marine sponge. *Antonie Leeuwenhoek* **2014**, *105*, 593–601. [CrossRef] [PubMed]
80. Vaca, I.; Faúndez, C.; Maza, F.; Paillavil, B.; Hernández, V.; Acosta, F.; Levicán, G.; Martínez, C.; Chávez, R. Cultivable psychrotolerant yeasts associated with Antarctic marine sponges. *World J. Microbiol. Biotechnol.* **2013**, *29*, 183–189. [CrossRef]
81. Kim, D.C.; Lee, H.S.; Ko, W.; Lee, D.S.; Sohn, J.H.; Yim, J.H.; Kim, Y.C.; Oh, H. Anti-inflammatory effect of methylpenicinoline from a marine isolate of *Penicillium* sp. (SF-5995): Inhibition of NF-κB and MAPK pathways in lipopolysaccharide-induced RAW264.7 macrophages and BV2 microglia. *Molecules* **2014**, *19*, 18073–18089. [CrossRef]
82. Henríquez, M.; Vergara, K.; Norambuena, J.; Beiza, A.; Maza, F.; Ubilla, P.; Araya, I.; Chávez, R.; San-Martín, A.; Darias, J.; et al. Diversity of cultivable fungi associated with Antarctic marine sponges and screening for their antimicrobial, antitumoral and antioxidant potential. *World J. Microbiol. Biotechnol.* **2014**, *30*, 65–76. [CrossRef]

83. Cui, X.; Zhu, G.; Liu, H.; Jiang, G.; Wang, Y.; Zhu, W. Diversity and function of the Antarctic krill microorganisms from *Euphausia superba*. *Sci. Rep.* **2016**, *6*, 36496. [CrossRef]
84. Wang, F.; Sheng, J.; Chen, Y.; Xu, J. Microbial diversity and dominant bacteria causing spoilage during storage and processing of the Antarctic krill, *Euphausia superba*. *Polar Biol.* **2021**, *44*, 163–171. [CrossRef]
85. Moreno-Pino, M.; Cristi, A.; Gillooly, J.F.; Trefault, N. Characterizing the microbiomes of Antarctic sponges: A functional metagenomic approach. *Sci. Rep.* **2020**, *10*, 645. [CrossRef]
86. Coda, R.; Cassone, A.; Rizzello, C.G.; Nionelli, L.; Cardinali, G.; Gobbetti, M. Antifungal activity of *Wickerhamomyces anomalus* and *Lactobacillus plantarum* during sourdough fermentation: Identification of novel compounds and long-term effect during storage of wheat bread. *Appl. Environ. Microbiol.* **2011**, *77*, 3484–3492. [CrossRef]
87. Walker, A.K.; Robicheau, B.M. Fungal diversity and community structure from coastal and barrier island beaches in the United States Gulf of Mexico. *Sci. Rep.* **2021**, *11*, 3889. [CrossRef]
88. Nagano, Y.; Nagahama, T.; Hatada, Y.; Nunoura, T.; Takami, H.; Miyazaki, J.; Takai, K.; Horikoshi, K. Fungal diversity in deep-sea sediments—The presence of novel fungal groups. *Fungal Ecol.* **2010**, *3*, 316–325. [CrossRef]
89. Rédou, V.; Ciobanu, M.C.; Pachiadaki, M.G.; Edgcomb, V.; Alain, K.; Barbier, G.; Burgaud, G. In-depth analyses of deep subsurface sediments using 454-pyrosequencing reveals a reservoir of buried fungal communities at record-breaking depths. *FEMS Microbiol. Ecol.* **2014**, *90*, 908–921. [CrossRef]
90. Singh, P.; Raghukumar, C.; Verma, P.; Shouche, Y. Fungal Community Analysis in the Deep-Sea Sediments of the Central Indian Basin by Culture-Independent Approach. *Microb. Ecol.* **2011**, *61*, 507–517. [CrossRef]
91. Gao, Y.; Du, X.; Xu, W.; Fan, R.; Zhang, X.; Yang, S.; Chen, X.; Lv, J.; Luo, Z. Fungal Diversity in Deep Sea Sediments from East Yap Trench and Their Denitrification Potential. *Geomicrobiol. J.* **2020**, *37*, 848–858. [CrossRef]
92. Vargas-Gastélum, L.; Riquelme, M. The Mycobiota of the Deep Sea: What Omics Can Offer. *Life* **2020**, *10*, 292. [CrossRef]
93. Barone, G.; Rastelli, E.; Corinaldesi, C.; Tangherlini, M.; Danovaro, R.; Dell'Anno, A. Benthic deep-sea fungi in submarine canyons of the Mediterranean Sea. *Prog. Oceanogr.* **2018**, *168*, 57–64. [CrossRef]
94. Yang, S.; Xu, W.; Gao, Y.; Chen, X.; Luo, Z.H. Fungal diversity in deep-sea sediments from Magellan seamounts environment of the western Pacific revealed by high-throughput Illumina sequencing. *J. Microbiol.* **2020**, *58*, 841–852. [CrossRef]
95. Alexander, E.; Stock, A.; Breiner, H.W.; Behnke, A.; Bunge, J.; Yakimov, M.M.; Stoeck, T. Microbial eukaryotes in the hypersaline anoxic L'Atalante deep-sea basin. *Environ. Microbiol.* **2009**, *11*, 360–381. [CrossRef]
96. Bernhard, J.M.; Kormas, K.; Pachiadaki, M.G.; Rocke, E.; Beaudoin, D.J.; Morrison, C.; Visscher, P.T.; Cobban, A.; Starczak, V.R.; Edgcomb, V.P. Benthic protists and fungi of Mediterranean deep hypsersaline anoxic basin redoxcline sediments. *Front. Microbiol.* **2014**, *5*, 605. [CrossRef]
97. Stock, A.; Breiner, H.W.; Pachiadaki, M.; Edgcomb, V.; Filker, S.; La Cono, V.; Yakimov, M.M.; Stoeck, T. Microbial eukaryote life in the new hypersaline deep-sea basin Thetis. *Extremophiles* **2012**, *16*, 21–34. [CrossRef]
98. Nagahama, T.; Takahashi, E.; Nagano, Y.; Abdel-Wahab, M.A.; Miyazaki, M. Molecular evidence that deep-branching fungi are major fungal components in deep-sea methane cold-seep sediments. *Environ. Microbiol.* **2011**, *13*, 2359–2370. [CrossRef]
99. Wang, Y.; Zhang, W.P.; Cao, H.L.; Shek, C.S.; Tian, R.M.; Wong, Y.H.; Batang, Z.; Al-Suwailem, A.; Qian, P.-Y.Y. Diversity and distribution of eukaryotic microbes in and around a brine pool adjacent to the Thuwal cold seeps in the Red Sea. *Front. Microbiol.* **2014**, *5*, 37. [CrossRef]
100. Nagano, Y.; Miura, T.; Nishi, S.; Lima, A.O.; Nakayama, C.; Pellizari, V.H.; Fujikura, K. Fungal diversity in deep-sea sediments associated with asphalt seeps at the Sao Paulo Plateau. *Deep Sea Res. Part II Top. Stud. Oceanogr.* **2017**, *146*, 59–67. [CrossRef]
101. Gadanho, M.; Sampaio, J.P. Occurrence and diversity of yeasts in the mid-atlantic ridge hydrothermal fields near the Azores Archipelago. *Microb. Ecol.* **2005**, *50*, 408–417. [CrossRef] [PubMed]
102. Burgaud, G.; Arzur, D.; Durand, L.; Cambon-Bonavita, M.-A.; Barbier, G. Marine culturable yeasts in deep-sea hydrothermal vents: Species richness and association with fauna. *FEMS Microbiol. Ecol.* **2010**, *73*, 121–133. [CrossRef] [PubMed]
103. Velez, P.; Gasca-Pineda, J.; Riquelme, M. Cultivable fungi from deep-sea oil reserves in the Gulf of Mexico: Genetic signatures in response to hydrocarbons. *Mar. Environ. Res.* **2020**, *153*, 104816. [CrossRef] [PubMed]
104. Gonçalves, V.N.; Carvalho, C.R.; Johann, S.; Mendes, G.; Alves, T.M.; Zani, C.L.; Junior, P.A.S.S.; Murta, S.M.F.F.; Romanha, A.J.; Cantrell, C.L.; et al. Antibacterial, antifungal and antiprotozoal activities of fungal communities present in different substrates from Antarctica. *Polar Biol.* **2015**, *38*, 1143–1152. [CrossRef]
105. Bochdansky, A.B.; Clouse, M.A.; Herndl, G.J. Eukaryotic microbes, principally fungi and labyrinthulomycetes, dominate biomass on bathypelagic marine snow. *ISME J.* **2017**, *11*, 362–373. [CrossRef]
106. Ogaki, M.B.; Coelho, L.C.; Vieira, R.; Neto, A.A.; Zani, C.L.; Alves, T.M.A.A.; Junior, P.A.S.S.; Murta, S.M.F.F.; Barbosa, E.C.; Oliveira, J.G.; et al. Cultivable fungi present in deep-sea sediments of Antarctica: Taxonomy, diversity, and bioprospecting of bioactive compounds. *Extremophiles* **2020**, *24*, 227–238. [CrossRef]
107. Gladfelter, A.S.; James, T.Y.; Amend, A.S. Marine fungi. *Curr. Biol.* **2019**, *29*, R191–R195. [CrossRef]
108. Harms, H.; Schlosser, D.; Wick, L.Y. Untapped potential: Exploiting fungi in bioremediation of hazardous chemicals. *Nat. Rev. Microbiol.* **2011**, *9*, 177–192. [CrossRef]
109. Barone, G.; Varrella, S.; Tangherlini, M.; Rastelli, E.; Dell'Anno, A.; Danovaro, R.; Corinaldesi, C.; Dell'Anno, A.; Danovaro, R.; Corinaldesi, C. Marine Fungi: Biotechnological Perspectives from Deep-Hypersaline Anoxic Basins. *Diversity* **2019**, *11*, 113. [CrossRef]

110. Grossart, H.P.; Rojas-Jimenez, K. Aquatic fungi: Targeting the forgotten in microbial ecology. *Curr. Opin. Microbiol.* **2016**, *31*, 140–145. [CrossRef]
111. Taylor, J.D.; Cunliffe, M. Multi-year assessment of coastal planktonic fungi reveals environmental drivers of diversity and abundance. *ISME J.* **2016**, *10*, 2118–2128. [CrossRef]
112. Grossart, H.P.; Van den Wyngaert, S.; Kagami, M.; Wurzbacher, C.; Cunliffe, M.; Rojas-Jimenez, K. Fungi in aquatic ecosystems. *Nat. Rev. Microbiol.* **2019**, *17*, 339–354. [CrossRef]
113. Kagami, M.; Miki, T.; Takimoto, G. Mycoloop: Chytrids in aquatic food webs. *Front. Microbiol.* **2014**, *5*, 166. [CrossRef]
114. Comeau, A.M.; Vincent, W.F.; Bernier, L.; Lovejoy, C. Novel chytrid lineages dominate fungal sequences in diverse marine and freshwater habitats. *Sci. Rep.* **2016**, *6*, 30120. [CrossRef]
115. De Vargas, C.; Audic, S.; Henry, N.; Decelle, J.; Mahé, F.; Logares, R.; Lara, E.; Berney, C.; Le Bescot, N.; Probert, I.; et al. Eukaryotic plankton diversity in the sunlit ocean. *Science* **2015**, *15*, 1261605. [CrossRef]
116. Frenken, T.; Alacid, E.; Berger, S.A.; Bourne, E.C.; Gerphagnon, M.; Grossart, H.P.; Gsell, A.S.; Ibelings, B.W.; Kagami, M.; Küpper, F.C.; et al. Integrating chytrid fungal parasites into plankton ecology: Research gaps and needs. *Environ. Microbiol.* **2017**, *19*, 3802–3822. [CrossRef]
117. Lepelletier, F.; Karpov, S.A.; Alacid, E.; Le Panse, S.; Bigeard, E.; Garcés, E.; Jeanthon, C.; Guillou, L. Dinomyces arenysensis gen. et sp. nov. (Rhizophydiales, Dinomycetaceae fam. nov.), a Chytrid Infecting Marine Dinoflagellates. *Protist* **2014**, *165*, 230–244. [CrossRef]
118. Sime-Ngando, T. Phytoplankton Chytridiomycosis: Fungal Parasites of Phytoplankton and Their Imprints on the Food Web Dynamics. *Front. Microbiol.* **2012**, *3*, 361. [CrossRef]
119. Scholz, B.; Küpper, F.C.; Vyverman, W.; Karsten, U. Eukaryotic pathogens (Chytridiomycota and Oomycota) infecting marine microphytobenthic diatoms—A methodological comparison. *J. Phycol.* **2014**, *50*, 1009–1019. [CrossRef]
120. Gleason, F.H.; Kagami, M.; Lefevre, E.; Sime-Ngando, T. The ecology of chytrids in aquatic ecosystems: Roles in food web dynamics. *Fungal Biol. Rev.* **2008**, *22*, 17–25. [CrossRef]
121. Gachon, C.M.M.; Sime-Ngando, T.; Strittmatter, M.; Chambouvet, A.; Kim, G.H. Algal diseases: Spotlight on a black box. *Trends Plant Sci.* **2010**, *15*, 633–640. [CrossRef]
122. Pusceddu, A.; Dell'Anno, A.; Fabiano, M.; Danovaro, R. Quantity and bioavailability of sediment organic matter as signatures of benthic trophic status. *Mar. Ecol. Prog. Ser.* **2009**, *375*, 41–52. [CrossRef]
123. Jebaraj, C.S.; Raghukumar, C.; Behnke, A.; Stoeck, T. Fungal diversity in oxygen-depleted regions of the Arabian Sea revealed by targeted environmental sequencing combined with cultivation. *FEMS Microbiol. Ecol.* **2010**, *71*, 399–412. [CrossRef] [PubMed]
124. Dethier, M.N.; Brown, A.S.; Burgess, S.; Eisenlord, M.E.; Galloway, A.W.E.; Kimber, J.; Lowe, A.T.; O'Neil, C.M.; Raymond, W.W.; Sosik, E.A.; et al. Degrading detritus: Changes in food quality of aging kelp tissue varies with species. *J. Exp. Mar. Bio. Ecol.* **2014**, *460*, 72–79. [CrossRef]
125. Crowther, T.W.; Grossart, H.P. The role of bottom-up and top-down interactions in determining microbial and fungal diversity and function. In *Trophic Ecology: Bottom-Up and Top-Down Interactions Across Aquatic and Terrestrial Systems*; Hanley, T.C., La Pierre, K.J., Eds.; Cambridge University Press: Cambridge, UK, 2015; pp. 260–287. ISBN 9781139924856.
126. Attermeyer, K.; Premke, K.; Hornick, T.; Hilt, S.; Grossart, H.P. Ecosystem-level studies of terrestrial carbon reveal contrasting bacterial metabolism in different aquatic habitats. *Ecology* **2013**, *94*, 2754–2766. [CrossRef] [PubMed]
127. Gomes, F.C.O.; Safar, S.V.B.; Marques, A.R.; Medeiros, A.O.; Santos, A.R.O.; Carvalho, C.; Lachance, M.-A.; Sampaio, J.P.; Rosa, C.A. The diversity and extracellular enzymatic activities of yeasts isolated from water tanks of *Vriesea minarum*, an endangered bromeliad species in Brazil, and the description of *Occultifur brasiliensis* f.a., sp. nov. *Antonie Leeuwenhoek* **2015**, *107*, 597–611. [CrossRef] [PubMed]
128. Krauss, G.J.; Solé, M.; Krauss, G.; Schlosser, D.; Wesenberg, D.; Bärlocher, F. Fungi in freshwaters: Ecology, physiology and biochemical potential. *FEMS Microbiol. Rev.* **2011**, *35*, 620–651. [CrossRef]
129. Bhadury, P.; Bik, H.; Lambshead, J.D.; Austen, M.C.; Smerdon, G.R.; Rogers, A.D. Molecular diversity of fungal phylotypes Co-Amplified alongside nematodes from coastal and Deep-Sea marine environments. *PLoS ONE* **2011**, *6*, e26445. [CrossRef]
130. Sapir, A.; Dillman, A.R.; Connon, S.A.; Grupe, B.M.; Ingels, J.; Mundo-Ocampo, M.; Levin, L.A.; Baldwin, J.G.; Orphan, V.J.; Sternberg, P.W. Microsporidia-nematode associations in methane seeps reveal basal fungal parasitism in the deep sea. *Front. Microbiol.* **2014**, *5*, 43. [CrossRef]
131. Dayton, P.K. Ecology of Kelp Communities. *Annu. Rev. Ecol. Syst.* **1985**, *16*, 215–245. [CrossRef]
132. Rateb, M.E.; Ebel, R. Secondary metabolites of fungi from marine habitats. *Nat. Prod. Rep.* **2011**, *28*, 290–344. [CrossRef]
133. Jones, E.B.G.; Pang, K.-L. *Marine Fungi: And Fungal-Like Organisms*; Walter de Gruyter: Berlin, Germany, 2012; ISBN 3110264064.
134. Richards, T.A.; Jones, M.D.M.; Leonard, G.; Bass, D. Marine fungi: Their ecology and molecular diversity. *Ann. Rev. Mar. Sci.* **2012**, *4*, 495–522. [CrossRef]
135. Zuccaro, A.; Mitchell, J.I. Fungal communities of seaweeds. In *The Fungal Community*; CRC Press: Boca Raton, USA, 2005; pp. 533–579.
136. Egan, S.; Harder, T.; Burke, C.; Steinberg, P.; Kjelleberg, S.; Thomas, T. The seaweed holobiont: Understanding seaweed–bacteria interactions. *FEMS Microbiol. Rev.* **2013**, *37*, 462–476. [CrossRef]
137. Singh, R.P.; Kumari, P.; Reddy, C.R.K. Antimicrobial compounds from seaweeds-associated bacteria and fungi. *Appl. Microbiol. Biotechnol.* **2015**, *99*, 1571–1586. [CrossRef]

138. Wichard, T.; Beemelmanns, C. Role of Chemical Mediators in Aquatic Interactions across the Prokaryote–Eukaryote Boundary. *J. Chem. Ecol.* **2018**, *44*, 1008–1021. [CrossRef] [PubMed]
139. Rämä, T.; Hassett, B.T.; Bubnova, E. Arctic marine fungi: From filaments and flagella to operational taxonomic units and beyond. *Bot. Mar.* **2017**, *60*, 433–452. [CrossRef]
140. Castro-Sowinski, S. *The Ecological Role of Micro-Organisms in the Antarctic Environment*; Springer: Berlin/Heidelberg, Germany, 2019; ISBN 3030027864.
141. D'Amico, S.; Collins, T.; Marx, J.C.; Feller, G.; Gerday, C. Psychrophilic microorganisms: Challenges for life. *EMBO Rep.* **2006**, *7*, 385–389. [CrossRef]
142. Cavicchioli, R.; Charlton, T.; Ertan, H.; Omar, S.M.; Siddiqui, K.S.; Williams, T.J. Biotechnological uses of enzymes from psychrophiles. *Microb. Biotechnol.* **2011**, *4*, 449–460. [CrossRef]
143. Corinaldesi, C. New perspectives in benthic deep-sea microbial ecology. *Front. Mar. Sci.* **2015**, *2*, 17. [CrossRef]
144. Chambergo, F.S.; Valencia, E.Y. Fungal biodiversity to biotechnology. *Appl. Microbiol. Biotechnol.* **2016**, *100*, 2567–2577. [CrossRef] [PubMed]
145. Coker, J.A. Extremophiles and biotechnology: Current uses and prospects. *F1000Research* **2016**, *5*, 396–403. [CrossRef] [PubMed]
146. Ul Arifeen, M.Z.; Ma, Y.N.; Xue, Y.R.; Liu, C.H. Deep-sea fungi could be the new arsenal for bioactive molecules. *Mar. Drugs* **2020**, *18*, 9. [CrossRef] [PubMed]
147. Buzzini, P.; Margesin, R. Cold-Adapted Yeasts: A Lesson from the Cold and a Challenge for the XXI Century. In *Cold-Adapted Yeasts: Biodiversity, Adaptation Strategies and Biotechnological Significance*; Buzzini, P., Margesin, R., Eds.; Springer: Berlin/Heidelberg, Germany, 2014; pp. 3–22. ISBN 978-3-642-39681-6.
148. Housman, G.; Byler, S.; Heerboth, S.; Lapinska, K.; Longacre, M.; Snyder, N.; Sarkar, S. Drug resistance in cancer: An overview. *Cancers* **2014**, *6*, 1769–1792. [CrossRef]
149. Ventola, C.L. The antibiotic resistance crisis: Part 1: Causes and threats. *Pharm. Ther.* **2015**, *40*, 277–283.
150. Newman, D.J.; Cragg, G.M. Natural Products as Sources of New Drugs from 1981 to 2014. *J. Nat. Prod.* **2016**, *79*, 629–661. [CrossRef]
151. Bernan, V.S.; Greenstein, M.; Maiese, W.M. Marine microorganisms as a source of new natural products. *Adv. Appl. Microbiol.* **1997**, *43*, 57–90.
152. Blunt, J.W.; Copp, B.R.; Keyzers, R.A.; Carroll, A.R.; Munro, M.M.H.G.; Prinsep, M.R.; Copp, B.R.; Davis, R.A.; Keyzers, R.A.; Prinsep, M.R. Marine natural products. *Nat. Prod. Rep.* **2016**, *35*, 8–53. [CrossRef]
153. Gerwick, W.H.; Moore, B.S. Lessons from the past and charting the future of marine natural products drug discovery and chemical biology. *Chem. Biol.* **2012**, *19*, 85–98. [CrossRef]
154. Bonugli-Santos, R.C.; dos Santos Vasconcelos, M.R.; Passarini, M.R.Z.; Vieira, G.A.L.; Lopes, V.C.P.; Mainardi, P.H.; dos Santos, J.A.; de Azevedo Duarte, L.; Otero, I.V.R.; da Silva Yoshida, A.M.; et al. Marine-derived fungi: Diversity of enzymes and biotechnological applications. *Front. Microbiol.* **2015**, *6*, 269. [CrossRef]
155. Imhoff, J.F. Natural products from marine fungi—Still an underrepresented resource. *Mar. Drugs* **2016**, *14*, 19. [CrossRef]
156. Demain, A.L. Importance of microbial natural products and the need to revitalize their discovery. *J. Ind. Microbiol. Biotechnol.* **2014**, *41*, 185–201. [CrossRef]
157. Gomes, N.; Lefranc, F.; Kijjoa, A.; Kiss, R. Can Some Marine-Derived Fungal Metabolites Become Actual Anticancer Agents? *Mar. Drugs* **2015**, *13*, 3950–3991. [CrossRef]
158. Varrella, S.; Tangherlini, M.; Corinaldesi, C. Deep Hypersaline Anoxic Basins as Untapped Reservoir of Polyextremophilic Prokaryotes of Biotechnological Interest. *Mar. Drugs* **2020**, *18*, 91. [CrossRef]
159. Deshmukh, S.K.; Prakash, V.; Ranjan, N. Marine Fungi: A Source of Potential Anticancer Compounds: A Source of Potential Anticancer Compounds. *Front. Microbiol.* **2018**, *8*, 2536. [CrossRef]
160. Núñez-Pons, L.; Shilling, A.; Verde, C.; Baker, B.J.; Giordano, D. Marine Terpenoids from Polar Latitudes and Their Potential Applications in Biotechnology. *Mar. Drugs* **2020**, *18*, 401. [CrossRef]
161. Frisvad, J.C. Cold-Adapted Fungi as a Source for Valuable Metabolites. In *Psychrophiles: From Biodiversity to Biotechnology*; Margesin, R., Schinner, F., Marx, J.-C., Gerday, C., Eds.; Springer: Berlin/Heidelberg, Germany, 2008; pp. 381–387. ISBN 978-3-540-74335-4.
162. Zucconi, L.; Canini, F.; Temporiti, M.E.; Tosi, S. Extracellular Enzymes and Bioactive Compounds from Antarctic Terrestrial Fungi for Bioprospecting. *Int. J. Environ. Res. Public Health* **2020**, *17*, 6459. [CrossRef]
163. Lin, A.; Wu, G.; Gu, Q.; Zhu, T.; Li, D. New eremophilane-type sesquiterpenes from an Antarctic deep-sea derived fungus, *Penicillium* sp. PR19 N-1. *Arch. Pharm. Res.* **2014**, *37*, 839–844. [CrossRef]
164. ChemSpider Search and Share Chemistry. Available online: https://www.chemspider.com/ (accessed on 7 May 2021).
165. Huang, J.-N.; Zou, Q.; Chen, J.; Xu, S.-H.; Luo, D.; Zhang, F.-G.; Lu, Y.-Y. Phenols and diketopiperazines isolated from Antarctic-derived fungi, *Penicillium citreonigrum* SP-6. *Phytochem. Lett.* **2018**, *27*, 114–118. [CrossRef]
166. Hu, Z.Y.; Li, Y.Y.; Huang, Y.J.; Su, W.J.; Shen, Y.M. Three new sesquiterpenoids from *Xylaria* sp. NCY2. *Helv. Chim. Acta* **2008**, *91*, 46–52. [CrossRef]
167. Figueroa, L.; Jiménez, C.; Rodríguez, J.; Areche, C.; Chávez, R.; Henríquez, M.; de la Cruz, M.; Díaz, C.; Segade, Y.; Vaca, I. 3-Nitroasterric Acid Derivatives from an Antarctic Sponge-Derived *Pseudogymnoascus* sp. Fungus. *J. Nat. Prod.* **2015**, *78*, 919–923. [CrossRef]

168. Li, Y.; Sun, B.; Liu, S.; Jiang, L.; Liu, X.; Zhang, H.; Che, Y. Bioactive Asterric Acid Derivatives from the Antarctic Ascomycete Fungus Geomyces sp. *J. Nat. Prod.* **2008**, *71*, 1643–1646. [CrossRef]
169. Ren, J.; Xue, C.; Tian, L.; Xu, M.; Chen, J.; Deng, Z.; Proksch, P.; Lin, W. Asperelines A−F, Peptaibols from the Marine-Derived Fungus *Trichoderma asperellum*. *J. Nat. Prod.* **2009**, *72*, 1036–1044. [CrossRef]
170. Gerday, C.; Aittaleb, M.; Bentahir, M.; Chessa, J.P.; Claverie, P.; Collins, T.; D'Amico, S.; Dumont, J.; Garsoux, G.; Georlette, D.; et al. Cold-adapted enzymes: From fundamentals to biotechnology. *Trends Biotechnol.* **2000**, *18*, 103–107. [CrossRef]
171. Nandanwar, S.K.; Borkar, S.B.; Lee, J.H.; Kim, H.J. Taking Advantage of Promiscuity of Cold-Active Enzymes. *Appl. Sci.* **2020**, *10*, 8128. [CrossRef]
172. Sočan, J.; Purg, M.; Åqvist, J. Computer simulations explain the anomalous temperature optimum in a cold-adapted enzyme. *Nat. Commun.* **2020**, *11*, 2644. [CrossRef] [PubMed]
173. Liu, X.; Kokare, C. Chapter 11—Microbial Enzymes of Use in Industry. In *Biotechnology of Microbial Enzymes Production, Biocatalysis and Industrial Applications*; Brahmachari, G., Ed.; Academic Press: Cambridge, MA, USA, 2017; pp. 267–298. ISBN 978-0-12-803725-6.
174. Singh, R.; Kumar, M.; Mittal, A.; Mehta, P.K. Microbial enzymes: Industrial progress in 21st century. *3 Biotech* **2016**, *6*, 174. [CrossRef]
175. Santiago, M.; Ramírez-Sarmiento, C.A.; Zamora, R.A.; Parra, L.P. Discovery, molecular mechanisms, and industrial applications of cold-active enzymes. *Front. Microbiol.* **2016**, *7*, 1408. [CrossRef]
176. De Maayer, P.; Anderson, D.; Cary, C.; Cowan, D.A. Some like it cold: Understanding the survival strategies of psychrophiles. *EMBO Rep.* **2014**, *15*, 508–517. [CrossRef]
177. Duarte, A.W.F.; dos Santos, J.A.; Vianna, M.V.; Vieira, J.M.F.; Mallagutti, V.H.; Inforsato, F.J.; Wentzel, L.C.P.; Lario, L.D.; Rodrigues, A.; Pagnocca, F.C.; et al. Cold-adapted enzymes produced by fungi from terrestrial and marine Antarctic environments. *Crit. Rev. Biotechnol.* **2018**, *38*, 600–619. [CrossRef]
178. Buzzini, P.; Branda, E.; Goretti, M.; Turchetti, B. Psychrophilic yeasts from worldwide glacial habitats: Diversity, adaptation strategies and biotechnological potential. *FEMS Microbiol. Ecol.* **2012**, *82*, 217–241. [CrossRef]
179. Flam, F. The chemistry of life at the margins. *Science* **1994**, *265*, 471–472. [CrossRef]
180. Dumorné, K.; Córdova, D.C.; Astorga-Eló, M.; Renganathan, P. Extremozymes: A potential source for industrial applications. *J. Microbiol. Biotechnol.* **2017**, *27*, 649–659. [CrossRef]
181. Bjelic, S.; Brandsdal, B.O.; Åqvist, J. Cold adaptation of enzyme reaction rates. *Biochemistry* **2008**, *47*, 10049–10057. [CrossRef]
182. Feller, G. Protein stability and enzyme activity at extreme biological temperatures. *J. Phys. Condens. Matter* **2010**, *22*, 323101. [CrossRef]
183. Feller, G. Psychrophilic Enzymes: From Folding to Function and Biotechnology. *Scientifica* **2013**, *2013*, 512840. [CrossRef]
184. Gerday, C. Psychrophily and Catalysis. *Biology* **2013**, *2*, 719–741. [CrossRef]
185. Sarmiento, F.; Peralta, R.; Blamey, J.M. Cold and Hot Extremozymes: Industrial Relevance and Current Trends. *Front. Bioeng. Biotechnol.* **2015**, *3*, 148. [CrossRef]
186. Martínez-Martínez, M.; Bargiela, R.; Ferrer, M. Metagenomics and the Search for Industrial Enzymes. In *Biotechnology of Microbial Enzymes: Production, Biocatalysis and Industrial Applications*; Brahmachari, G., Demain, A.L., Adrio, J.L., Eds.; Academic Press: Cambridge, MA, USA, 2016; pp. 167–184.
187. Jin, M.; Gai, Y.; Guo, X.; Hou, Y.; Zeng, R. Properties and Applications of Extremozymes from Deep-Sea Extremophilic Microorganisms: A Mini Review. *Mar. Drugs* **2019**, *17*, 656. [CrossRef]
188. Castilla, I.A.; Woods, D.F.; Reen, F.J.; O'Gara, F. Harnessing marine biocatalytic reservoirs for green chemistry applications through metagenomic technologies. *Mar. Drugs* **2018**, *16*, 227. [CrossRef]
189. Martorell, M.M.; Ruberto, L.A.M.; de Figueroa, L.I.C.; Mac Cormack, W.P. Antarctic Yeasts as a Source of Enzymes for Biotechnological Applications. In *Fungi of Antarctica*; Springer International Publishing: Berlin/Heidelberg, Germany, 2019; pp. 285–304.
190. Arora, N.K.; Panosyan, H. Extremophiles: Applications and roles in environmental sustainability. *Environ. Sustain.* **2019**, *2*, 217–218. [CrossRef]
191. Burhan, H.; Ravinder, S.R.; Deepak, C.; Poonam, S.; Fayaz, A.M.; Sanjay, S.; Ishfaq, A. Psychrophilic yeasts and their biotechnological applications—A review. *Afr. J. Biotechnol.* **2014**, *13*, 2188–2197. [CrossRef]
192. Feller, G. Molecular adaptations to cold in psychrophilic enzymes. *Cell. Mol. Life Sci.* **2003**, *60*, 648–662. [CrossRef]
193. Feller, G.; Gerday, C. Psychrophilic enzymes: Hot topics in cold adaptation. *Nat. Rev. Microbiol.* **2003**, *1*, 200–208. [CrossRef]
194. Gomes, J.; Steiner, W. The biocatalytic potential of extremophiles and extremozymes. *Food Technol. Biotechnol.* **2004**, *42*, 223–235.
195. Dalmaso, G.Z.L.; Ferreira, D.; Vermelho, A.B.; Zamith, G.; Dalmaso, L.; Ferreira, D.; Vermelho, A.B.; Dalmaso, G.Z.L.; Ferreira, D.; Vermelho, A.B. Marine Extremophiles: A Source of Hydrolases for Biotechnological Applications. *Mar. Drugs* **2015**, *13*, 1925–1965. [CrossRef]
196. Kuddus, M. Cold-active enzymes in food biotechnology: An updated mini review. *J. Appl. Biol. Biotechnol.* **2018**, *6*, 58–63.
197. Blamey, J.M.; Fischer, F.; Meyer, H.P.; Sarmiento, F.; Zinn, M. Enzymatic Biocatalysis in Chemical Transformations: A Promising and Emerging Field in Green Chemistry Practice. In *Biotechnology of Microbial Enzymes: Production, Biocatalysis and Industrial Applications*; Brahmachari, G., Demain, A.L., Adrio, J.L., Eds.; Academic Press: Cambridge, MA, USA, 2017; pp. 347–403. ISBN 9780128037461.

198. Fenice, M. The Psychrotolerant Antarctic Fungus *Lecanicillium muscarium* CCFEE 5003: A Powerful Producer of Cold-Tolerant Chitinolytic Enzymes. *Molecules* **2016**, *21*, 447. [CrossRef]
199. Barghini, P.; Moscatelli, D.; Garzillo, A.M.V.; Crognale, S.; Fenice, M. High production of cold-tolerant chitinases on shrimp wastes in bench-top bioreactor by the Antarctic fungus *Lecanicillium muscarium* CCFEE 5003: Bioprocess optimization and characterization of two main enzymes. *Enzym. Microb. Technol.* **2013**, *53*, 331–338. [CrossRef]
200. Ramli, A.N.M.; Mahadi, N.M.; Shamsir, M.S.; Rabu, A.; Joyce-Tan, K.H.; Murad, A.M.A.; Md. Illias, R. Structural prediction of a novel chitinase from the psychrophilic *Glaciozyma antarctica* PI12 and an analysis of its structural properties and function. *J. Comput. Aided Mol. Des.* **2012**, *26*, 947–961. [CrossRef]
201. Teoh, C.P.; Koh, S.P.; Ling, C.M.W.V. Characterisation of an Antarctic Yeast, *Glaciozyma antarctica* PI12. *Borneo Int. J. Biotechnol.* **2020**, *1*, 89. [CrossRef]
202. Bharudin, I.; Abu Bakar, M.F.; Hashim, N.H.F.; Mat Isa, M.N.; Alias, H.; Firdaus-Raih, M.; Md Illias, R.; Najimudin, N.; Mahadi, N.M.; Abu Bakar, F.D.; et al. Unravelling the adaptation strategies employed by *Glaciozyma antarctica* PI12 on Antarctic sea ice. *Mar. Environ. Res.* **2018**, *137*, 169–176. [CrossRef] [PubMed]
203. Mohammadi, S.; Hashim, N.H.F.; Mahadi, M.N.; Murad, A.M. The Cold-Active Endo-β-1,3(4)-Glucanase from a Marine Psychrophilic Yeast, *Glaciozyma antarctica* PI12: Heterologous Expression, Biochemical Characterisation, and Molecular Modeling. *Int. J. Appl. Biol. Pharm. Technol.* **2021**, *12*, 279–300. [CrossRef]
204. Mohammadi, S.; Parvizpour, S.; Razmara, J.; Abu Bakar, F.D.; Illias, R.M.; Mahadi, N.M.; Murad, A.M.A. Structure Prediction of a Novel Exo-β-1,3-Glucanase: Insights into the Cold Adaptation of Psychrophilic Yeast *Glaciozyma antarctica* PI12. *Interdiscip. Sci. Comput. Life Sci.* **2018**, *10*, 157–168. [CrossRef]
205. Hashim, N.H.F.; Mahadi, N.M.; Illias, R.M.; Feroz, S.R.; Abu Bakar, F.D.; Murad, A.M.A. Biochemical and structural characterization of a novel cold-active esterase-like protein from the psychrophilic yeast *Glaciozyma antarctica*. *Extremophiles* **2018**, *22*, 607–616. [CrossRef]
206. Duarte, A.W.F.; Barato, M.B.; Nobre, F.S.; Polezel, D.A.; de Oliveira, T.B.; dos Santos, J.A.; Rodrigues, A.; Sette, L.D. Production of cold-adapted enzymes by filamentous fungi from King George Island, Antarctica. *Polar Biol.* **2018**, *41*, 2511–2521. [CrossRef]
207. Duarte, A.W.F.; Lopes, A.M.; Molino, J.V.D.; Pessoa, A.; Sette, L.D. Liquid-liquid extraction of lipase produced by psychrotrophic yeast *Leucosporidium scottii* L117 using aqueous two-phase systems. *Sep. Purif. Technol.* **2015**, *156*, 215–225. [CrossRef]
208. Carrasco, M.; Rozas, J.M.; Barahona, S.; Alcaíno, J.; Cifuentes, V.; Baeza, M. Diversity and extracellular enzymatic activities of yeasts isolated from King George Island, the sub-Antarctic region. *BMC Microbiol.* **2012**, *12*, 251. [CrossRef]
209. Tsuji, M.; Yokota, Y.; Shimohara, K.; Kudoh, S.; Hoshino, T. An Application of Wastewater Treatment in a Cold Environment and Stable Lipase Production of Antarctic Basidiomycetous Yeast *Mrakia blollopis*. *PLoS ONE* **2013**, *8*, e59376. [CrossRef]
210. Poveda, G.; Gil-Durán, C.; Vaca, I.; Levicán, G.; Chávez, R. Cold-active pectinolytic activity produced by filamentous fungi associated with Antarctic marine sponges. *Biol. Res.* **2018**, *51*, 28. [CrossRef]
211. Yu, P.; Wang, X.-T.; Liu, J.-W. Purification and characterization of a novel cold-adapted phytase from *Rhodotorula mucilaginosa* strain JMUY14 isolated from Antarctic. *J. Basic Microbiol.* **2015**, *55*, 1029–1039. [CrossRef]
212. Lario, L.D.; Chaud, L.; das Graças Almeida, M.; Converti, A.; Durães Sette, L.; Pessoa, A.; Duraes Sette, L.; Pessoa, A.; Durães Sette, L.; Pessoa, A. Production, purification, and characterization of an extracellular acid protease from the marine Antarctic yeast *Rhodotorula mucilaginosa* L7. *Fungal Biol.* **2015**, *119*, 1129–1136. [CrossRef]
213. Chaud, L.C.S.; Lario, L.D.; Bonugli-Santos, R.C.; Sette, L.D.; Pessoa Junior, A.; Felipe, M.d.G.d.A. Improvement in extracellular protease production by the marine antarctic yeast *Rhodotorula mucilaginosa* L7. *New Biotechnol.* **2016**, *33*, 807–814. [CrossRef]
214. Turkiewicz, M.; Pazgier, M.; Kalinowska, H.; Bielecki, S. A cold-adapted extracellular serine proteinase of the yeast *Leucosporidium antarcticum*. *Extremophiles* **2003**, *7*, 435–442. [CrossRef]
215. Alias, N.; Ahmad Mazian, M.; Salleh, A.B.; Basri, M.; Rahman, R.N.Z.R.A. Molecular Cloning and Optimization for High Level Expression of Cold-Adapted Serine Protease from Antarctic Yeast *Glaciozyma antarctica* PI12. *Enzym. Res.* **2014**, *2014*, 197938. [CrossRef]
216. Glodowsky, A.P.; Ruberto, L.A.; Martorell, M.M.; Mac Cormack, W.P.; Levin, G.J. Cold active transglutaminase from antarctic *Penicillium chrysogenum*: Partial purification, characterization and potential application in food technology. *Biocatal. Agric. Biotechnol.* **2020**, *29*, 101807. [CrossRef]
217. Gil-Durán, C.; Ravanal, M.-C.; Ubilla, P.; Vaca, I.; Chávez, R. Heterologous expression, purification and characterization of a highly thermolabile endoxylanase from the Antarctic fungus *Cladosporium* sp. *Fungal Biol.* **2018**, *122*, 875–882. [CrossRef]
218. Del-Cid, A.; Ubilla, P.; Ravanal, M.-C.M.-C.C.; Medina, E.; Vaca, I.; Levican, G.; Eyzaguirre, J.; Chavez, R.; Levicán, G.; Eyzaguirre, J.; et al. Cold-Active Xylanase Produced by Fungi Associated with Antarctic Marine Sponges. *Appl. Biochem. Biotechnol.* **2014**, *172*, 524–532. [CrossRef]
219. Azhar, M. Cloning, Expression and Analysis of α-Amylase Gene from Psychrophilic Yeast *Leucosporidium antarcticum* PI12. Ph.D. Thesis, Universiti Teknologi Malaysia, Skudai, Malaysia, 2012.
220. Song, C.; Liu, G.L.; Xu, J.L.; Chi, Z.M. Purification and characterization of extracellular β-galactosidase from the psychrotolerant yeast *Guehomyces pullulans* 17-1 isolated from sea sediment in Antarctica. *Process Biochem.* **2010**, *45*, 954–960. [CrossRef]
221. Erich, S.; Kuschel, B.; Schwarz, T.; Ewert, J.; Böhmer, N.; Niehaus, F.; Eck, J.; Lutz-Wahl, S.; Stressler, T.; Fischer, L. Novel high-performance metagenome β-galactosidases for lactose hydrolysis in the dairy industry. *J. Biotechnol.* **2015**, *210*, 27–37. [CrossRef]

222. Ugidos-Rodríguez, S.; Matallana-González, M.C.; Sánchez-Mata, M.C. Lactose malabsorption and intolerance: A review. *Food Funct.* **2018**, *9*, 4056–4068. [CrossRef] [PubMed]
223. Collins, T.; Gerday, C.; Feller, G. Xylanases, xylanase families and extremophilic xylanases. *FEMS Microbiol. Rev.* **2005**, *29*, 3–23. [CrossRef] [PubMed]
224. Collins, T.; Gerday, C.; Feller, G.; Polizeli, M.L.T.M.; Rizzatti, A.C.S.; Monti, R.; Terenzi, H.F.; Jorge, J.A.; Amorim, D.S. Xylanases from fungi: Properties and industrial applications. *Appl. Microbiol. Biotechnol.* **2005**, *67*, 577–591.
225. Bhardwaj, N.; Kumar, B.; Verma, P. A detailed overview of xylanases: An emerging biomolecule for current and future prospective. *Bioresour. Bioprocess.* **2019**, *6*, 40. [CrossRef]
226. Guerrand, D. Lipases industrial applications: Focus on food and agroindustries. *OCL* **2017**, *24*, D403. [CrossRef]
227. Gurung, N.; Ray, S.; Bose, S.; Rai, V. A broader view: Microbial enzymes and their relevance in industries, medicine, and beyond. *Biomed. Res. Int.* **2013**, *2013*, 329121. [CrossRef]
228. Chandra, P.; Enespa; Singh, R.; Arora, P.K. Microbial lipases and their industrial applications: A comprehensive review. *Microb. Cell Fact.* **2020**, *19*, 169. [CrossRef]
229. Al-Ghanayem, A.A.; Joseph, B. Current prospective in using cold-active enzymes as eco-friendly detergent additive. *Appl. Microbiol. Biotechnol.* **2020**, *104*, 2871–2882. [CrossRef]
230. Kavitha, M. Cold active lipases—An update. *Front. Life Sci.* **2016**, *9*, 226–238. [CrossRef]
231. Vakhlu, J.; Kour, A. Yeast lipases: Enzyme purification, biochemical properties and gene cloning. *Electron. J. Biotechnol.* **2006**, *9*, 717–3458. [CrossRef]
232. Domínguez De María, P.; Carboni-Oerlemans, C.; Tuin, B.; Bargeman, G.; Van Der Meer, A.; Van Gemert, R. Biotechnological applications of *Candida antarctica* lipase A: State-of-the-art. *J. Mol. Catal. B Enzym.* **2005**, *37*, 36–46. [CrossRef]
233. Szczesna-Antczak, M.; Kamińska, J.; Florczak, T.; Turkiewicz, M. Cold-active yeast lipases: Recent issues and future prospects. In *Cold-Adapted Yeasts: Biodiversity, Adaptation Strategies and Biotechnological Significance*; Springer: Berlin/Heidelberg, Germany, 2013; pp. 353–375. ISBN 9783642396816.
234. Joseph, B.; Ramteke, P.W.; Thomas, G. Cold active microbial lipases: Some hot issues and recent developments. *Biotechnol. Adv.* **2008**, *26*, 457–470. [CrossRef]
235. Martorell, M.M.; Ruberto, L.A.M.; Fernández, P.M.; Castellanos de Figueroa, L.I.; Mac Cormack, W.P. Bioprospection of cold-adapted yeasts with biotechnological potential from Antarctica. *J. Basic Microbiol.* **2017**, *57*, 504–516. [CrossRef]
236. Cabrera, M.Á.; Blamey, J.M. Biotechnological applications of archaeal enzymes from extreme environments. *Biol. Res.* **2018**, *51*, 37. [CrossRef]
237. Furhan, J. Adaptation, production, and biotechnological potential of cold-adapted proteases from psychrophiles and psychrotrophs: Recent overview. *J. Genet. Eng. Biotechnol.* **2020**, *18*, 36. [CrossRef]
238. Razzaq, A.; Shamsi, S.; Ali, A.; Ali, Q.; Sajjad, M.; Malik, A.; Ashraf, M. Microbial proteases applications. *Front. Bioeng. Biotechnol.* **2019**, *7*, 110. [CrossRef]
239. Sharma, K.M.; Kumar, R.; Panwar, S.; Kumar, A. Microbial alkaline proteases: Optimization of production parameters and their properties. *J. Genet. Eng. Biotechnol.* **2017**, *15*, 115–126. [CrossRef]
240. Grossart, H.P.; Wurzbacher, C.; James, T.Y.; Kagami, M. Discovery of dark matter fungi in aquatic ecosystems demands a reappraisal of the phylogeny and ecology of zoosporic fungi. *Fungal Ecol.* **2016**, *19*, 28–38. [CrossRef]
241. Overy, D.P.; Rämä, T.; Oosterhuis, R.; Walker, A.K.; Pang, K.L. The neglected marine fungi, sensu stricto, and their isolation for natural products' discovery. *Mar. Drugs* **2019**, *17*, 42. [CrossRef]
242. Vester, J.K.; Glaring, M.A.; Stougaard, P. Improved cultivation and metagenomics as new tools for bioprospecting in cold environments. *Extremophiles* **2015**, *19*, 17–29. [CrossRef] [PubMed]
243. Amann, R.I.; Ludwig, W.; Schleifer, K.-H. Phylogenetic Identification and In Situ Detection of Individual Microbial Cells without Cultivation. *Microbiol. Rev.* **1995**, *59*, 143–169. [CrossRef] [PubMed]
244. Solden, L.; Lloyd, K.; Wrighton, K. The bright side of microbial dark matter: Lessons learned from the uncultivated majority. *Curr. Opin. Microbiol.* **2016**, *31*, 217–226. [CrossRef] [PubMed]
245. Ambrosino, L.; Tangherlini, M.; Colantuono, C.; Esposito, A.; Sangiovanni, M.; Miralto, M.; Sansone, C.; Chiusano, M.L. Bioinformatics for Marine Products: An Overview of Resources, Bottlenecks, and Perspectives. *Mar. Drugs* **2019**, *17*, 576. [CrossRef] [PubMed]
246. Barone, R.; De Santi, C.; Palma Esposito, F.; Tedesco, P.; Galati, F.; Visone, M.; Di Scala, A.; De Pascale, D. Marine metagenomics, a valuable tool for enzymes and bioactive compounds discovery. *Front. Mar. Sci.* **2014**, *1*, 38. [CrossRef]
247. Madhavan, A.; Sindhu, R.; Parameswaran, B.; Sukumaran, R.K.; Pandey, A. Metagenome Analysis: A Powerful Tool for Enzyme Bioprospecting. *Appl. Biochem. Biotechnol.* **2017**, *183*, 636–651. [CrossRef]
248. Hug, J.J.; Bader, C.D.; Remškar, M.; Cirnski, K.; Müller, R. Concepts and methods to access novel antibiotics from actinomycetes. *Antibiotics* **2018**, *7*, 44. [CrossRef]
249. Zhang, M.M.; Qiao, Y.; Ang, E.L.; Zhao, H. Using natural products for drug discovery: The impact of the genomics era. *Expert Opin. Drug Discov.* **2017**, *12*, 475–487. [CrossRef]
250. Skellam, E. Strategies for Engineering Natural Product Biosynthesis in Fungi. *Trends Biotechnol.* **2019**, *37*, 416–427. [CrossRef]

251. Goyal, D.; Swaroop, S.; Pandey, J. Harnessing the Genetic Diversity and Metabolic Potential of Extremophilic Microorganisms through the Integration of Metagenomics and Single-Cell Genomics. In *Extremophilic Microbes and Metabolites—Diversity, Bioprespecting and Biotechnological Applications*; IntechOpen: London, UK, 2021.
252. Milshteyn, A.; Schneider, J.S.; Brady, S.F. Mining the metabiome: Identifying novel natural products from microbial communities. *Chem. Biol.* **2014**, *21*, 1211–1223. [CrossRef]

Journal of
Fungi

Article

Local Environmental Conditions Promote High Turnover Diversity of Benthic Deep-Sea Fungi in the Ross Sea (Antarctica)

Giulio Barone [1,2], Cinzia Corinaldesi [3], Eugenio Rastelli [4], Michael Tangherlini [5], Stefano Varrella [3], Roberto Danovaro [2,6] and Antonio Dell'Anno [2,*]

1 Institute for Marine Biological Resources and Biotechnology, National Research Council, Largo Fiera della Pesca 2, 60125 Ancona, Italy; giulio.barone@irbim.cnr.it
2 Department of Life and Environmental Sciences, Polytechnic University of Marche, Via Brecce Bianche, 60131 Ancona, Italy; r.danovaro@univpm.it
3 Department of Materials, Environmental Sciences and Urban Planning, Polytechnic University of Marche, Via Brecce Bianche, 60131 Ancona, Italy; c.corinaldesi@staff.univpm.it (C.C.); s.varrella@univpm.it (S.V.)
4 Department of Marine Biotechnology, Stazione Zoologica Anton Dohrn, Fano Marine Centre, Viale Adriatico 1-N, 61032 Fano, Italy; eugenio.rastelli@szn.it
5 Department of Research Infrastructures for Marine Biological Resources, Stazione Zoologica Anton Dohrn, Fano Marine Centre, Viale Adriatico 1-N, 61032 Fano, Italy; michael.tangherlini@szn.it
6 Stazione Zoologica Anton Dohrn, Villa Comunale, 80121 Naples, Italy
* Correspondence: a.dellanno@univpm.it

Abstract: Fungi are a ubiquitous component of marine systems, but their quantitative relevance, biodiversity and ecological role in benthic deep-sea ecosystems remain largely unexplored. In this study, we investigated fungal abundance, diversity and assemblage composition in two benthic deep-sea sites of the Ross Sea (Southern Ocean, Antarctica), characterized by different environmental conditions (i.e., temperature, salinity, trophic availability). Our results indicate that fungal abundance (estimated as the number of 18S rDNA copies g^{-1}) varied by almost one order of magnitude between the two benthic sites, consistently with changes in sediment characteristics and trophic availability. The highest fungal richness (in terms of Amplicon Sequence Variants—ASVs) was encountered in the sediments characterized by the highest organic matter content, indicating potential control of trophic availability on fungal diversity. The composition of fungal assemblages was highly diverse between sites and within each site (similarity less than 10%), suggesting that differences in environmental and ecological characteristics occurring even at a small spatial scale can promote high turnover diversity. Overall, this study provides new insights on the factors influencing the abundance and diversity of benthic deep-sea fungi inhabiting the Ross Sea, and also paves the way for a better understanding of the potential responses of benthic deep-sea fungi inhabiting Antarctic ecosystems in light of current and future climate changes.

Keywords: deep-sea sediments; fungal diversity; trophic conditions; Antarctica; Ross Sea

Citation: Barone, G.; Corinaldesi, C.; Rastelli, E.; Tangherlini, M.; Varrella, S.; Danovaro, R.; Dell'Anno, A. Local Environmental Conditions Promote High Turnover Diversity of Benthic Deep-Sea Fungi in the Ross Sea (Antarctica). *J. Fungi* **2022**, *8*, 65. https://doi.org/10.3390/jof8010065

Academic Editor: Federico Baltar

Received: 20 December 2021
Accepted: 5 January 2022
Published: 8 January 2022

Publisher's Note: MDPI stays neutral with regard to jurisdictional claims in published maps and institutional affiliations.

1. Introduction

The Southern Ocean, surrounding Antarctica, plays a key role in global ocean circulation and biogeochemical cycles [1,2]. Here, primary productivity and carbon export to the seafloor are highly variable in space and time, with the highest rates of primary production occurring during the austral summer in the coastal polynyas (regions of open water surrounded by sea ice; [3,4]), marginal ice zone [5] and continental shelf [6–8]. Despite the extreme environmental conditions (e.g., low temperature, highly variable nutrient availability), the Southern Ocean hosts rich and diverse benthic deep-sea assemblages, including several endemic species [9,10].

Microbial assemblages in benthic deep-sea ecosystems play an important role in C and nutrient cycling and transfer of energy and material to the higher trophic levels [11]. Besides prokaryotes, fungi are widespread in deep-sea environments spanning from hypersaline anoxic basins [12–14] to cold seeps [15,16], from hydrothermal vents [17–19] to surface and subsurface sediments [13,20–24], including benthic Antarctic systems [25–28] and references therein). Theoretical estimates suggest that fungi can be the most diversified component of eukaryotes on Earth, with more than 5 million species of which only 5% have been described [29,30]. This gap applies in particular to deep-sea ecosystems, where a significant fraction of fungal diversity is still unknown [24,31,32]. Recent studies suggest that a variety of environmental factors (e.g., temperature, salinity, nutrient availability) can influence the diversity and assemblage composition of fungi in marine ecosystems [33,34]. However, factors controlling the distribution and diversity of fungi in benthic deep-sea ecosystems remain largely unexplored to date [24] and even less is known of fungi inhabiting the Southern Ocean. In deep-sea ecosystems, fungi are not only highly diversified, but they are likely involved in the degradation and cycling of organic matter [13,18,35–38]. In benthic deep-sea ecosystems, organic matter mainly consists of refractory organic compounds [39,40], and fungi are known to be efficient degraders of complex organic molecules not suitable for other heterotrophic microbes [41,42]. However, their role in C and nutrient cycling in benthic deep-sea ecosystems remains poorly understood [24].

Global climate change is altering marine biodiversity and food web dynamics, and such effects are particularly pronounced at high latitudes [43–45]. Changes in environmental conditions (e.g., temperature, salinity, nutrient availability) due to climate changes in polar regions can induce a domino effect that could impact biodiversity and ecosystem functioning from the continental shelf down to the deep seafloor [46,47]. Nevertheless, information on benthic deep-sea fungal assemblages of Antarctic ecosystems is scant and insufficient to understand and predict how these components will respond to the expected changes in environmental conditions.

In this study, we investigated the abundance, diversity and assemblage composition of fungi in two benthic deep-sea sites of the Ross Sea (one of the most productive sectors of the whole Southern Ocean; [48]), characterized by different environmental conditions in terms of trophic availability and thermohaline regime. This work aims at shedding light on the ecology of benthic deep-sea fungi and factors shaping their distribution at different spatial scales (i.e., between stations of the same site and between sites). This information is crucial for better comprehension of the responses of benthic fungal assemblages inhabiting Antarctic ecosystems, and also in light of climate change scenarios.

2. Materials and Methods

2.1. Study Area and Sampling Strategy

Sediment samples were collected in the Ross Sea, Southern Ocean (Figure 1), during the austral summer 2017 onboard the research vessel *M/N Italica* in the framework of the XXXII Italian Antarctic Expedition. Samples were collected at two different sites, named B and C, located about 170 km from each other. Site B (average depth of ca. 580 m) is located in the cross-shelf valley in the northern part of the Joides basin, and it is characterized by bio-siliceous olive-gray mud sediments. Site C is located at ca. 430 m depth close to the shelf break on the northern flank of the Mawson Bank, and it is characterized by sand, gravel, pebbles and coarse biogenic carbonate debris and high near-bottom current velocities (up to 20 cm s^{-1}; [49]). At both sites, two stations located at ca. 2 km from each other were selected to investigate spatial variability within the same site (hereafter defined B1 and B2 and C1 and C2). At each station, the main physical–chemical characteristics of the bottom waters were acquired by CTD casts along with the collection of undisturbed sediment samples by three independent box corer deployments. Once on board, sediment subsamples of the top 1 cm were collected and stored at −20 °C until laboratory analyses for the determination of organic matter quantity and quality (used as a proxy of trophic conditions [50,51]), fungal abundance, diversity and assemblage composition. All samples were processed within six months of their collection.

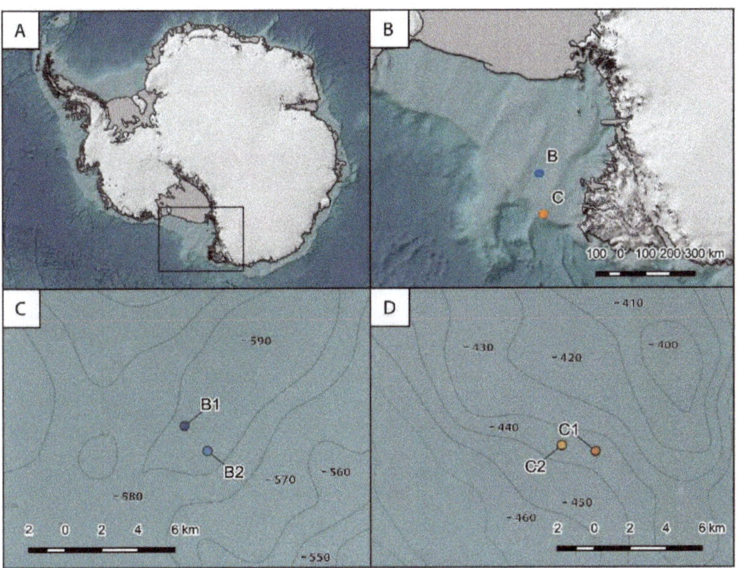

Figure 1. Location of the study area (**A**) and sampling locations of the two benthic deep-sea sites investigated (**B**) and of the stations within sites B (**C**) and C (**D**). The map was generated upon freely available layers within QGIS 3.22 environment (http://www.qgis.org; accessed on 7 January 2022).

2.2. Trophic Conditions

Trophic conditions of benthic systems were assessed on the basis of the quantity and biochemical composition of organic matter [50,51]. Chloroplastic pigments (chlorophyll-a and phaeopigments) were analyzed fluorometrically [52]. Pigments were extracted with 90% acetone (12 h in the dark at 4 °C). After centrifugation, the supernatant was used to determine the functional chlorophyll-a and then acidified with 0.1N HCl to estimate phaeopigments. Total phytopigment concentration (CPE) was defined as the sum of chlorophyll-a and phaeopigment concentrations.

Protein, carbohydrate and lipid concentrations in the sediment were determined according to previously described protocols [52]. Briefly, protein concentration was assessed by a colorimetric method, based on the reaction of proteins with copper tartrate and the Folin–Ciocalteau in a basic environment (pH 10), which provides a stable blue coloration with an intensity proportional to protein concentration. Carbohydrate concentration was determined spectrophotometrically based on the reaction between carbohydrates and phenol in the presence of sulfuric acid, which provides a coloration whose intensity is proportional to carbohydrate concentration. Lipids were extracted by direct elution with chloroform and methanol followed by reaction with sulfuric acid and determination by a colorimetric method. Protein, carbohydrate and lipid concentrations were expressed as albumin, glucose and tripalmitin equivalents, respectively. All analyses were carried out in three replicates. Protein, carbohydrate and lipid concentrations were converted to carbon equivalents (conversion factors: 0.49, 0.40 and 0.75 gC g^{-1}, respectively) to determine biopolymeric C content (BPC) in the sediments [50]. The protein to carbohydrate ratio (P:C) was used as a proxy of organic matter quality [51].

2.3. DNA Extraction and Purification for Molecular Analysis

DNA was extracted and purified from sediment samples using PowerSoil DNA isolation kit (QIAGEN), following the manufacturer's instruction with slight modifications to remove extracellular DNA (based on three subsequent washing steps), before DNA extraction [52].

2.4. Fungal Abundance Estimated by Quantitative Real-Time PCR (qPCR)

To estimate fungal abundance, DNA aliquots were used for quantitative real-time PCR (qPCR) analysis of the fungal-specific 18S rRNA gene [53]. Briefly, fungi-specific primers (FR1 5′-AIC CAT TCA ATC GGT AIT-3′) and FF390 (5′-CGA TAA CGA ACG AGA CCT-3′), which amplify a 18S rRNA gene fragment of about 350 bp [54], were used with the Sensi-FAST SYBR Q-PCR kit (Bioline, London, UK). The 15 μL reactions contained 8 μl Sensi-FAST master mix, 1 μL of each primer (final concentration 1 μM), 1μL of DNA template and 5 μL nuclease-free molecular-grade water [53]. A Bio-Rad iQ5 instrument was used to perform qPCR analyses using the following thermal protocol: 94 °C for 3 min., then 40 cycles of 94 °C for 10 s, annealing at 50 °C for 15 s, elongation at 72 °C for 20 s and acquisition of fluorescence data at 82 °C. The CFX Manager™ (v3.1) software was used to calculate Cq, efficiency (E) and R^2 values of standard curves for each plate and to quantify 18S rDNA copy numbers present in the samples analyzed. Standard curves were generated using known concentrations of *Aspergillus niger* 18S rDNA. The number of fungal 18S rDNA copies was standardized per gram of dry sediment.

2.5. Fungal Diversity and Assemblage Composition

DNA extracted from two sediment samples collected at each station by independent box corer deployments was amplified using the primer set ITS1F (5′-GGAAGTAAAAGTCGTAACAAGG-3′) and ITS2 (5′- GCTGCGTTCTTCATCGATGC-3′), which amplify the internal transcribed spacer-1 (ITS1) region of the fungal rRNA gene [55,56]. Amplicons were sequenced on an Illumina MiSeq platform by the LGC group (Berlin, Germany), following the Earth Microbiome Project protocols (http://www.earthmicrobiome.org/emp-standard-protocols/; accessed on 15 September 2018). Paired-end sequences were analyzed within the QIIME2 environment [57]. First, the ITSxpress plugin was used to trim sequences targeting the ITS1 region [58]; then, trimmed paired-end sequences were analyzed through the DADA2 procedure [59], and the resulting biologically significant Amplicon Sequence Variants (ASVs) were compared against the UNITE database (Version: 8.3; Last updated: 11 December 2020) for taxonomic affiliation [58]. Taxonomic affiliation was performed through the USEARCH SINTAX procedure [60] using three different thresholds: 0.8 (default), 0.6 and 0.5 to evaluate potential distant affiliations. To allow for a proper comparison among samples, the ASV table was then rarefied to 900 randomly selected sequences, corresponding to the lowest read count obtained in our samples [61,62].

2.6. Data Analyses

Differences in environmental and trophic variables, fungal abundance (as 18S rDNA copy number) and ASV richness between and within sites were tested by permutational two-way nested analysis of variance (2-way nested PERMANOVA; [63,64], considering the two factors Site (2 levels: B and C) and Station (nested in Site, 2 levels: 1 and 2). P-values were calculated with unrestricted permutation of raw data (perm.: 9999) with adonis function in *vegan* package. To investigate the relationships between fungal abundance and ASV richness and environmental and trophic variables, Spearman Rank correlation analyses were carried out.

The rarefied ASV table was used to assess the number of either "core" ASVs (i.e., at least one ASV present in all samples) and "exclusive" ASVs (i.e., ASVs found only in a single sample) and the output was visualized by network analysis through the Gephi package [65]. To determine similarities of the fungal assemblage composition between stations and sites, a similarity percentage analysis (SIMPER; [66]) was carried out. To identify potential factors influencing fungal assemblage composition, DistLM routine and distance-based redundancy analyses (dbRDA) were carried out with Primer+PERMANOVA (v7) [67]. In particular, temperature, salinity and dissolved oxygen concentrations of the bottom waters and quantity (i.e., biopolymeric C concentrations and total phytopigment content) and quality (protein to carbohydrate ratio) of organic matter in the sediments were used as predictor variables.

3. Results and Discussion

The thermohaline conditions of bottom waters of the benthic systems investigated in the present study changed widely, with temperature values ranging from −1.880 °C to −0.046 °C and salinity values ranging from 34.650 to 34.756 (Table 1). Stations at Site B were characterized by colder, saltier and more oxygenated waters than stations at Site C, reflecting differences in water mass characteristics. In particular, cold waters with temperatures below the surface freezing point observed at stations of Site B were associated with the Ice Shelf Water (ISW) overflowing on the continental slope of the Ross Sea [68]. The two benthic sites were also characterized by differences in terms of sediment grain size, which was mainly represented by silt–clay particles at stations of Site B and by coarse particles, including carbonate debris, at stations of Site C.

Table 1. Values of temperature, salinity and dissolved oxygen concentrations of the bottom waters and biopolymeric carbon and total phytopigment concentrations and protein to carbohydrate ratios (P:C) in the surface sediments of the different stations investigated. Mean values and standard deviations (±) are reported.

Site	Station	Depth (m)	Latitude EPSG: 4326	Longitude EPSG: 4326	Temperature (°C)	Salinity	Dissolved Oxygen (mg L^{-1})
B	B1	585	−74.03951	175.07945	−1.88 ± 0.0002	34.752 ± 0.0002	10.96 ± 0.05
B	B2	587	−74.01603	175.04279	−1.878 ± 0.0002	34.756 ± 0.0001	11 ± 0.05
C	C1	433	−72.49527	174.94336	−0.5 ± 0.001	34.65 ± 0.0002	6.25 ± 0.003
C	C2	434	−72.49967	174.99696	−0.046 ± 0.0024	34.667 ± 0.0001	9.92 ± 0.005
		Biopolymeric Carbon (mg g^{-1})		Total phytopigments (μg g^{-1})		P:C	
B	B1	3.23 ± 0.23		3.28 ± 1.27		0.39 ± 0.1	
B	B2	2.32 ± 0.61		3.44 ± 0.84		0.19 ± 0.1	
C	C1	0.28 ± 0.08		0.33 ± 0.08		0.7 ± 0.19	
C	C2	0.23 ± 0.08		0.24 ± 0.11		0.46 ± 0.06	

There is evidence that the distribution and accumulation of organic matter in surface sediments of deep-sea systems of the Ross Sea are influenced not only by vertical inputs from the upper water column, but also by lateral advection processes [69,70]. The formation and cascading of the Ice Shelf Water can represent an important process of C supply to the sea bottom, which may be responsible for the high organic matter content observed at the benthic deep-sea stations of Site B (Table 1). Values of total phytopigment concentrations, a proxy of the organic material produced by photosynthesis in surface waters and settling on the seafloor, and biopolymeric C concentrations in surface sediments of stations of Site B were, indeed, 10 times higher than those of Site C stations (Table 1).

There is a general consensus that trophic availability controls the abundance and distribution of benthic deep-sea standing stocks from prokaryotes to meio- and macrofauna [71,72], but information on its relevance in influencing the distribution of deep-sea fungi is still largely lacking [24]. In this study, fungal abundance, expressed as fungal 18S rDNA copies per gram of dry sediment, ranged from $1.3 \pm 0.7 \times 10^6$ copies g^{-1} to $1.6 \pm 0.5 \times 10^7$ copies g^{-1} (Figure 2A,B). Our results fall within previously reported ranges for deep-sea sediments of the Pacific Ocean (from 3.5×10^6 to 5.2×10^7 28S rDNA copies g^{-1} [73]) and the Mediterranean Sea (from 1.4×10^6 to 5.1×10^7 18S rDNA copies g^{-1} [24]) and provide the first evidence of the quantitative importance of fungi in benthic deep-sea ecosystems of the Southern Ocean. Fungal abundance changed between and within sites (Site: Pseudo-$F_{2,8}$ = 5.27, $p < 0.05$; Station (Site): $F_{1,8}$ = 15.86, $p < 0.05$; Figure 2A,B), with values up to 1 order of magnitude higher at stations of Site B than at stations of Site C. Significant positive relationships were found between fungal abundance and total phytopigment and biopolymeric C concentrations in the sediment (Figure 3A,B). In particular, biopolymeric C concentrations alone explained 87% of the total variation in fungal abundance (t = 8.1, $p < 0.001$, Spearman's ρ = 0.87). Overall, these findings suggest that benthic deep-sea fungi, besides prokaryotes, can be actively involved in the decomposition and utilization of organic matter settling on the seafloor, thus contributing to its cycling.

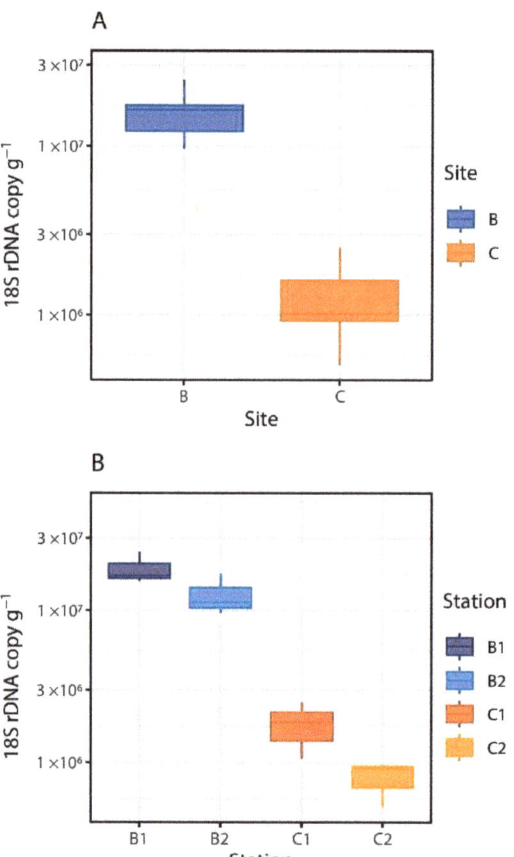

Figure 2. Boxplots showing fungal abundance, expressed as the number of 18S rDNA copies per gram of dry sediment, of the two benthic deep-sea sites investigated (**A**) and of the stations within sites B and C (**B**).

The clustering of the 113,635 fungal ITS sequences obtained in the present study after trimming allowed us to identify a total of 1251 fungal ASVs. Rarefaction curves indicated that the sequencing effort was sufficient to describe the fungal diversity present in the benthic systems investigated, even after rarefaction to the lowest sequencing depth (Figure 4A). We found a high variability of fungal ASV richness between the two sites (Figure 4B), and also between the stations, in particular of Site C (Figure 4C). Fungal ASV richness was higher at Site B, where a higher trophic availability was also found, compared to Site C ($F_{1,4}$ = 108.2; $p < 0.01$). However, the values of fungal richness reported in the present study fell within the range previously reported for other benthic deep-sea ecosystems [22,24], including Southern Ocean sediments [26].

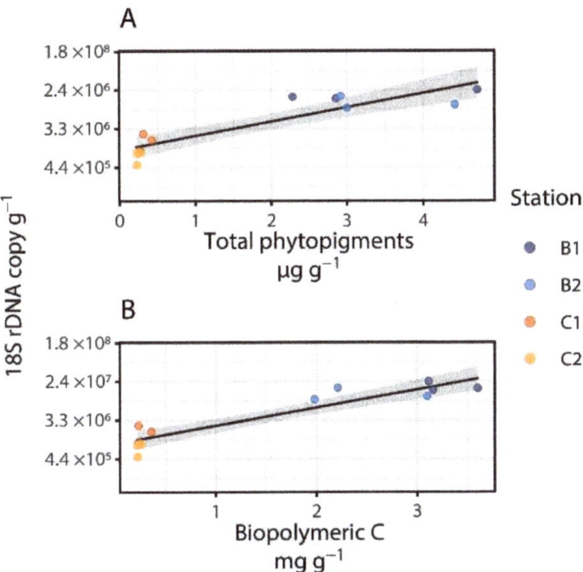

Figure 3. Relationships between fungal abundance (expressed as the number of 18S rDNA copies per gram of dry sediment) and total phytopigment (**A**) and biopolymeric C (**B**) concentrations in the study area. Gray shade represents the 95% confidence interval of the linear model interpolating the observations.

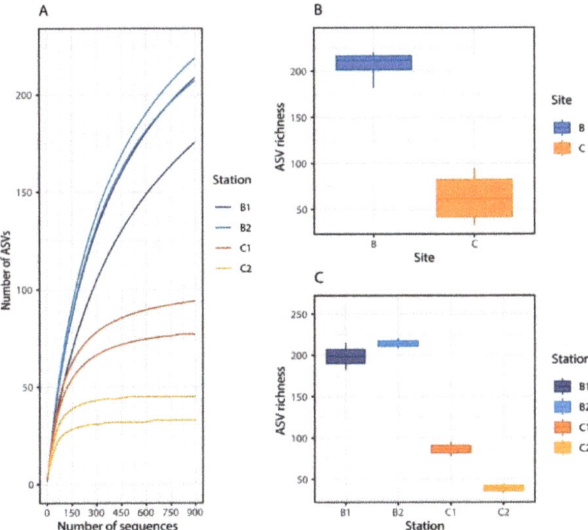

Figure 4. (**A**) Rarefaction curves from random subsampling (100 iterations) of 900 reads corresponding to the lowest read number obtained from the different sediment samples analyzed; (**B**) boxplot showing the fungal ASV richness in the two benthic deep-sea sites investigated, and (**C**) boxplot showing the fungal ASV richness in the four benthic deep-sea stations investigated.

Taxonomic analysis showed that using a default confidence threshold (cutoff of 0.8), most of the fungal ASVs could not be assigned to known fungal taxa (on average ca. 90%, Figures 5 and S1). Relaxing the confidence thresholds, the number of unknown fungal ASVs decreased (68% with a 0.6 cutoff and 58% with a 0.5 cutoff), but with a less-reliable classification (Figure S1). This result indicates that benthic deep-sea Antarctic sediments can harbor a large number of novel fungal lineages, while fungal ASVs affiliating to known fungal taxa included members affiliated to Ascomycota and Basidiomycota, which typically represent the main phyla reported in different benthic deep-sea ecosystems worldwide [24,26,27,32].

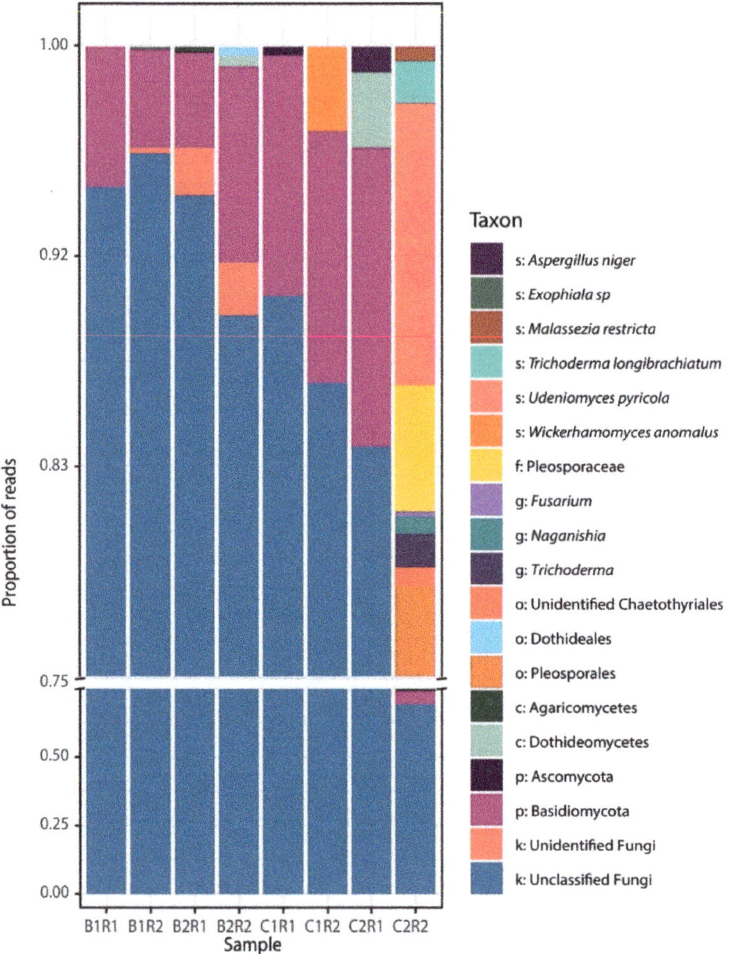

Figure 5. Stacked bar plot showing the relative abundance of fungal taxa at the lowest taxonomic levels identified in each deep-sea sediment sample collected in the Ross Sea. Taxonomic affiliation was assigned using a confidence threshold cutoff of 0.8.

Only a few ASVs could be affiliated to known fungal genera. In particular, we found 12 ASVs affiliating to nine genera, including genera commonly encountered in a variety of benthic deep-sea ecosystems (e.g., *Aspergillus*; [24,32]) and polar systems (e.g., *Naganishia, Dothideomycetes* and *Agaricomycetes*; [27,74,75]). Fungi belonging to the genera *Trichoderma* found in this study were already reported and isolated from lake and sediments of the Penguin Island in Antarctica [76,77], while other genera commonly found in Antarctic sediments, such as *Metschnikowia*, *Galciozyma* and *Psedogymnoascus*, were not encountered (for a more detailed list see [28]). Furthermore, other fungal taxa, including members belonging to *Fusarium* and *Wickerhamomyces*, have been reported to be associated with Antarctic sponges and macroalgae [78–80], while taxa affiliated with *Exophiala* and *Aspergillus* have been previously isolated from different Antarctic marine samples [28,81]. Such a comparison suggests that Antarctic deep-sea sediments can host profoundly different fungal assemblages depending on specific environmental and ecological settings.

SIMPER analysis revealed a very low similarity between the fungal assemblage compositions of the two sites and within them, as highlighted by the network plot (Figure 6). In particular, the average similarity between stations of Site B were higher than those between stations of Site C (7.7 vs. 1.6%), while the similarity between Site B and C was on average < 1%. Such very low similarity values were due to the presence of a large fraction of exclusive ASVs of each sample (accounting for 76–94% of the total ASVs; Figure 6). Overall, these findings suggest that differences in ecological and environmental conditions occurring even at spatial scales of a few meters (i.e., between replicates) can have a major role in shaping fungal assemblage composition, thus contributing to increase fungal turnover diversity.

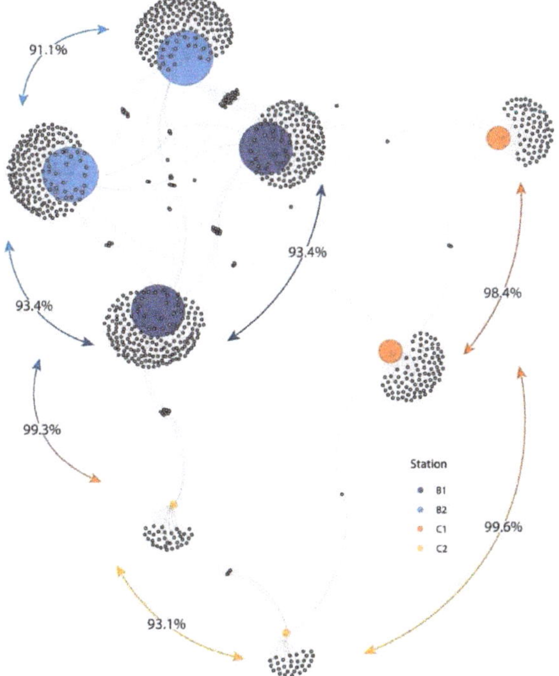

Figure 6. Network plot based on fungal assemblage composition found in the different sediment samples analyzed. Colored nodes represent sampling stations, while gray nodes represent the ASVs belonging to one or more samples to which are connected by gray edges. Arrows connecting samples indicate the Bray–Curtis dissimilarity calculated upon the log-transformed rarefied ASV table.

Previous studies suggested that environmental factors and trophic availability can influence fungal assemblage composition [24,82–84]. The distance-based redundancy analysis (dbRDA) allowed us to identify significant relationships between fungal assemblage composition and trophic (total phytopigment and biopolymeric C concentrations, protein to carbohydrate ratio) and environmental variables (temperature, salinity and oxygen concentrations). Altogether, the environmental and trophic variables explained 87% of the observed variation in fungal assemblage composition, but only temperature significantly explained 18% of the total variance (Figure 7). Thus, other factors acting at the local scale, such as habitat heterogeneity, competition and predation processes [85], may have an additional role in promoting a high diversification of benthic deep-sea fungi. Overall, results of the present study indicate that Antarctic deep-sea sediments host abundant and highly diversified fungal assemblages most of which still unidentified and suggest that fungi inhabiting Antarctic benthic deep-sea ecosystems can be sensitive to an interplay of environmental and ecological factors, whose variations, potentially induced also by climate changes, can profoundly influence their assemblage composition.

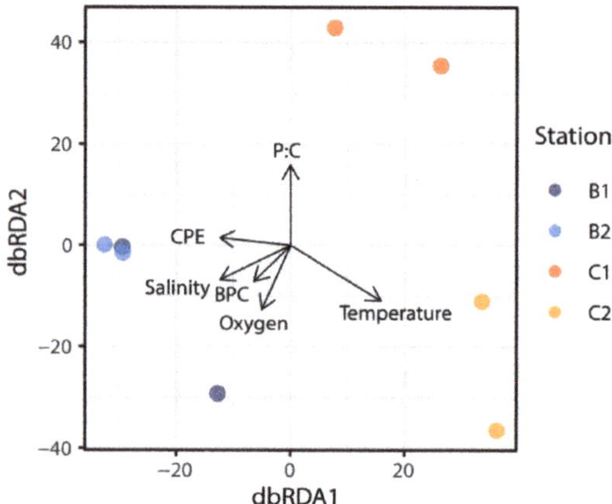

Figure 7. Output of the distance-based redundancy analysis (dbRDA) showing the relationships between the fungal assemblage composition and environmental and trophic variables analyzed in this study.

4. Conclusions

This study provides new insights into the quantitative relevance and diversity of benthic deep-sea fungi in the Ross Sea. Our findings reveal that the distribution of fungal abundance and richness is primarily driven by trophic availability, whereas an interplay of factors shapes fungal assemblage composition. Our findings also suggest that the spatial variability even at a small scale can promote important differences in deep-sea fungal assemblages, thus allowing for the maintenance of overall high fungal diversity. Results reported in this study could be relevant for a better understanding of the potential impact of thermohaline and trophic modifications due to climate changes on Antarctic deep-sea ecosystems. Modifications of ice coverage and thermohaline conditions affecting the planktonic food web structure could, indeed, profoundly influence organic carbon export to the seafloor, with cascading effects on benthic deep-sea biodiversity and ecosystem functioning. Although altered freezing and melting cycles of Antarctic pack ice are expected to drastically change ecosystem functioning, we still have a limited knowledge of biogeochemical cycles and ecological processes in which fungi are involved. Therefore, our results highlight the need to improve our understanding of the ecological role of benthic deep-sea fungi for better comprehension and prediction of the potential effects of climate changes on Antarctic deep-sea ecosystem functioning.

Supplementary Materials: The following supporting information can be downloaded at: https://www.mdpi.com/article/10.3390/jof8010065/s1: Figure S1. Taxonomic analysis of fungal ASVs obtained using 3 different confidence thresholds through the USEARCH SINTAX command.

Author Contributions: Conceptualization, A.D., C.C. and R.D.; methodology, G.B. and M.T.; data curation, G.B., M.T. and S.V.; writing—original draft preparation, G.B. and M.T.; writing—review and editing, A.D., C.C., S.V. and E.R.; funding acquisition, A.D. and R.D. All authors have read and agreed to the published version of the manuscript.

Funding: This research was carried out in the framework of the BEDROSE project financially supported by PNRA and the VIRIDE project (n. 2017EKFA98) financially supported by the Ministry of University and Research.

Institutional Review Board Statement: Not applicable.

Informed Consent Statement: Not applicable.

Data Availability Statement: Raw sequencing reads have been deposited in the NCBI Sequence Read Archive (SRA) under the accession numbers PRJNA791191.

Acknowledgments: The authors thank all the crew members for supporting sample collection.

Conflicts of Interest: The authors declare no conflict of interest.

References

1. Arrigo, K.R.; van Dijken, G.L.; Bushinsky, S. Primary production in the Southern Ocean, 1997–2006. *J. Geophys. Res. Ocean* **2008**, *113*, 1–27. [CrossRef]
2. Arrigo, K.R.; van Dijken, G.; Long, M. Coastal Southern Ocean: A strong anthropogenic CO_2 sink. *Geophys. Res. Lett.* **2008**, *35*, 1–6.
3. Arrigo, K.R.; van Dijken, G.L. Phytoplankton dynamics within 37 Antarctic coastal polynya systems. *J. Geophys. Res. Ocean* **2003**, *108*, 3271. [CrossRef]
4. Arrigo, K.R.; Van Dijken, G.L. Interannual variation in air-sea CO_2 flux in the Ross Sea, Antarctica: A model analysis. *J. Geophys. Res. Ocean* **2007**, *112*, 1–16.
5. Smith, W.O.; Nelson, D.M. Importance of ice edge phytoplankton production in the Southern Ocean. *Bioscience* **1986**, *36*, 251–257. [CrossRef]
6. Smith, W.O.; Gordon, L.I. Hyperproductivity of the Ross Sea (Antarctica) polynya during austral spring. *Geophys. Res. Lett.* **1997**, *24*, 233–236. [CrossRef]
7. Sweeney, C. The annual cycle of surface water CO_2 and O_2 in the Ross Sea: A model for gas exchange on the continental shelves of Antarctica. *Biogeochem. Ross Sea* **2003**, *78*, 295–312.
8. Arrigo, K.R.; Van Dijken, G.L. Annual changes in sea-ice, chlorophyll α, and primary production in the Ross Sea, Antarctica. *Deep. Res. Part II Top. Stud. Oceanogr.* **2004**, *51*, 117–138. [CrossRef]
9. Brandt, A.; De Broyer, C.; De Mesel, I.; Ellingsen, K.E.; Gooday, A.J.; Hilbig, B.; Linse, K.; Thomson, M.R.A.; Tyler, P.A. The biodiversity of the deep Southern Ocean benthos. *Philos. Trans. R. Soc. B Biol. Sci.* **2007**, *362*, 39–66. [CrossRef]
10. Griffiths, H.J. Antarctic Marine Biodiversity—What Do We Know About the Distribution of Life in the Southern Ocean? *PLoS ONE* **2010**, *5*, e11683. [CrossRef]
11. Danovaro, R.; Snelgrove, P.V.R.; Tyler, P. Challenging the paradigms of deep-sea ecology. *Trends Ecol. Evol.* **2014**, *29*, 465–475. [CrossRef]
12. Bernhard, J.M.; Kormas, K.; Pachiadaki, M.G.; Rocke, E.; Beaudoin, D.J.; Morrison, C.; Visscher, P.T.; Cobban, A.; Starczak, V.R.; Edgcomb, V.P. Benthic protists and fungi of Mediterranean deep hypsersaline anoxic basin redoxcline sediments. *Front. Microbiol.* **2014**, *5*, 1–13. [CrossRef] [PubMed]
13. Edgcomb, V.P.; Pachiadaki, M.G.; Mara, P.; Kormas, K.A.; Leadbetter, E.R.; Bernhard, J.M. Gene expression profiling of microbial activities and interactions in sediments under haloclines of E. Mediterranean deep hypersaline anoxic basins. *ISME J.* **2016**, *10*, 2643–2657. [CrossRef]
14. Barone, G.; Varrella, S.; Tangherlini, M.; Rastelli, E.; Dell'Anno, A.; Danovaro, R.; Corinaldesi, C. Marine fungi: Biotechnological perspectives from deep-hypersaline anoxic basins. *Diversity* **2019**, *11*, 113. [CrossRef]
15. Nagahama, T.; Takahashi, E.; Nagano, Y.; Abdel-Wahab, M.A.; Miyazaki, M. Molecular evidence that deep-branching fungi are major fungal components in deep-sea methane cold-seep sediments. *Environ. Microbiol.* **2011**, *13*, 2359–2370. [CrossRef]
16. Thaler, A.D.; Van Dover, C.L.; Vilgalys, R. Ascomycete phylotypes recovered from a Gulf of Mexico methane seep are identical to an uncultured deep-sea fungal clade from the Pacific. *Fungal Ecol.* **2012**, *5*, 270–273. [CrossRef]
17. Burgaud, G.; Arzur, D.; Durand, L.; Cambon-Bonavita, M.A.; Barbier, G. Marine culturable yeasts in deep-sea hydrothermal vents: Species richness and association with fauna. *FEMS Microbiol. Ecol.* **2010**, *73*, 121–133. [CrossRef] [PubMed]
18. Burgaud, G.; Le Calvez, T.; Arzur, D.; Vandenkoornhuyse, P.; Barbier, G. Diversity of culturable marine filamentous fungi from deep-sea hydrothermal vents. *Environ. Microbiol.* **2009**, *11*, 1588–1600. [CrossRef] [PubMed]
19. Xu, W.; Guo, S.; Pang, K.L.; Luo, Z.H. Fungi associated with chimney and sulfide samples from a South Mid-Atlantic Ridge hydrothermal site: Distribution, diversity and abundance. *Deep. Res. Part I Oceanogr. Res. Pap.* **2017**, *123*, 48–55. [CrossRef]
20. Edgcomb, V.P.; Beaudoin, D.; Gast, R.; Biddle, J.F.; Teske, A. Marine subsurface eukaryotes: The fungal majority. *Environ. Microbiol.* **2011**, *13*, 172–183. [CrossRef]

21. Orsi, W.; Biddle, J.F.; Edgcomb, V. Deep Sequencing of Subseafloor Eukaryotic rRNA Reveals Active Fungi across Marine Subsurface Provinces. *PLoS ONE* **2013**, *8*, e56335. [CrossRef]
22. Zhang, T.; Wang, N.F.; Zhang, Y.Q.; Liu, H.Y.; Yu, L.Y. Diversity and distribution of fungal communities in the marine sediments of Kongsfjorden, Svalbard (High Arctic). *Sci. Rep.* **2015**, *5*, 1–11. [CrossRef]
23. Pachiadaki, M.G.; Rédou, V.; Beaudoin, D.J.; Burgaud, G.; Edgcomb, V.P. Fungal and prokaryotic activities in the Marine subsurface biosphere at Peru Margin and canterbury basin inferred from RNA-based analyses and microscopy. *Front. Microbiol.* **2016**, *7*, 846. [CrossRef] [PubMed]
24. Barone, G.; Rastelli, E.; Corinaldesi, C.; Tangherlini, M.; Danovaro, R.; Dell'Anno, A. Benthic deep-sea fungi in submarine canyons of the Mediterranean Sea. *Prog. Oceanogr.* **2018**, *168*, 57–64. [CrossRef]
25. Ogaki, M.B.; Coelho, L.C.; Vieira, R.; Neto, A.A.; Zani, C.L.; Alves, T.M.A.; Junior, P.A.S.; Murta, S.M.F.; Barbosa, E.C.; Oliveira, J.G.; et al. Cultivable fungi present in deep-sea sediments of Antarctica: Taxonomy, diversity, and bioprospecting of bioactive compounds. *Extremophiles* **2020**, *24*, 227–238. [CrossRef]
26. Ogaki, M.B.; Pinto, O.H.B.; Vieira, R.; Neto, A.A.; Convey, P.; Carvalho-Silva, M.; Rosa, C.A.; Câmara, P.E.A.S.; Rosa, L.H. Fungi Present in Antarctic Deep-Sea Sediments assessed Using DNA Metabarcoding. *Microb. Ecol.* **2021**, *82*, 157–164. [CrossRef]
27. Rosa, L.H. *Fungi of Antarctica Diversity, Ecology and Biotechnological Applications*; Springer International Publishing: Cham, Germany, 2019; ISBN 978-3-030-18367-7.
28. Varrella, S.; Barone, G.; Tangherlini, M.; Rastelli, E.; Dell'anno, A.; Corinaldesi, C. Diversity, ecological role and biotechnological potential of antarctic marine fungi. *J. Fungi* **2021**, *7*, 391. [CrossRef] [PubMed]
29. Blackwell, M. The fungi: 1, 2, 3 … 5.1 million species? *Am. J. Bot.* **2011**, *98*, 426–438. [CrossRef]
30. Hawksworth, D.L.; Rossman, A.Y. Where are all the undescribed fungi? *Phytopathology* **1997**, *87*, 888–891. [CrossRef] [PubMed]
31. Jeffries, T.C.; Curlevski, N.J.; Brown, M.V.; Harrison, D.P.; Doblin, M.A.; Petrou, K.; Ralph, P.J.; Seymour, J.R. Partitioning of fungal assemblages across different marine habitats. *Environ. Microbiol. Rep.* **2016**, *8*, 235–238. [CrossRef]
32. Vargas-Gastélum, L.; Riquelme, M. The mycobiota of the deep sea: What omics can offer. *Life* **2020**, *10*, 292. [CrossRef]
33. Tisthammer, K.H.; Cobian, G.M.; Amend, A.S. Global biogeography of marine fungi is shaped by the environment. *Fungal Ecol.* **2016**, *19*, 39–46. [CrossRef]
34. Li, W.; Wang, M.M.; Wang, X.G.; Cheng, X.L.; Guo, J.J.; Bian, X.M.; Cai, L. Fungal communities in sediments of subtropical Chinese seas as estimated by DNA metabarcoding. *Sci. Rep.* **2016**, *6*, 1–9. [CrossRef]
35. Orsi, W.D.; Vuillemin, A.; Coskun, Ö.K.; Rodriguez, P.; Oertel, Y.; Niggemann, J.; Mohrholz, V.; Gomez-Saez, G.V. Carbon assimilating fungi from surface ocean to subseafloor revealed by coupled phylogenetic and stable isotope analysis. *ISME J.* **2021**, *15*, 1–17. [CrossRef]
36. Hyde, K.D.; Jones, E.B.G.; Leaño, E.; Pointing, S.B.; Poonyth, A.D.; Vrijmoed, L.L.P. Role of fungi in marine ecosystems. *Biodivers. Conserv.* **1998**, *7*, 1147–1161. [CrossRef]
37. Cathrine, S.J.; Raghukumar, C. Anaerobic denitrification in fungi from the coastal marine sediments off Goa, India. *Mycol. Res.* **2009**, *113*, 100–109. [CrossRef]
38. Jebaraj, C.S.; Raghukumar, C.; Behnke, A.; Stoeck, T. Fungal diversity in oxygen-depleted regions of the Arabian Sea revealed by targeted environmental sequencing combined with cultivation. *FEMS Microbiol. Ecol.* **2010**, *71*, 399–412. [CrossRef]
39. Dell'Anno, A.; Danovaro, R. Ecology: Extracellular DNA plays a key role in deep-sea ecosystem functioning. *Science* **2005**, *309*, 2179. [CrossRef]
40. Dell'Anno, A.; Pusceddu, A.; Corinaldesi, C.; Canals, M.; Heussner, S.; Thomsen, L.; Danovaro, R. Trophic state of benthic deep-sea ecosystems from two different continental margins off Iberia. *Biogeosciences* **2013**, *10*, 2945–2957. [CrossRef]
41. Grinhut, T.; Hadar, Y.; Chen, Y. Degradation and transformation of humic substances by saprotrophic fungi: Processes and mechanisms. *Fungal Biol. Rev.* **2007**, *21*, 179–189. [CrossRef]
42. Raghukumar, S. Animals in Coastal Benthic Ecosystem and Aquaculture Systems. In *Fungi in Coastal and Oceanic Marine Ecosystems*; Raghukumar, S., Ed.; Springer International Publishing: Cham, Germany, 2017; pp. 163–183. ISBN 978-3-319-54304-8.
43. Doney, S.C.; Ruckelshaus, M.; Emmett Duffy, J.; Barry, J.P.; Chan, F.; English, C.A.; Galindo, H.M.; Grebmeier, J.M.; Hollowed, A.B.; Knowlton, N.; et al. Climate change impacts on marine ecosystems. *Ann. Rev. Mar. Sci.* **2012**, *4*, 11–37. [CrossRef]
44. Garciá Molinos, J.; Halpern, B.S.; Schoeman, D.S.; Brown, C.J.; Kiessling, W.; Moore, P.J.; Pandolfi, J.M.; Poloczanska, E.S.; Richardson, A.J.; Burrows, M.T. Climate velocity and the future global redistribution of marine biodiversity. *Nat. Clim. Chang.* **2016**, *6*, 83–88. [CrossRef]
45. Silvano, A.; Foppert, A.; Rintoul, S.R.; Holland, P.R.; Tamura, T.; Kimura, N.; Castagno, P.; Falco, P.; Budillon, G.; Haumann, F.A.; et al. Recent recovery of Antarctic Bottom Water formation in the Ross Sea driven by climate anomalies. *Nat. Geosci.* **2020**, *13*, 780–786. [CrossRef]
46. Learman, D.R.; Henson, M.W.; Thrash, J.C.; Temperton, B.; Brannock, P.M.; Santos, S.R.; Mahon, A.R.; Halanych, K.M. Biogeochemical and microbial variation across 5500 km of Antarctic surface sediment implicates organic matter as a driver of benthic community structure. *Front. Microbiol.* **2016**, *7*, 284. [CrossRef]
47. Sweetman, A.K.; Thurber, A.R.; Smith, C.R.; Levin, L.A.; Mora, C.; Wei, C.L.; Gooday, A.J.; Jones, D.O.B.; Rex, M.; Yasuhara, M.; et al. Major impacts of climate change on deep-sea benthic ecosystems. *Elementa* **2017**, *5*, 4. [CrossRef]
48. Smith, W.O.; Ainley, D.G.; Arrigo, K.R.; Dinniman, M.S. The oceanography and ecology of the ross sea. *Ann. Rev. Mar. Sci.* **2014**, *6*, 469–487. [CrossRef]

49. Fabiano, M.; Danovaro, R. Enzymatic activity, bacterial distribution, and organic matter composition in sediments of the Ross Sea (Antarctica). *Appl. Environ. Microbiol.* **1998**, *64*, 3838–3845. [CrossRef]
50. Dell'Anno, A.; Mei, M.L.; Pusceddu, A.; Danovaro, R. Assessing the trophic state and eutrophication of coastal marine systems: A new approach based on the biochemical composition of sediment organic matter. *Mar. Pollut. Bull.* **2002**, *44*, 611–622. [CrossRef]
51. Pusceddu, A.; Dell'Anno, A.; Fabiano, M.; Danovaro, R.; Dell'Anno, A.; Fabiano, M.; Danovaro, R. Quantity and bioavailability of sediment organic matter as signatures of benthic trophic status. *Mar. Ecol. Prog. Ser.* **2009**, *375*, 41–52. [CrossRef]
52. Danovaro, R. *Methods for the Study of Deep-Sea Sediments, Their Functioning and Biodiversity*; Danovaro, R., Ed.; CRC Press: Boca Raton, FL, USA, 2010; ISBN 9781439811382.
53. Taylor, J.D.; Cunliffe, M. Multi-year assessment of coastal planktonic fungi reveals environmental drivers of diversity and abundance. *ISME J.* **2016**, *10*, 2118–2128. [CrossRef] [PubMed]
54. Chemidlin Prévost-Bouré, N.; Christen, R.; Dequiedt, S.; Mougel, C.; Lelièvre, M.; Jolivet, C.; Shahbazkia, H.R.; Guillou, L.; Arrouays, D.; Ranjard, L. Validation and application of a PCR primer set to quantify fungal communities in the soil environment by real-time quantitative PCR. *PLoS ONE* **2011**, *6*, e24166. [CrossRef]
55. White, T.J.; Bruns, T.; Lee, S.; Taylor, J. Amplification and direct sequencing of fungal ribosomal RNA genes for phylogenetics. In *PCR Protocols: A Guide to Methods and Applications*; Innis, M.A., Gelfand, D.H., Sninsky, J.J., White, T.J., Eds.; Academic Press: New York, NY, USA, 1990; pp. 315–322.
56. Gardes, M.; Bruns, T.D. ITS primers with enhanced specificity for basidiomycetes–application to the identification of mycorrhizae and rusts. *Mol. Ecol.* **1993**, *2*, 113–118. [CrossRef]
57. Bolyen, E.; Rideout, J.R.; Dillon, M.R.; Bokulich, N.A.; Abnet, C.C.; Al-Ghalith, G.A.; Alexander, H.; Alm, E.J.; Arumugam, M.; Asnicar, F.; et al. Reproducible, interactive, scalable and extensible microbiome data science using QIIME 2. *Nat. Biotechnol.* **2019**, *37*, 852–857. [CrossRef] [PubMed]
58. Rivers, A.R.; Weber, K.C.; Gardner, T.G.; Liu, S.; Armstrong, S.D. ITSxpress: Software to rapidly trim internally transcribed spacer sequences with quality scores for marker gene analysis. *F1000Research* **2018**, *7*, 1418. [CrossRef] [PubMed]
59. Callahan, B.J.; McMurdie, P.J.; Rosen, M.J.; Han, A.W.; Johnson, A.J.A.; Holmes, S.P. DADA2: High-resolution sample inference from Illumina amplicon data. *Nat. Methods* **2016**, *13*, 581–583. [CrossRef] [PubMed]
60. Edgar, R.C. Accuracy of taxonomy prediction for 16S rRNA and fungal ITS sequences. *PeerJ* **2018**, *2018*, e4652. [CrossRef]
61. Hughes, J.B.; Hellmann, J.J. The application of rarefaction techniques to molecular inventories of microbial diversity. *Methods Enzymol.* **2005**, *397*, 292–308. [PubMed]
62. Gihring, T.M.; Green, S.J.; Schadt, C.W. Massively parallel rRNA gene sequencing exacerbates the potential for biased community diversity comparisons due to variable library sizes. *Environ. Microbiol.* **2012**, *14*, 285–290. [CrossRef] [PubMed]
63. Anderson, M.J. A new method for non-parametric multivariate analysis of variance. *Austral Ecol.* **2001**, *26*, 32–46.
64. Anderson, M.J. Permutation tests for univariate or multivariate analysis of variance and regression. *Can. J. Fish. Aquat. Sci.* **2001**, *58*, 626–639. [CrossRef]
65. Bastian, M.; Heymann, S.; Jacomy, M. Gephi: An open source software for exploring and manipulating networks. In Proceedings of the Third International AAAI Conference on Weblogs and Social Media, San Jose, CA, USA, 17–20 May 2009.
66. Clarke, K.R. Non-parametric multivariate analyses of changes in community structure. *Aust. J. Ecol.* **1993**, *18*, 117–143. [CrossRef]
67. McArdle, B.H.; Anderson, M.J. Fitting Multivariate Models to Community Data: A Comment on Distance-Based Redundancy Analysis. *Ecology* **2001**, *82*, 290. [CrossRef]
68. Bergamasco, A.; Defendi, V.; Zambianchi, E.; Spezie, G. Evidence of dense water overflow on the Ross Sea shelf-break. *Antarct. Sci.* **2002**, *14*, 271–277. [CrossRef]
69. Gardner, W.D.; Richardson, M.J.; Smith, W.O. Seasonal patterns of water column particulate organic carbon and fluxes in the ross sea, antarctica. *Deep. Res. Part II Top. Stud. Oceanogr.* **2000**, *47*, 3423–3449. [CrossRef]
70. Frignani, M.; Giglio, F.; Accornero, A.; Langone, L.; Ravaioli, M. Sediment characteristics at selected sites of the Ross Sea continental shelf: Does the sedimentary record reflect water column fluxes? *Antarct. Sci.* **2003**, *15*, 133–139. [CrossRef]
71. Danovaro, R.; Molari, M.; Corinaldesi, C.; Dell'Anno, A. Macroecological drivers of archaea and bacteria in benthic deep-sea ecosystems. *Sci. Adv.* **2016**, *2*, 1–12. [CrossRef] [PubMed]
72. Wei, C.L.; Rowe, G.T.; Briones, E.E.; Boetius, A.; Soltwedel, T.; Caley, M.J.; Soliman, Y.; Huettmann, F.; Qu, F.; Yu, Z.; et al. Global patterns and predictions of seafloor biomass using random forests. *PLoS ONE* **2010**, *5*, e15323. [CrossRef]
73. Xu, W.; Pang, K.L.; Luo, Z.H. High Fungal Diversity and Abundance Recovered in the Deep-Sea Sediments of the Pacific Ocean. *Microb. Ecol.* **2014**, *68*, 688–698. [CrossRef]
74. Rämä, T.; Hassett, B.T.; Bubnova, E. Arctic marine fungi: From filaments and flagella to operational taxonomic units and beyond. *Bot. Mar.* **2017**, *60*, 433–452. [CrossRef]
75. Duarte, A.W.F.; dos Santos, J.A.; Vianna, M.V.; Vieira, J.M.F.; Mallagutti, V.H.; Inforsato, F.J.; Wentzel, L.C.P.; Lario, L.D.; Rodrigues, A.; Pagnocca, F.C.; et al. Cold-adapted enzymes produced by fungi from terrestrial and marine Antarctic environments. *Crit. Rev. Biotechnol.* **2018**, *38*, 600–619. [CrossRef]
76. Ren, J.; Xue, C.; Tian, L.; Xu, M.; Chen, J.; Deng, Z.; Proksch, P.; Lin, W. Asperelines A−F, Peptaibols from the Marine-Derived Fungus *Trichoderma asperellum*. *J. Nat. Prod.* **2009**, *72*, 1036–1044. [CrossRef]
77. Gonçalves, V.N.; Vaz, A.B.M.; Rosa, C.A.; Rosa, L.H. Diversity and distribution of fungal communities in lakes of Antarctica. *FEMS Microbiol. Ecol.* **2012**, *82*, 459–471. [CrossRef]

78. Godinho, V.M.; Furbino, L.E.; Santiago, I.F.; Pellizzari, F.M.; Yokoya, N.S.; Pupo, D.; Alves, T.M.A.; S Junior, P.A.; Romanha, A.J.; Zani, C.L.; et al. Diversity and bioprospecting of fungal communities associated with endemic and cold-adapted macroalgae in Antarctica. *ISME J.* **2013**, *7*, 1434–1451. [CrossRef] [PubMed]
79. Furbino, L.E.; Godinho, V.M.; Santiago, I.F.; Pellizari, F.M.; Alves, T.M.A.A.; Zani, C.L.; Junior, P.A.S.S.; Romanha, A.J.; Carvalho, A.G.O.O.; Gil, L.H.V.G.V.G.; et al. Diversity Patterns, Ecology and Biological Activities of Fungal Communities Associated with the Endemic Macroalgae Across the Antarctic Peninsula. *Microb. Ecol.* **2014**, *67*, 775–787. [CrossRef]
80. Wang, F.; Sheng, J.; Chen, Y.; Xu, J. Microbial diversity and dominant bacteria causing spoilage during storage and processing of the Antarctic krill, *Euphausia superba*. *Polar Biol.* **2021**, *44*, 163–171. [CrossRef]
81. Gonçalves, V.N.; Vitoreli, G.A.; de Menezes, G.C.A.; Mendes, C.R.B.; Secchi, E.R.; Rosa, C.A.; Rosa, L.H. Taxonomy, phylogeny and ecology of cultivable fungi present in seawater gradients across the Northern Antarctica Peninsula. *Extremophiles* **2017**, *21*, 1005–1015. [CrossRef] [PubMed]
82. Richards, T.A.; Talbot, N.J. Horizontal gene transfer in osmotrophs: Playing with public goods. *Nat. Rev. Microbiol.* **2013**, *11*, 720–727. [CrossRef] [PubMed]
83. Richards, T.A.; Leonard, G.; Mahé, F.; Del Campo, J.; Romac, S.; Jones, M.D.M.; Maguire, F.; Dunthorn, M.; De Vargas, C.; Massana, R.; et al. Molecular diversity and distribution of marine fungi across 130 european environmental samples. *Proc. R. Soc. B Biol. Sci.* **2015**, *282*, 20152243. [CrossRef] [PubMed]
84. Couturier, M.; Bennati-Granier, C.; Urio, M.B.; Ramos, L.P.; Berrin, J.G. Fungal enzymatic degradation of cellulose. In *Green Fuels Technology*; Green Energy and Technology; Soccol, C., Brar, S., Faulds, C., Ramos, L., Eds.; Springer: Cham, Switzerland, 2016; pp. 133–146.
85. Fabiano, M.; Danovaro, R. Meiofauna distribution and mesoscale variability in two sites of the Ross Sea (Antarctica) with contrasting food supply. *Polar Biol.* **1999**, *22*, 115–123. [CrossRef]

Journal of
Fungi

MDPI

Article

Insights on *Lulworthiales* Inhabiting the Mediterranean Sea and Description of Three Novel Species of the Genus *Paralulworthia*

Anna Poli, Valeria Prigione *, Elena Bovio, Iolanda Perugini and Giovanna Cristina Varese

Mycotheca Universitatis Taurinensis, Department of Life Sciences and Systems Biology, University of Torino, Viale Mattioli 25, 10125 Torino, Italy; anna.poli@unito.it (A.P.); elena.bovio@inrae.fr (E.B.); iolanda.perugini@unito.it (I.P.); cristina.varese@unito.it (G.C.V.)
* Correspondence: valeria.prigione@unito.it

Abstract: The order *Lulworthiales*, with its sole family *Lulworthiaceae*, consists of strictly marine genera found on a wide range of substrates such as seagrasses, seaweeds, and seafoam. Twenty-one unidentified *Lulworthiales* were isolated in previous surveys aimed at broadening our understanding of the biodiversity hosted in the Mediterranean Sea. Here, these organisms, mostly found in association with *Posidonia oceanica* and with submerged woods, were examined using thorough multi-locus phylogenetic analyses and morphological observations. Maximum-likelihood and Bayesian phylogeny based on nrITS, nrSSU, nrLSU, and four protein-coding genes led to the introduction of three novel species of the genus *Paralulworthia*: *P. candida*, *P. elbensis*, and *P. mediterranea*. Once again, the marine environment is a confirmed huge reservoir of novel fungal lineages with an under-investigated biotechnological potential waiting to be explored.

Keywords: marine fungi; novel lineages; phylogeny; genetic markers

Citation: Poli, A.; Prigione, V.; Bovio, E.; Perugini, I.; Varese, G.C. Insights on *Lulworthiales* Inhabiting the Mediterranean Sea and Description of Three Novel Species of the Genus *Paralulworthia*. *J. Fungi* **2021**, *7*, 940. https://doi.org/10.3390/jof7110940

Academic Editor: Federico Baltar

Received: 8 September 2021
Accepted: 3 November 2021
Published: 6 November 2021

1. Introduction

The increasing interest in marine fungi continues to widen our knowledge of marine biodiversity. So far, more than 1900 species inhabiting the oceans have been described (www.marinefungi.org); however, most of the fungal diversity, estimated to exceed 10,000 taxa [1], is yet to be uncovered. Marine habitats and substrates, both biotic and abiotic, are continuously being explored worldwide, leading to the discovery of new marine fungal lineages.

Sordariomycetes are one of the classes mostly detected in the sea (www.marinefungi.org) and include several orders, namely, *Coronophorales*, *Chaetosphaeriales*, *Diaporthales*, *Hypocreales*, *Koralionastetales*, *Lulworthiales*, *Magnaporthales*, *Microascales*, *Ophiostomatales*, *Phyllachorales*, *Savoryellales*, *Sordariales*, *Tirisporellales*, *Torpedosporales*, and *Xylariales* [2,3]. *Lulworthiales* and *Koralionastetales*, recently placed in the new subclass *Lulworthiomycetidae* [4,5], consist of exclusively marine taxa. The order *Lulworthiales*, with its sole family *Lulworthiaceae*, was introduced on the basis of morphological characters and phylogenetic analyses built upon nrSSU and nrLSU partial sequences to accommodate the polyphyletic genera *Lulworthia* and *Lindra* [6,7]. The family *Lulworthiaceae* consists of strictly marine genera—including *Cumulospora*, *Halazoon*, *Hydea*, *Kohlmeyerella*, *Lulwoana*, *Lulwoidea*, *Lulworthia*, *Lindra*, *Matsusporium*, *Moleospora*, *Rostrupiella*, *Sammeyersia*, and the recently described genus *Paralulworthia*—that are distributed worldwide and found on a variety of substrates such as submerged wood, seaweeds, seagrasses, seafoam, and aquatic plants [4,8,9]. Members of this family are well-known cellulase producers and can break down complex lignocellulose compounds, thus contributing to the recycling of nutrients [10]. Morphologically, they are characterized by ascomata subglobose to cylindrical, 8-spored asci, cylindrical to fusiform and filamentous ascospores with end chambers filled with mucus (the latter character is missing in *Lindra*) [6,11].

Twenty-one unidentified *Lulworthiales* were isolated in previous surveys aimed at broadening our knowledge on the underwater fungal diversity of the Mediterranean Sea: sixteen isolates were obtained from the seagrass *Posidonia oceanica* [12], three from submerged wood [13], and two from seawater contaminated by oil spills. Traditionally, the identification of fungi at species level is based on the description of sexual and/or asexual reproductive structures. However, it is not unusual to deal with marine fungi that neither sporulate nor develop reproductive structures in axenic culture. Therefore, the identification of sterile mycelia must rely on molecular data [4,14–17]. In light of valuable biotechnological exploitations of marine fungi, correct taxonomic placement of sterile mycelia is necessary.

With this study, we tried to provide a better phylogenetic placement of the Mediterranean *Lulworthiales* by applying a combined multi-locus molecular phylogeny. Following phylogenetic inference and morphological insights, the three new species, *Paralulworthia candida*, *Paralulworthia elbensis*, and *Paralulworthia mediterranea*, were hereunder proposed.

2. Materials and Methods

2.1. Fungal Isolates

The isolates analysed in this study were recovered during previous surveys from the Mediterranean Sea in Italy. In detail, two isolates were derived from a site chronically contaminated by an oil spill in Gela (Caltanissetta, Italy) [17], three from submerged woods sampled in the Marine Protected Areas Island of Bergeggi (Savona, Italy) [13], and sixteen from twelve plants of *P. oceanica* collected in the coastal waters off the Elba Island (Livorno, Italy) from two sampling sites, Ghiaie and Margidore [12] (Table 1). The strains were isolated on Corn Meal Agar medium supplemented with sea salts (CMASS; 3.5% *w*/*v* sea salt mix, Sigma-Aldrich, Saint Louis, MO, USA, in ddH_2O), and are currently preserved at the Mycotheca Universitatis Taurinensis (MUT), Italy.

Table 1. Dataset used for phylogenetic analysis. Genbank sequences include newly generated nrITS, nrLSU, and nrSSU amplicons relative to the novel species *Paralulworthia candida*, *P. elbensis*, and *P. mediterranea* (in bold).

Species	Strain	Source	nrITS	nrSSU	nrLSU
Lulworthiales					
Lulworthiaceae					
Cumulospora marina Schmidt	MF46	Submerged wood	–	GU252136	GU252135
	GC53	Submerged wood	–	GU256625	GU256626
C. varia Chatmala and Somrithipol	GR78	Submerged wood	–	EU848593	EU848578
	IT 152	Submerged wood	EU848579	EU848579	
Halazoon mehlae Abdel-Aziz, Abdel-Wahab and Nagah.	MF819 [T]	Drift stems of *Phragmites australis*	–	GU252144	GU252143
H. fuscus (Schmidt) Abdel-Wahab, Pang, Nagah., Abdel-Aziz and Jones	NBRC 105256	Driftwood	–	GU252148	GU252147
Hydea pigmaea (Kohlm) Pang and Jones	NBRC 33069	Driftwood	–	GU252134	GU252133
	IT081	Driftwood	–	GU256632	GU256633
Kohlmeyeriella crassa (Nakagiri) Kohlm., Volkm.–Kohlm., Campb., Spatafora and Gräfenhan	NBRC 32133 [T]	Sea foam	LC146741	AY879005	LC146742
K. tubulata (Kohlm.) Jones, Johnson and Moss	PP115	Marine environment	–	AY878998	AF491265
	PP0989	Marine environment	–	AY878997	AF491264
Lindra marinera Meyers	JK 5091	Marine environment	–	AY879000	AY878958
L. obtusa Nakagiri and Tubaki	NRBC 31317 [T]	Sea foam	LC146744	AY879002	AY878960
	AFTOL 5012	Marine environment	–	FJ176847	FJ176902
	CBS 113030	n.d.		AY879001	AY878959

Table 1. *Cont.*

Species	Strain	Source	nrITS	nrSSU	nrLSU
L. thalassiae Orpurt, Meyers, Boral and Simms	JK 5090A	Marine environment	–	U46874	U46891
	AFTOL 413	Marine environment	DQ491508	DQ470994	DQ470947
	JK 5090	Marine environment	–	AF195634	AF195635
	JK 4322	*Thalassia testudinum* leaves	–	AF195632	AF195633
Lulwoana uniseptata (Nakagiri) Kohlmeyer et al.	NBRC 32137 T	Submerged wood	LC146746	LC146746	LC146746
	CBS 16760	Driftwood	–	AY879034	AY878991
Zalerion maritima (Linder) Anastasiou	FCUL280207CP1	Sea water	KT347216	KT347203	JN886806
	FCUL010407SP2	Sea water	KT347217	KT347204	JN886805
Lulworthia atlantica Azevedo, Caeiro and Barata	FCUL210208SP4	Sea water	KT347205	KT347193	JN886843
	FCUL190407CF4	Sea water	KT347207	KT347198	JN886816
	FCUL061107CP3	Sea water	KT347208	KT347196	JN886825
L. fucicola Sutherl.	ATCC 64288 T	Intertidal wood	–	AY879007	AY878965
	PP1249	Marine environment	–	AY879008	AY878966
L. grandispora Meyers	AFTOL 424	Dead *Rhizophora* sp. branch	–	DQ522855	DQ522856
	NTOU3841	Driftwood	–	KY026044	KY026048
	NTOU3847	Decayed mangrove wood	–	KY026046	KY026049
	NTOU3849	Decayed mangrove wood	–	KY026047	KY026050
Lulworthia lignoarenaria (Koch and Jones) Kohlm., Volkm.–Kohlm., Campb., Spatafora and Gräfenhan	AFTOL 5013	Marine environment	–	FJ176848	FJ176903
L. medusa (Ellis and Everh.) Cribb and Cribb	JK 5581 T	Spartina	–	AF195636	AF195637
L. opaca (Linder) Cribb and J.W. Cribb	CBS 218.60	Driftwood in seawater	–	AY879003	AY87896
L. cf. *purpurea* (Wilson) Johnson	FCUL170907CP5	Sea water	KT347219	KT347201	JN886824
	FCUL280207CF9	Sea water	KT347218	KT347202	JN886808
Matsusporium tropicale (Kohlm.) Jones and Pang	NBRC 32499	Submerged wood	–	GU252142	GU252141
Moleospora maritima Abdel-Wahab, Abdel-Aziz and Nagah.	MF 836 T	Drift stems of *Phragmites australis*	–	GU252138	GU252137
***Paralulworthia candida* sp. nov.**	MUT 5430	*P. oceanica*	**MZ357724**	**MZ357767**	**MZ357746**
***Paralulworthia elbensis* sp. nov.**	MUT 377	*P. oceanica*	**MZ357710**	**MZ357753**	**MZ357732**
	MUT 5422	*P. oceanica*	**MZ357723**	**MZ357766**	**MZ357745**
	MUT 5438	*P. oceanica*	**MZ357712**	**MZ357755**	**MZ357734**
	MUT 5461	*P. oceanica*	**MZ357725**	**MZ357768**	**MZ357747**
Paralulworthia gigaspora Prigione, Poli, Bovio and Varese	MUT 435 T	*P. oceanica*	MN649242	MN649246	MN649250
	MUT 5413	*P. oceanica*	MN649243	MN649247	MN649251
	MUT 263	Oil-contaminated sea water	**MZ357729**	**MZ357772**	**MZ357751**

Table 1. *Cont.*

Species	Strain	Source	nrITS	nrSSU	nrLSU
	MUT 465	*P. oceanica*	**MZ357726**	**MZ357769**	**MZ357748**
	MUT 1753	Oil-contaminated sea water	**MZ357730**	**MZ357773**	**MZ357752**
	MUT 5085	*P. oceanica*	**MZ357715**	**MZ357758**	**MZ357737**
	MUT 5086	*P. oceanica*	**MZ357716**	**MZ357759**	**MZ357738**
	MUT 5093	*P. oceanica*	**MZ357718**	**MZ357761**	**MZ357740**
	MUT 5094	*P. oceanica*	**MZ357719**	**MZ357762**	**MZ357741**
Paralulworthia halima (Anastasiou) Gonçalves, Abreu and Alves	CMG 68	Submerged wood	MT235736	MT235712	MT235753
	CMG 69	Submerged wood	MT235737	MT235713	MT235754
	MUT 1483	Submerged wood	**MZ357727**	**MZ357770**	**MZ357749**
	MUT 2919	Submerged wood	**MZ357713**	**MZ357756**	**MZ357735**
	MUT 3347	Submerged wood	**MZ357728**	**MZ357771**	**MZ357750**
Paralulworthia posidoniae Poli, Prigione, Bovio and Varese	MUT 5261 T	*P. oceanica*	MN649245	MN649249	MN649253
	MUT 5092	*P. oceanica*	**MZ357717**	**MZ357760**	**MZ357739**
	MUT 5110	*P. oceanica*	**MZ357720**	**MZ357763**	**MZ357742**
	MUT 5419	*P. oceanica*	**MZ357722**	**MZ35776**	**MZ357744**
***Paralulworthia mediterranea* sp. nov.**	MUT 654	*P. oceanica*	**MZ357711**	**MZ357754**	**MZ357733**
	MUT 5080	*P. oceanica*	**MZ357714**	**MZ357757**	**MZ357736**
	MUT 5417 T	*P. oceanica*	**MZ357721**	**MZ357764**	**MZ357743**
Pisorisporiales					
Pisorisporiaceae					
Achroceratosphaeria potamia Réblová, Fourn. and Hyde	JF 08139 T	Submerged wood of *Platanus* sp.	–	GQ996541	GQ996538
Pleosporales					
Melanommataceae					
Bimuria novae-zelandiae Hawksw., Chea and Sheridan	CBS 107.79 T	Soil	MH861181	FJ190605	MH872950
Pleosporaceae					
Setosphaeria monoceras Alcorn	CBS 154.26	n.d.	DQ337380	DQ238603	AY016368
Dydimosphaeriaceae					
Letendraea helminthicola (Berk. and Broome) Weese ex Petch	CBS 884.85	Yerba mate	MK404145	AY016345	AY016362

T = Type Strain.

2.2. Morphological Analysis

The strains were grown on Malt Extract Agar seawater (MEASW; 20 g malt extract, 20 g glucose, 2 g peptone, 20 g agar—Sigma-Aldrich, Saint Louis, MO, USA—in 1 L of seawater) for one month at 21 °C prior to inoculation in triplicate onto new MEASW Petri dishes (9 cm Ø). Plates were incubated at 15 and 21 °C. The colony growth was monitored periodically for 28 days, while macroscopic and microscopic features were assessed at the end of the incubation period.

Efforts to induce sporulation were carried out by applying sterile pieces of *Quercus ruber* cork and *Pinus pinaster* wood (species autochthonous to the Mediterranean area) on three-week-old fungal colonies [18]. Plates were further incubated for four weeks at 21 °C. Cork and wood specimens were transferred to 50 mL tubes containing 20 mL of sterile seawater. Samples were incubated at 21 °C for a minimum of three months up to nine months.

Morphological structures were observed, and images captured using an optical microscope (Leica DM4500B, Leica microsystems GmbH, Wetzlar, Germany) equipped with a camera (Leica DFC320, Leica microsystems GmbH, Wetzlar, Germany).

2.3. DNA Extraction, PCR Amplification, and Data Assembling

Fresh mycelium carefully scraped from MEASW plates was transferred to a 2 mL Epperndorf tube and disrupted by a MM400 tissue lyzer (Retsch GmbH, Haan, Germany). Genomic DNA was extracted following the manufacturer's instructions of a NucleoSpin kit (Macherey Nagel GmbH, Duren, DE, USA). The quality and quantity of DNA were measured spectrophotometrically (Infinite 200 PRO NanoQuant; Tecan, Männedorf, Switzerland), and DNA samples were stored at −20 °C.

The partial sequences of seven genetic markers were amplified by PCR. Primer pairs ITS1/ITS4 [19], LR0R/LR7 [20], and NS1/NS4 [19] were used to amplify the internal transcribed spacers, including the 5.8S rDNA gene (nrITS), 28S large ribosomal subunit (nrLSU), and 18S small ribosomal subunit (nrSSU). The translation elongation factor (TEF-1α), the β-tubulin (β-TUB), and the largest and second-largest subunits of RNA polymerase II (RPB1 and RPB2) were amplified by using the following primer pairs: EF-dF/EF-2218R [21], Bt2a/Bt2b [22], RPB1Af/RPB1Cr [23], and fRPB2-5F/fPB2-7R [24]. Reaction mixtures consisted of 20–40 ng DNA template, 10× PCR Buffer (15 mM $MgCl_2$, 500 mM KCl, 100 mM Tris-HCl, pH 8.3), 200 μM each dNTP, 1 μM each primer, and 2.5 U Taq DNA Polymerase (Qiagen, Chatsworth, CA, USA) in 50 μL final volume. Negative controls with no DNA template were included. For problematic cases, additional $MgCl_2$, BSA, and/or 2.5% DMSO were supplied. Amplifications were run in a T100 Thermal Cycler (Bio-Rad, Hercules, CA, USA) programmed as described in Table 2.

Table 2. Genetic markers, primers, and thermocycler conditions used in this study.

	Forward and Reverse Primers	Thermocycler Conditions	References
ITS	ITS1–ITS4	95 °C: 5 min (95 °C: 40 s, 55 °C: 50 s, 72 °C: 50 s) × 35 cycles; 72 °C: 8 min; 4 °C: ∞	[19]
LSU	LR0R–LR7	95 °C: 5 min (95 °C: 1 min, 50 °C: 1 min, 72 °C: 2 min) × 35 cycles; 72 °C: 10 min; 4 °C: ∞	[20]
SSU	NS1–NS4	95 °C: 5 min (95 °C: 1 min, 50 °C: 1 min, 72 °C: 2 min) × 35 cycles; 72 °C: 10 min; 4 °C: ∞	[19]
TEF-1α	EF-dF/EF-2218R	95 °C: 5 min (95 °C: 1 min, 50 °C: 1 min; 72 °C: 2 min) × 40 cycles, 72 °C: 10 min; 4 °C: ∞	[21]
βTUB	Bt2a–Bt2b	94 °C: 4 min (94 °C: 35 s, 58 °C: 35 s, 72 °C: 50 s) × 35 cycles; 72 °C: 5 min; 4 °C: ∞	[22]
RPB1	RPB1Af–RPB1Cr	96 °C: 5 min (94 °C: 30 s, 52 °C: 30 s, 72 °C: 1 min) × 40 cycles; 72 °C: 8 min; 4 °C: ∞	[23]
RPB2	fRPB2-5F/fPB2-7cR	94 °C: 3 min (94 °C: 30 s; 55 °C: 30 s; 72 °C: 1 min) × 40 cycles, 72 °C: 10 min; 4 °C: ∞	[24]

Amplicons and a GelPilot 1 kb, plus a DNA Ladder, were visualized on a 1.5% agarose gel stained with 5 mL 100 mL^{-1} ethidium bromide. PCR products were purified and sequenced at the Macrogen Europe Laboratory (Madrid, Spain). The resulting Applied Biosystem (ABI) chromatograms were inspected, trimmed, and assembled to obtain consensus sequences using Sequencer 5.0 (GeneCodes Corporation, Ann Arbor, MI, USA; http://www.genecodes.com). Newly generated sequences were deposited in GenBank with the accession numbers reported in Table 1 and Table S1.

2.4. Sequence Alignment and Phylogenetic Analysis

A dataset consisting of nrSSU, nrITS, and nrLSU was assembled on the basis of BLASTn results and of the available phylogenetic studies focused on *Lulworthiales*, *Lulworthiaceae*, and *Lulworthia* [5–7,9,11,25,26]. Reference sequences were obtained from GenBank (Table 1). Sequences were aligned using MUSCLE (default conditions for gap openings and gap extension penalties), implemented in MEGA X (Molecular Evolutionary Genetics Analysis), visually inspected, and manually trimmed to delimit and discard ambiguously aligned regions. Alignments were concatenated into a single data matrix with SequenceMatrix [27] since no incongruence was observed among single-loci phylogenetic trees. The best evolutionary model under the Akaike Information Criterion (AIC) was determined with jModelTest 2 [28]. Phylogenetic inference was estimated using Maximum Likehood (ML) and Bayesian Inference (BI) criteria. The ML analysis was generated using RAxML v. 8.1.2 [29] under GTR + I + G evolutionary model and 1000 bootstrap replicates. Support values from bootstrapping runs (BS) were mapped on the global best tree using the "-f a" option of RAxML and "-x 12345" as a random seed to invoke the novel rapid bootstrapping algorithm. BI was performed with MrBayes 3.2.2 [30] with the same substitution model (GTR + I + G). The alignment was run for 10 million generations with two independent runs each, containing four Markov Chains Monte Carlo (MCMC) and sampling every 100 iterations. The first 25% of generated trees were discarded as "burn-in". A consensus tree was generated using the "sumt" function of MrBayes and Bayesian posterior probabilities (BYPP) were calculated. Consensus trees were visualized in FigTree v. 1.4.2 (http://tree.bio.ed.ac.uk/software/figtree). Three species of Pleosporales, namely, *Bimuria novae-zelandiae*, *Letendraea helminthicola*, and *Setosphaeria monoceras*, were used as an outgroup, as indicated in previous studies [26]. Due to a topological similarity of the two resulting trees, only Bayesian analysis with BS and BYPP values was reported (Figure 1).

Following, a new phylogenetic analysis was conducted, focusing only on the under investigation, whose relationships were unclear. To this aim, TEF-1α, β-TUB, RPB1, and RPB2 sequences were added to the restricted dataset (Table S1). Alignments and multi-loci phylogeny were conducted as described above. *Lulworthia* cf. *purpurea*, *Halazoon melhae*, *Lulworthia medusa*, and *Cirrenalia fusca* were used as the outgroup.

Sequence alignments and phylogenetic trees were deposited in TreeBASE (http://www.treebase.org, submission number S28658 and S28660).

3. Results

3.1. Phylogenetic Inference

Preliminary analyses carried out individually with nrITS, nrSSU, and nrLSU revealed no incongruence in the topology of the single-loci trees. The combined three-markers dataset—built on the basis of the BLASTn results and available phylogenetic studies [5–7,9,11,25,26]—consisted of 69 taxa (including MUT strains) that represented 15 genera and 29 species (Table 1). A total of 148 sequences (21 nrITS, 21 nrSSU, 21 nrLSU, 20 TEF-1α, 24 β-TUB, 17 RPB1, and 24 RPB2) were newly generated, whereas 115 were obtained from GenBank.

The dataset combining nrSSU, nrITS, and nrLSU had an aligned length of 2166 characters, of which 1130 were conserved, 355 were parsimony-uninformative, and 681 parsimony-informative (TL = 2511, CI = 0.530999, RI = 0.776449, RC = 0.412294, HI = 0.469001). The strains investigated formed a monophyletic lineage (BYPP = 0.99; BS = 65%), with its closest relatives being *Lulworthia* cf. *purpurea*, *Halazoon mehlae*, *Lulworthia medusa*, and *H. fuscus* (Figure 1). Within this new group, five clades could be observed, as follows: MUT 5092, MUT 5110, and MUT 5419 clustered together with *Paralulworthia posidoniae*; MUT 1483 MUT 2919 and MUT 3347 were identified as *Paralulworthia halima* by performing BLASTn analysis of nrITS, nr SSU, and nrLSU relative to the three strains (nucleotide similarity between 99% and 100%); MUT 263, MUT 465, MUT 1753, MUT 5085, MUT 5086, MUT 5093, and MUT 5094, grouped together with *Paralulworthia gigaspora*; the fourth clade included MUT 654, MUT 5080, and MUT 5417 and appeared to support a new species of *Paralulworthia*; likewise, MUT 377, MUT 5422, MUT 5430, MUT 5438, and MUT 5461 formed the fifth clade.

Figure 1. Phylogenetic inference based on a combined nrITS, nrSSU, and nrLSU dataset. The tree is rooted to three species of Pleosporales. Different colors indicate the belonging to different clades; in bold the strains analyzed in this study. Branch numbers indicate BYPP and BS values; T = Type Strain; Bar = expected changes per site (0.04).

The supplemental dataset, implemented with the addition of TEF-1α, RPB1, RPB2, and β-TUB sequence data relative to the strains investigated, had an aligned length of 4623 characters, of which 4322 were conserved, 144 were parsimony-uninformative, and 157 were parsimony-informative (TL = 351, CI = 0.800971, RI = 0.909292, RC = 0. 728316, HI = 0.199029). The segregation of the strains was more evident, confirming the conclusions previously drawn. In detail, by inspecting the tree, rooted to the group consisting of *L.* cf. *purpurea*, *H. mehlae*, *L. medusa*, and *H. fuscus*, it was possible to distinguish two groups in the genus *Paralulworthia sensu lato*: (a) the cluster (BYPP = 1.00; BS = 98%) that included the new species, *Paralulworthia mediterranea* (MUT 654, MUT 5080, and MUT 5417); and (b) the cluster consisting of MUT 377, MUT 5422, MUT 5430, MUT 5438, and MUT 5461 (BYPP = 0.83; BS = 70%) that represented two additional novel species of the same genus (Figure S1), namely, *Paralulworthia elbensis* (MUT 377, MUT 5422, MUT 5438) and *Paralulworthia candida* (MUT 5430).

3.2. Taxonomy

3.2.1. *Paralulworthia mediterranea* sp. nov. A. Poli, E. Bovio, G.C. Varese and V. Prigione

- MYCOBANK: MB841118
- Type: Italy, Tuscany, the Mediterranean Sea, Elba Island (Livorno), Ghiaie, 3–5 m depth, 42°49′04″ N, 10°19′20″ E, from *Posidonia oceanica* roots, March 2010, R. Mussat-Sartor and N. Nurra, MUT 5417 holotype, living culture permanently preserved in metabolically inactive state by deep-freezing at MUT.
- Additional material examined: Italy, Tuscany, the Mediterranean Sea, Elba Island (Livorno), Ghiaie, 3–5 m depth, 42°49′04″ N, 10°19′20″ E from *Posidonia oceanica* rhizomes, March 2010, R. Mussat-Sartor and N. Nurra, MUT 654. Italy, Tuscany, the Mediterranean Sea, Elba Island (Livorno), Ghiaie, 3–5 m depth, 42°49′04″ N, 10°19′20″ E from *Posidonia oceanica* rhizomes, March 2010, R. Mussat-Sartor and N. Nurra, MUT 5080.
- Etymology: In reference to the Mediterranean Sea.
- Description: Growing actively on *Pinus pinaster* wood and *Quercus ruber* cork, more markedly on the first. *Hyphae* 2.4–4 μm wide, septate, from hyaline to dematiaceous. *Chlamydospores* light brown 4–5 × 5–6 μm, unicellular or two-celled often present. *Bulbils* on the colony surface single or in group, pale yellow or cream colored, becoming ochre with age, nearly spherical, 150–400 μm diameter, formed by swollen cells (10–15 μm diameter) (Figure 2).
- Sexual morph not observed. Asexual morph with differentiated conidiogenesis not observed.
- Colony description: Colony growing on MEASW, reaching 57–70 mm diameter after 14 days at 21 °C, mycelium feltrose, becoming granular with age due to the presence of bulbils, with irregular edges, beige, sometimes with greyish shades at the edges; reverse from amber to dark orange. A yellowish brown colored diffusible pigment was often present (Figure 2).

3.2.2. *Paralulworthia candida* sp. nov. A. Poli, E. Bovio, V. Prigione and G.C. Varese

- MYCOBANK: 841116
- Type: Italy, Tuscany, the Mediterranean Sea, Elba Island (Livorno), Ghiaie, 3–5 m depth, 42°49′04″ N, 10°19′20″ E, from *Posidonia oceanica* roots, March 2010, R. Mussat-Sartor and N. Nurra, MUT 5430 holotype, living culture permanently preserved in metabolically inactive state by deep-freezing at MUT.
- Etymology: In reference to the colony color.
- Description: Poor colonization of *Pinus pinaster* wood and *Quercus ruber* cork. *Hyphae* 2.2–4.2 μm wide, septate, hyaline. *Chlamydospores* abundant, brown, globose, or subglobose, from unicellular (5–7 × 5–8 μm) to eight-cellular (8–13 μm diameter), in the shape of a sarcina (Figure 3).
- Sexual morph not observed. Asexual morph with differentiated conidiogenesis not observed.
- Colony description. Growing on MEASW, reaching 27–32 mm diameter after 14 days at 21 °C, mycelium floccose, white with yellowish shades in the center, submerged edges giving a beige halo to the colony; reverse light orange. A pinkish colored diffusible pigment present (Figure 3).

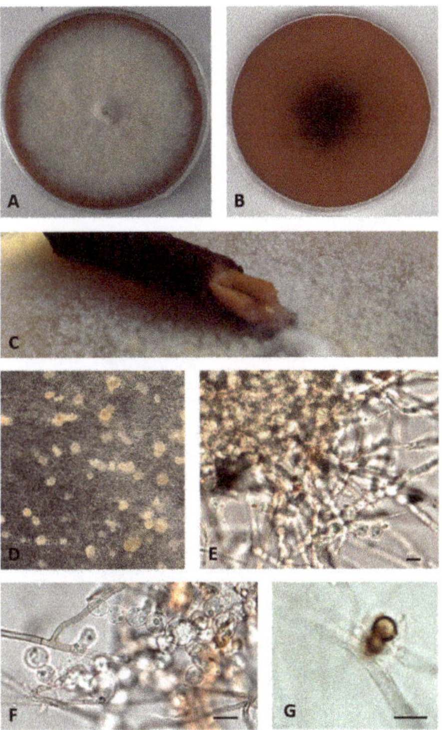

Figure 2. *Paralulworthia mediterranea* sp. nov. MUT 5417. 28-day-old colony at 21 °C on MEASW (**A**) and reverse (**B**); early colonization of *Pinus pinaster* wood (**C**); mycelium with bulbils (**D**); particular of a bulbil (**E**); swollen hyphae (**F**); two-celled chlamydospore (**G**). Scale bar: 10 μm.

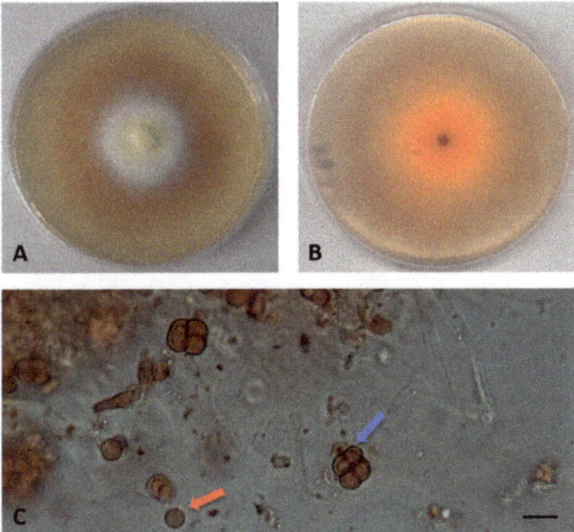

Figure 3. *Paralulorthia candida* sp. nov. MUT 5430. 28-day-old colony at 21 °C on MEASW (**A**) and reverse (**B**); unicellular (red arrow) and eight-cellular chlamydospores—blue arrow, (**C**). Scale bar: 10 μm.

3.2.3. *Paralulworthia elbensis* sp. nov. A. Poli, E. Bovio, V. Prigione and G.C. Varese

- MYCOBANK: MB841117
- Type: Italy, Tuscany, the Mediterranean Sea, Elba Island (Livorno), Margidore,14–15 m depth, 42°45′29″ N, 10°18′24″ E, from *Posidonia oceanica* roots, March 2010, R. Mussat-Sartor and N. Nurra, MUT 5422 holotype, living culture permanently preserved in metabolically inactive state by deep-freezing at MUT.
- Additional material examined: Italy, Tuscany, the Mediterranean Sea, Elba Island (Livorno), Ghiaie, 3–5 m depth, 42°49′04″ N, 10°19′20″ E, from *Posidonia oceanica* roots, March 2010, R. Mussat-Sartor and N. Nurra, MUT 377. Italy, Tuscany, the Mediterranean Sea, Elba Island (Livorno), Margidore,14–15 m depth, 42°45′29″ N, 10°18′24″ E, from *Posidonia oceanica* roots, March 2010, R. Mussat-Sartor and N. Nurra, MUT 5438. Italy, Tuscany, the Mediterranean Sea, Elba Island (Livorno), Margidore, 14–15 m depth, 42°45′29″ N, 10°18′24″ E, from *Posidonia oceanica* roots, March 2010, R. Mussat-Sartor and N. Nurra, MUT 5461.
- Etymology: In reference to the location of isolation.
- Description: Poor colonization of *Pinus pinaster* wood and *Quercus ruber* cork. *Hyphae* 2.6–4.5 μm wide, septate, hyaline. *Chlamydospores* abundant, brown, single or in chains, from globose to ellipsoidal, unicellular (5–7 × 6–7 μm) or multicellular (8–11 × 9–12 μm diameter) (Figure 4).

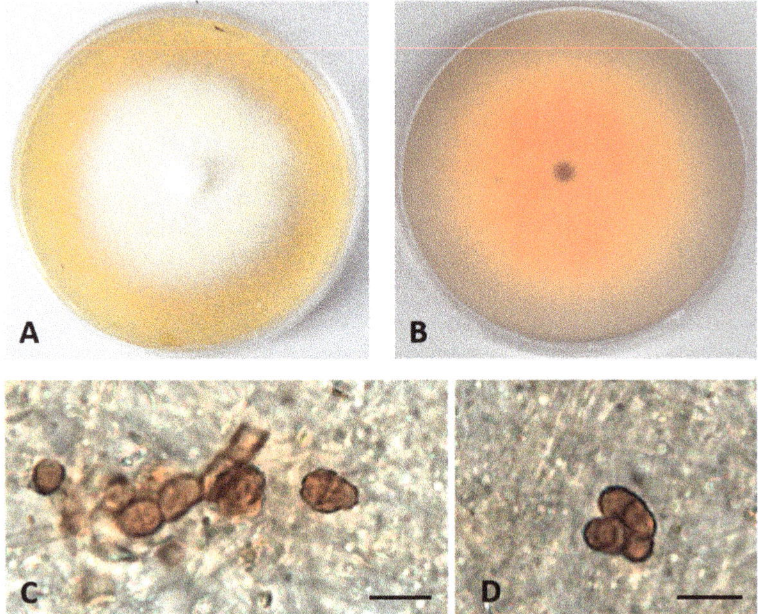

Figure 4. *Paraulworthia elbensis* sp. nov. MUT 5422. 28-day-old colony at 21 °C on MEASW (**A**) and reverse (**B**); chlamidospores in chain (**C**); multicellular chlamydospores (**D**). Scale bar: 10 μm.

- Sexual morph not observed. Asexual morph with differentiated conidiogenesis not observed.
- Colony description. Growing on MEASW, reaching 35–37 mm diameter after 14 days at 21 °C, mycelium feltrose, white with yellowish shades, submerged edges; reverse light orange (Figure 4).

4. Discussion

The morphological description of the strains object of this study was complicated by the absence of reproductive structures in pure cultures, thus making the description of diagnostic features amongst the newly recognized lineages impossible. For the same reason, the morphological comparison between the strains in analysis and the recently accepted species was unfeasible.

For a better characterization of these fungi, we used a culture medium supplemented with seawater to mimic their natural environment. Indeed, it is known that only the addition of seawater supports a measurable growth of vegetative mycelium [16,31]. Placing wood and cork specimens on the colonies' surface, followed by their transfer into seawater to induce sporulation, was not successful. In fact, despite wood colonization, only chlamydospores were produced. Strictly vegetative growth is now a recognized feature of a relatively high percentage of marine fungi isolated from different substrates [31–36]. This may be due to the lack of appropriate environmental conditions these organisms are adapted to (e.g., high salinity, low temperature, high hydrostatic pressure, etc.) or simply to the fact that the dispersal of sterile marine fungi relies on hyphal fragments and/or resistance structures. In addition, it must be considered that, in fungi, sexual reproduction is controlled by the mating-type (MAT1) locus and that, contrary to homothallic self-fertile filamentous ascomycetes, mating in heterothallic self-sterile species is possible only between strains morphologically indistinguishable with a different idiomorph at the MAT1 locus [37–39]. As a consequence of fungal sterility, the identification of the 21 *Lulworthiales* was achieved with the help of molecular and phylogenetic data. Notwithstanding, the lack of sporulation in the strains identified as *P. gigaspora* and *P. posidoniae* puzzled us since sexual structures had previously been observed and described [8]. A possible reason for this behavior may be found in the strain-specific behavior of a homothallic species: homothallism, may, in fact, be a necessary but not sufficient condition for self-fertility to occur. Strains may be more or less sensitive to a range of conditions such as light, temperature, or salinity. Alternatively, an idiomorph may be eliminated via homologous recombination, as demonstrated in *Chromocrea spinulosa*—which exhibits both homothallic and heterothallic behaviour [37]—or, as seen in some species such as *Thielaviopsis cerberus* [40], it may display a unidirectional mating-type switching.

The inspection of the phylogenetic tree, based on the three ribosomal genes (nrITS, nrLSU, and nrSSU), highlights the presence of five hypothetical clades (Figure 1). In detail, three strains grouped with *P. posidoniae*, three clustered with *P. halima*, and seven seemed affiliated with *P. gigaspora*. The remaining formed two well-supported clusters that did not encompass any known fungus, indicating the presence of new lineages. To clarify and be certain of the relations among the species, a new dataset focusing on the five clades and their closest relatives (*L.* cf. *purpurea*, *H. mehlae*, *L. medusa*, and *H. fuscus*) was built with the addition of four protein-coding genes, namely, TEF-1α, RPB1, RPB2, and βTUB (Figure S1). Given the presence of intron regions, which can evolve at a faster rate compared to ribosomal regions, protein-coding genes are more informative and can be employed to improve phylogenetic accuracy, providing a clear species-level identification [23,41]. Indeed, the phylogenetic tree constructed upon seven markers points out the presence of two groups: (a) the cluster that includes the *P. gigaspora*, *P. halima*, *P. posidoniae*, and the newly found *P. mediterranea* clades; and (b) the cluster that consists of two additional new species, *P. candida* and *P. elbensis*. Besides the ultimate scope of our investigation, all the newly generated protein-coding sequences greatly enrich the public databases, thus increasing the availability of molecular data for researchers dealing with this group of fungi.

One could argue and contest the fact the introduction of novel species is based on molecular and phylogenetic data only. However, we followed the key recommendations outlined by Jeewon and Hyde [42]. As indicated by the authors, all the ITS sequences (including 5.8S) analyzed are longer than the minimum requirement of 450 base pairs; the tree is based on genes with strong phylogenetic signals and is statistically supported and includes the minimum number of closely related taxa of the same genus (Figure 1). Finally, reliable statistical support for each new clade (at least 60% BS or 0.9 BYPP) confirms taxa distinctiveness (Figure 1 and Figure S1).

The order Lulworthiales, with its sole family Lulworthiaceae, was erected to accommodate the genera *Lulworthia* and *Lindra*, once considered part of *Halosphaeriales* (fam. *Halosphaeriaceae*) [6,7], and was then moved to the new subclass *Lulworthiomycetidae* [5]. The polyphyletic nature of these two genera initially confused taxonomists, although nowadays, following a number of revisions [7,8,25,26], it is broadly accepted and is once more demonstrated in our investigation (Figure 1).

Lulworthiaceae are found in cold, temperate, and tropical waters in association with woods, seaweeds, seagrasses, and seafoam [4,8,9]. Likewise Goncalves et al. [9], the strains of *P. halima*, produced only chlamydospores and derived from submerged woods, indicating a preference of this species for such a substrate. Two strains of *P. gigaspora* were isolated from an oil spill, while the rest were associated with the seagrass *P. oceanica*. Members of *Lulworthiaceae* are known saprobes (http://www.funguild.org) and cellulases producers [10]. Considering the substrates of isolation and the production of lignocellulosic enzymes, we can hypothesize a lignicolous nature of the newly identified species. The retrieval of two strains of *P. gigaspora* from an oil spill reinforces the idea that these organisms can break down complex lignocellulose and recalcitrant compounds, thus contributing to the recycling of nutrients and possibly degrading contaminants such as polycyclic aromatic hydrocarbons (PAH). Interestingly, Paço and collaborators demonstrated the ability of a strain of *Zalerion maritimum* to degrade polyethylene [43], suggesting a key role of *Lulworthiaceae* in offering a solution to microplastic pollution. Further experiments will be necessary to assess the full degradative potential of these organisms that could be harnessed for bioremediation purposes.

5. Conclusions

In conclusion, the retrieval of fungi affiliated with *Lulworthiales*—together with the introduction of the novel species *Paralulworthia candida*, *Paralulworthia elbensis*, and *Paralulworthia mediterranea*—greatly contributes to improving our knowledge on this strictly marine order and to step-by-step unveiling the fungal communities hosted in the Mediterranean Sea.

Due to the extraordinary biotechnological potential demonstrated by marine fungi, a few strains described in this paper are currently being investigated for the production of novel bioactive molecules. However, we must bear in mind that the applicative value of these organisms depends on their identification at the species level, safe long-term preservation, and on the accessibility guaranteed by the public collections of biological resources.

Supplementary Materials: The following are available online at https://www.mdpi.com/article/10.3390/jof7110940/s1: Figure S1, Phylogenetic inference based on a combined nrITS, nrSSU, nrLSU, RPB1, RPB2, TEF-1α, and βTUB dataset; Table S1, Dataset based on nrITS, nrLSU, nrSSU, RPB1, RPB2, TEF-1α, and βTUB used for phylogenetic analysis.

Author Contributions: Conceptualization, A.P., E.B. and V.P.; methodology, A.P., G.C.V., E.B. and V.P.; software, A.P.; validation, A.P., I.P. and V.P.; formal analysis, A.P., E.B. and V.P.; investigation, A.P., E.B. and V.P.; resources, V.P. and G.C.V.; data curation, A.P., I.P. and V.P.; writing—original draft preparation, A.P. and V.P.; writing—review and editing, A.P., V.P. and G.C.V.; visualization, A.P. and V.P.; supervision, V.P. and G.C.V.; project administration, A.P., V.P. and G.C.V.; funding acquisition, G.C.V. All authors have read and agreed to the published version of the manuscript.

Funding: This project has received funding from the European Union's Horizon 2020 research and innovation programme under grant agreement No. 871129.

Institutional Review Board Statement: Not applicable.

Informed Consent Statement: Not applicable.

Data Availability Statement: Sequence data are available at Genbank (NCBI) under the accession numbers reported in the manuscript.

Acknowledgments: The authors would like to acknowledge JRU MIRRI-IT (http://www.mirri-it.it/) and IS_MIRRI21 (https://ismirri21.mirri.org/) for scientific support.

Conflicts of Interest: The authors declare no conflict of interest.

References

1. Jones, E.B.G.; Pang, K.-L.; Abdel-Wahab, M.A.; Scholz, B.; Hyde, K.D.; Boekhout, T.; Ebel, R.; Rateb, M.E.; Henderson, L.; Sakayaroj, J.; et al. An online resource for marine fungi. *Fungal Divers.* **2019**, *96*, 347–433. [CrossRef]
2. Jones, E.B.G.; Suetrong, S.; Sakayaroj, J.; Bahkali, A.H.; Abdel-Wahab, M.A.; Boekhout, T.; Pang, K.L. Classification of marine Ascomycota, Basidiomycota, Blastocladiomycota and Chytridiomycota. *Fungal Divers.* **2015**, *73*, 1–72. [CrossRef]
3. Maharachchikumbura, S.S.N.; Hyde, K.D.; Jones, E.B.G.; McKenzie, E.H.C.; Bhat, J.D.; Dayarathne, M.C.; Huang, S.K.; Norphanphoun, C.; Senanayake, I.C.; Perera, R.H.; et al. Families of Sordariomycetes. *Fungal Divers.* **2016**, *79*, 1–317. [CrossRef]
4. Jones, E.G.; Pang, K.-L. (Eds.) *Marine Fungi and Fungal-like Organisms*; Walter de Gruyter: Berlin, Germany, 2012.
5. Maharachchikumbura, S.S.N.; Hyde, K.D.; Jones, E.B.G.; McKenzie, E.H.C.; Huang, S.K.; Abdel-Wahab, M.A.; Daranagama, D.A.; Dayarathne, M.; D'Souza, M.J.; Goonasekara, I.D.; et al. Towards a natural classification and backbone tree for Sordariomycetes. *Fungal Divers.* **2015**, *72*, 199–301. [CrossRef]
6. Kohlmeyer, J.; Spatafora, J.W.; Volkmann-Kohlmeyer, B. Lulworthiales, a new order of marine Ascomycota. *Mycologia* **2000**, *92*, 453–458. [CrossRef]
7. Campbell, J.; Volkmann-Kohlmeyer, B.; Grafenhan, T.; Spatafora, J.W.; Kohlmeyer, J. A re-evaluation of Lulworthiales: Relationships based on 18S and 28S rDNA. *Mycol. Res.* **2005**, *109*, 556–568. [CrossRef]
8. Poli, A.; Bovio, E.; Ranieri, L.; Varese, G.C.; Prigione, V. Fungal Diversity in the Neptune Forest: Comparison of the Mycobiota of *Posidonia oceanica*, *Flabellia petiolata*, and *Padina pavonica*. *Front. Microbiol.* **2020**, *11*, 933. [CrossRef] [PubMed]
9. Goncalves, M.F.M.; Abreu, A.C.; Hilario, S.; Alves, A. Diversity of marine fungi associated with wood baits in the estuary Ria de Aveiro, with descriptions of *Paralulworthia halima*, comb. nov., *Remispora submersa*, sp. nov., and *Zalerion pseudomaritima*, sp. nov. *Mycologia* **2021**, *113*, 664–683. [CrossRef]
10. Raghukumar, S. *Fungi in Coastal and Oceanic Marine Ecosystems: Marine Fungi*; Springer: Cham, Switzerland, 2017.
11. Campbell, J.; Inderbitzin, P.; Kohlmeyer, J.; Volkmann-Kohlmeyer, B. Koralionastetales, a new order of marine Ascomycota in the Sordariomycetes. *Mycol. Res.* **2009**, *113*, 373–380. [CrossRef]
12. Panno, L.; Bruno, M.; Voyron, S.; Anastasi, A.; Gnavi, G.; Miserere, L.; Varese, G.C. Diversity, ecological role and potential biotechnological applications of marine fungi associated to the seagrass *Posidonia oceanica*. *New Biotechnol.* **2013**, *30*, 685–694. [CrossRef]
13. Garzoli, L.; Gnavi, G.; Tamma, F.; Tosi, S.; Varese, G.C.; Picco, A.M. Sink or swim: Updated knowledge on marine fungi associated with wood substrates in the Mediterranean Sea and hints about their potential to remediate hydrocarbons. *Prog. Oceanogr.* **2015**, *137*, 140–148. [CrossRef]
14. Abdel-Wahab, M.A.; Hodhod, M.S.; Bahkali, A.H.A.; Jones, E.B.G. Marine fungi of Saudi Arabia. *Bot. Mar.* **2014**, *57*, 323–335. [CrossRef]
15. Dayarathne, M.C.; Wanasinghe, D.N.; Devadatha, B.; Abeywickrama, P.; Jones, E.B.G.; Chomnunti, P.; Sarma, V.V.; Hyde, K.D.; Lumyong, S.; McKenzie, E.H.C. Modern taxonomic approaches to identifying diatrypaceous fungi from marine habitats, with a novel genus Halocryptovalsa Dayarathne & KD Hyde, gen. nov. *Cryptogam. Mycol.* **2020**, *41*, 21–67. [CrossRef]
16. Poli, A.; Bovio, E.; Ranieri, L.; Varese, G.C.; Prigione, V. News from the Sea: A New Genus and Seven New Species in the Pleosporalean Families Roussoellaceae and Thyridariaceae. *Diversity* **2020**, *12*, 144. [CrossRef]
17. Bovio, E.; Gnavi, G.; Prigione, V.; Spina, F.; Denaro, R.; Yakimov, M.; Calogero, R.; Crisafi, F.; Varese, G.C. The culturable mycobiota of a Mediterranean marine site after an oil spill: Isolation, identification and potential application in bioremediation. *Sci. Total Environ.* **2017**, *576*, 310–318. [CrossRef]
18. Panebianco, C.; Tam, W.Y.; Jones, E.B.G. The effect of pre-inoculation of balsa wood by selected marine fungi and their effect on subsequent colonisation in the sea. *Fungal Divers.* **2002**, *10*, 77–88.
19. White, T.J.; Bruns, T.; Lee, S.; Taylor, J. Amplification and direct sequencing of fungal ribosomal RNA genes for phylogenetics. In *PCR Protocols: A Guide to Methods and Applications*; Academic Press: London, UK, 1990; pp. 315–322.
20. Vilgalys, R.; Hester, M. Rapid genetic identification and mapping of enzymatically amplified ribosomal DNA from several Cryptococcus species. *J. Bacteriol.* **1990**, *172*, 4238–4246. [CrossRef] [PubMed]
21. Matheny, P.B.; Wang, Z.; Binder, M.; Curtis, J.M.; Lim, Y.W.; Nilsson, R.H.; Hughes, K.W.; Hofstetter, V.; Ammirati, J.F.; Schoch, C.L.; et al. Contributions of rpb2 and tef1 to the phylogeny of mushrooms and allies (Basidiomycota, Fungi). *Mol. Phylogenet. Evol.* **2007**, *43*, 430–451. [CrossRef] [PubMed]
22. Glass, N.L.; Donaldson, G.C. Development of primer sets designed for use with the PCR to amplify conserved genes from filamentous ascomycetes. *Appl. Environ. Microbiol.* **1995**, *61*, 1323–1330. [CrossRef] [PubMed]
23. Raja, H.A.; Baker, T.R.; Little, J.G.; Oberlies, N.H. DNA barcoding for identification of consumer-relevant mushrooms: A partial solution for product certification? *Food Chem.* **2017**, *214*, 383–392. [CrossRef] [PubMed]

24. Liu, Y.J.; Whelen, S.; Hall, B.D. Phylogenetic relationships among ascomycetes: Evidence from an RNA polymerse II subunit. *Mol. Biol. Evol.* **1999**, *16*, 1799–1808. [CrossRef] [PubMed]
25. Abdel-Wahab, M.A.; Pang, K.L.; Nagahama, T.; Abdel-Aziz, F.A.; Jones, E.B.G. Phylogenetic evaluation of anamorphic species of *Cirrenalia* and *Cumulospora* with the description of eight new genera and four new species. *Mycol. Prog.* **2010**, *9*, 537–558. [CrossRef]
26. Azevedo, E.; Barata, M.; Marques, M.I.; Caeiro, M.F. *Lulworthia atlantica*: A new species supported by molecular phylogeny and morphological analysis. *Mycologia* **2017**, *109*, 287–295. [CrossRef] [PubMed]
27. Vaidya, G.; Lohman, D.J.; Meier, R. SequenceMatrix: Concatenation software for the fast assembly of multi-gene datasets with character set and codon information. *Cladistics* **2011**, *27*, 171–180. [CrossRef]
28. Darriba, D.; Taboada, G.L.; Doallo, R.; Posada, D. jModelTest 2: More models, new heuristics and parallel computing. *Nat. Methods* **2012**, *9*, 772. [CrossRef]
29. Stamatakis, A. RAxML version 8: A tool for phylogenetic analysis and post-analysis of large phylogenies. *Bioinformatics* **2014**, *30*, 1312–1313. [CrossRef]
30. Ronquist, F.; Teslenko, M.; van der Mark, P.; Ayres, D.L.; Darling, A.; Hohna, S.; Larget, B.; Liu, L.; Suchard, M.A.; Huelsenbeck, J.P. MrBayes 3.2: Efficient Bayesian Phylogenetic Inference and Model Choice Across a Large Model Space. *Syst. Biol.* **2012**, *61*, 539–542. [CrossRef]
31. Poli, A.; Bovio, E.; Perugini, I.; Varese, G.C.; Prigione, V. *Corollospora mediterranea*: A Novel Species Complex in the Mediterranean Sea. *Appl. Sci.* **2021**, *11*, 5452. [CrossRef]
32. Poli, A.; Vizzini, A.; Prigione, V.; Varese, G.C. Basidiomycota isolated from the Mediterranean Sea—Phylogeny and putative ecological roles. *Fungal Ecol.* **2018**, *36*, 51–62. [CrossRef]
33. Garzoli, L.; Poli, A.; Prigione, V.; Gnavi, G.; Varese, G.C. Peacock's tail with a fungal cocktail: First assessment of the mycobiota associated with the brown alga *Padina pavonica*. *Fungal Ecol.* **2018**, *35*, 87–97. [CrossRef]
34. Gnavi, G.; Garzoli, L.; Poli, A.; Prigione, V.; Burgaud, G.; Varese, G.C. The culturable mycobiota of *Flabellia petiolata*: First survey of marine fungi associated to a Mediterranean green alga. *PLoS ONE* **2017**, *12*, e0175941. [CrossRef] [PubMed]
35. Zuccaro, A.; Schulz, B.; Mitchell, J.A. Molecular detection of ascomycetes associated with *Fucus serratus*. *Mycol. Res.* **2003**, *107*, 1451–1466. [CrossRef] [PubMed]
36. Zuccaro, A.; Schoch, C.L.; Spatafora, J.W.; Kohlmeyer, J.; Draeger, S.; Mitchell, J.A. Detection and identification of fungi intimately associated with the brown seaweed *Fucus serratus*. *Appl. Environ. Microbiol.* **2008**, *74*, 931–941. [CrossRef]
37. Bennett, R.J.; Turgeon, B.G. Fungal Sex: The Ascomycota. *Microbiol. Spectr.* **2016**, *4*, 4–5. [CrossRef] [PubMed]
38. Kumar, A.; Sorensen, J.L.; Hansen, F.; Arvas, M.; Syed, M.F.; Hassan, L.; Benz, J.P.; Record, E.; Henrissat, B.; Poggeler, S.; et al. Genome sequencing and analyses of two marine fungi from the North Sea unraveled a plethora of novel biosynthetic gene clusters. *Sci. Rep.* **2018**, *8*, 10187. [CrossRef] [PubMed]
39. Yun, S.H.; Kim, H.K.; Lee, T.; Turgeon, B.G. Self-fertility in *Chromocrea spinulosa* is a consequence of direct repeat-mediated loss of MAT1-2, subsequent imbalance of nuclei differing in mating type, and recognition between unlike nuclei in a common cytoplasm. *Plos Genet.* **2017**, *13*, 27. [CrossRef] [PubMed]
40. Kramer, D.; Lane, F.A.; Steenkamp, E.T.; Wingfield, B.D.; Wilken, P.M. Unidirectional mating-type switching confers self-fertility to *Thielaviopsis cerberus*, the only homothallic species in the genus. *Fungal Biol.* **2021**, *125*, 427–434. [CrossRef]
41. Chethana, K.W.T.; Jayawardena, R.S.; Hyde, K.D. Hurdles in fungal taxonomy: Effectiveness of recent methods in discriminating taxa. *Megataxa* **2020**, *1*, 114–122.
42. Jeewon, R.; Hyde, K.D. Establishing species boundaries and new taxa among fungi: Recommendations to resolve taxonomic ambiguities. *Mycosphere* **2016**, *7*, 1669–1677. [CrossRef]
43. Paco, A.; Duarte, K.; da Costa, J.P.; Santos, P.S.M.; Pereira, R.; Pereira, M.E.; Freitas, A.C.; Duarte, A.C.; Rocha-Santos, T.A.P. Biodegradation of polyethylene microplastics by the marine fungus *Zalerion maritimum*. *Sci. Total Environ.* **2017**, *586*, 10–15. [CrossRef]

Journal of
Fungi

MDPI

Review

Fungal Biodiversity in Salt Marsh Ecosystems

Mark S. Calabon [1,2], E. B. Gareth Jones [3], Itthayakorn Promputtha [4,5] and Kevin D. Hyde [1,2,4,6,*]

1 Center of Excellence in Fungal Research, Mae Fah Luang University, Chiang Rai 57100, Thailand; mscalabon@up.edu.ph
2 School of Science, Mae Fah Luang University, Chiang Rai 57100, Thailand
3 Department of Botany and Microbiology, College of Science, King Saud University, P.O. Box 2455, Riyadh 11451, Saudi Arabia; torperadgj@gmail.com
4 Department of Biology, Faculty of Science, Chiang Mai University, Chiang Mai 50200, Thailand; itthayakorn.p@cmu.ac.th
5 Environmental Science Research Center, Faculty of Science, Chiang Mai University, Chiang Mai 50200, Thailand
6 Innovative Institute of Plant Health, Zhongkai University of Agriculture and Engineering, Guangzhou 510225, China
* Correspondence: kdhyde3@gmail.com

Abstract: This review brings together the research efforts on salt marsh fungi, including their geographical distribution and host association. A total of 486 taxa associated with different hosts in salt marsh ecosystems are listed in this review. The taxa belong to three phyla wherein Ascomycota dominates the taxa from salt marsh ecosystems accounting for 95.27% (463 taxa). The Basidiomycota and Mucoromycota constitute 19 taxa and four taxa, respectively. Dothideomycetes has the highest number of taxa, which comprises 47.12% (229 taxa), followed by Sordariomycetes with 167 taxa (34.36%). Pleosporales is the largest order with 178 taxa recorded. Twenty-seven genera under 11 families of halophytes were reviewed for its fungal associates. *Juncus roemerianus* has been extensively studied for its associates with 162 documented taxa followed by *Phragmites australis* (137 taxa) and *Spartina alterniflora* (79 taxa). The highest number of salt marsh fungi have been recorded from Atlantic Ocean countries wherein the USA had the highest number of species recorded (232 taxa) followed by the UK (101 taxa), the Netherlands (74 taxa), and Argentina (51 taxa). China had the highest number of salt marsh fungi in the Pacific Ocean with 165 taxa reported, while in the Indian Ocean, India reported the highest taxa (16 taxa). Many salt marsh areas remain unexplored, especially those habitats in the Indian and Pacific Oceans areas that are hotspots of biodiversity and novel fungal taxa based on the exploration of various habitats.

Keywords: halophytes; marine fungi; marine mycology; salt marsh fungi; worldwide distribution

Citation: Calabon, M.S.; Jones, E.B.G.; Promputtha, I.; Hyde, K.D. Fungal Biodiversity in Salt Marsh Ecosystems. *J. Fungi* **2021**, *7*, 648. https://doi.org/10.3390/jof7080648

Academic Editor: Wei Li

Received: 26 June 2021
Accepted: 30 July 2021
Published: 9 August 2021

Publisher's Note: MDPI stays neutral with regard to jurisdictional claims in published maps and institutional affiliations.

1. Introduction

Salt marsh ecosystems are known for their high productivity, exceeding primary production estimates of species rich ecosystems (e.g., tropical rainforests, coral reefs) [1]. The flora in salt marsh ecosystems is mainly composed of grasses, herbs, and shrubs and these are terrestrial organisms variously adapted to, or tolerant of, a semi-marine environment. Halophytes are a diverse group of plants that have a worldwide distribution, and grow in different climatic regions, wherein soils have high salinity levels [2]. Halophytes are common in temperate and Mediterranean climates, and fewer both in the tropics and at high latitudes [3–6]. The vegetation in these ecosystems shows the vertical zonation of different communities as tidal submergence decreases with increasing elevation, and species tolerance to changing gradient conditions. Salt marsh vegetation generally increases the attenuation of both tidal currents and waves as they pass over the vegetated area and immobilize elements with their sediments. Furthermore, halophytes serve as a natural buffer, protecting other shoreline ecosystems from human impacts and disturbances. The area provides a habitat and nursery for marine organisms [7]. Worldwide, salt marshes cover an area of 5,495,089 hectare in 43 countries [8].

There are over 500 species of salt marsh plants worldwide [9]. The families Amaranthaceae (subfamilies Chenopodiaceae, Salicornieae), Poaceae, Juncaceae, and Cyperaceae are the major vegetation in salt marsh ecosystems, while the minor components are Plumbaginaceae and Frankeniaceae [3], and are represented in Figures 1 and 2. Salinity, latitude, region of the world, the frequency and duration of tidal flooding, substrate, oxygen and nutrient availability, surface elevation, competition among species, disturbance by wrack deposition are interacting factors that influence the species of halophytes in the salt marshes [10,11]. For example, *Spartina alterniflora* is a dominant grass from mid-tide to high-tide levels in temperate Eastern North America, while *Puccinellia* dominates in boreal and arctic marshes [10,11].

Figure 1. Salt marsh ecosystems in UK (**a–d**) and Thailand (**e–f**). (**b–d**) Tidal grasses, *Spartina townsendii* (Poaceae) and *Phragmites* (Poaceae), dominate the salt marsh in UK (50°49′55.4″ N 0°58′25.1″ W; 51°43′03.1″ N 5°10′24.8″ W); (**e**) *Spartina (Poaceae)* (12°22′4.0″ N 99°59′6.7″ E) (**f**) and *Suaeda* (Amaranthaceae) (12°10′19.6″ N 99°58′20.3″ E) in tidal marsh areas in southern Thailand.

Figure 2. Halophytes in salt marsh ecosystems: (**a**) flowering inflorescence of *Spartina*, (**b**) *Phragmites*, (**c**) *Salicornia*, (**d**) *Typha*, (**e**,**f**) *Atriplex*, and (**g**,**h**) *Suaeda*.

Major studies on halophytes focus on ecology and conservation [12–14]. One of these is the decomposition of vascular plant material wherein the detritus breakdown was reviewed in Pomeroy and Wiegert [15], Howarth and Hobbie [16], and Long and Mason [17]. The active decomposition processes in salt marsh ecosystems reflects to the relatively high rates of primary production. Three phases of plant decomposition were noted by Valiela et al. [18]. The early phase involves the leaching of soluble compounds, resulting in a fast rate of weight loss lasting for less than a month. Organic matter breakdown by microorganisms and continuous leaching of decayed products occurs in the second phase that lasts for a year. The last phase lasts for another year when there is a slow decay of refractory materials such as humates and fulvates [19].

The continuous breakdown of detritus into smaller fragments increases the surface-to-volume ratio and this is exposed to further microbial degradation. Bacteria and fungi are key decomposers in the salt marsh ecosystem that are essential for the transformation and recycling of nutrients through the environment. The colonization of fungi on standing dead halophytes commences during the early stages of decomposition before leaf fall to the salt marsh sediment surface [20,21]. The decomposition of the senescent tissues of halophytes by salt marsh fungi is brought about by the direct penetration of the host cell wall and the production of enzymes active in degrading lignocellulosic compounds, such as lignin, cellulose, and hemicellulose [22–26]. Bacterial communities are the major decomposers in the latter stage of decomposition [27,28]. Studies in salt marsh ecosystems not only consider microbial activity and the recycling of nutrients, but also bacterial [29,30] and fungal diversity [20,31,32].

The present review compiles the published data of fungi from halophytes, including their geographical distribution and host association. When compared to other fungal groups, salt marsh fungi are underexplored, and this review brings together the research efforts on these undiscovered habitats and plants. The pertinent literature from bibliographic databases (e.g., Scopus, Web of Science, Google Scholar) and published resources on salt marsh fungi documenting halophytes were compiled. Published works, wherein the documented fungal taxa were observed directly from halophytic substrates, are included (Table 1). The different host parts, living and dead, that are either partly or wholly submerged are documented, as well as drift plant portions washed up in salt marsh areas. Salt marsh fungi isolated using cultivation-dependent techniques were not included since it is not known if these fungi were actively growing and reproducing on the halophytes. The taxa were listed based on the recent outline of fungi and fungus-like taxa by Wijayawardene et al. [33]. Since previous works only listed the taxa and the hosts [34–36], here we include the plant parts where the fungus was observed, the location (country: state/province) where the host was collected, the life mode of the fungus, and the pertinent literature citations are included (Table 1). The accepted name of the host was based on the webpage of the World Flora Online consortium (http://www.worldfloraonline.org/; accessed on 10 May 2021), GrassBase (https://www.kew.org/data/grasses-db/sppindex.htm; accessed on 10 May 2021) and CRC World Dictionary of Grasses by Quattrocchi [37]. The graphs presented in the next sections summarizes the information from Table 1 and was developed using data visualization tools (Excel Office 365, Tableau Desktop Professional Edition 19.2.2).

Table 1. Geographical distribution of salt marsh fungi recorded from various halophytes.

Taxon	Host Part	Life Mode	Hosts	Distribution	References
ASCOMYCOTA					
DOTHIDEOMYCETES					
Acrospermales					
Acrospermaceae					
Acrospermum graminum Lib.	–	–	*Elymus pungens*	UK	[38]
Asterinales					
Morenoinaceae					
Morenoina phragmitis J.P. Ellis	Living/decomposing leaf sheaths and stems	Saprobic	*Phragmites australis*	Netherlands: Zeeland	[39,40]
Botryosphaeriales					
Botryosphaeriaceae					
Botryosphaeria festucae (Lib.) Arx and E. Müll.	Living/decomposing leaf sheaths and stems	Saprobic	*Phragmites australis*	Netherlands: Zeeland	[39,40]
Macrophomina sp.	Decaying stems and leaf sheaths	Saprobic	*Phragmites australis*	China: Hong Kong	[41]
Tiarosporella halmyra Kohlm. and Volkm.-Kohlm.	Senescent culms	Saprobic	*Juncus roemerianus*	USA: North Carolina	[42]
Phyllostictaceae					
Guignardia spp.	Senescent leaves	Saprobic	*Juncus roemerianus*	USA: Florida	[43]
Phyllosticta sp.	–	Pathogenic	*Spartina cynosuroides*	USA: Maryland	[44]
Phyllosticta spartinae Brunaud	–	–	*Spartina maritima*	France	[45]
Phyllosticta suaedae Lobik	Leaves	–	*Suaeda maritima*	Russia	[46]
Capnodiales					
Cladosporiaceae					
Cladosporium algarum Cooke and Massee	–	–	*Spergularia marina*	–	[35]
	–	–	*Suaeda maritima*	–	[35]
Cladosporium allicinum (Fr.) Bensch, U. Braun and Crous	–	–	*Elymus pungens*	UK	[38]
Cladosporium cladosporioides (Fresen.) G.A. de Vries	Leaves	Saprobic	*Distichlis spicata*	Argentina: Buenos Aires	[47]
	Living, senescent, and decaying leaves	Saprobic	*Juncus roemerianus*	USA: Florida	[43]
	Leaves and roots	Saprobic	*Spartina* sp.	Canada: Bay of Fundy	[48]
Cladosporium herbarum (Pers.) Link	Leaves	Saprobic	*Distichlis spicata*	Argentina: Buenos Aires	[47]
	Stem	Saprobic	*Spartina townsendii*	UK: England	[49]
	Leaves, stems, and roots	Saprobic	*Spartina* sp.	Canada: Bay of Fundy	[48]

Table 1. *Cont.*

Taxon	Host Part	Life Mode	Hosts	Distribution	References
Cladosporium macrocarpum Preuss	Leaves	Saprobic	*Spartina* sp.	Canada: Bay of Fundy	[48]
Cladosporium sphaerospermum Penz.	Living, senescent, and decaying leaves	Saprobic	*Juncus roemerianus*	USA: Florida	[43]
	Living/decomposing leaf sheaths and blades	Saprobic	*Phragmites australis*	Netherlands: Zeeland	[39,41,50]
	–	Saprobic	*Spartina patens*	USA: Rhode Island	[36]
	–	Saprobic	*Spartina* sp.	Canada	[36]
Capnodiales genera *incertae sedis*					
Mucomycosphaerella eurypotami (Kohlm., Volkm.-Kohlm. and O.E. Erikss.) Quaedvl. and Crous	Senescent leaves	Saprobic	*Juncus roemerianus*	USA: North Carolina	[51]
Mycosphaerellaceae					
Fulvia fulva (Cooke) Cif.	Leaves and stems	Saprobic	*Spartina* sp.	Canada: Bay of Fundy	[48]
Micronectriella agropyri Apinis and Chesters	–	–	*Elymus pungens*	UK	[38]
	–	–	*Puccinellia maritima*	UK	[38]
	–	–	*Spartina townsedii*	UK	[38]
Mycosphaerella lineolata (Roberge ex Desm.) J. Schröt.	Living/decomposing leaf sheaths and stems	Saprobic	*Phragmites australis*	Netherlands: Zeeland	[39,40]
	–	–	*Elymus pungens*	UK	[38]
Mycosphaerella salicorniae (Auersw.) Lindau	–	–	*Arthrocnemum subterminale*	–	[35]
	–	–	*Limonium* sp.	–	[35]
	–	–	*Sarcocornia perennis*	–	[35]
	–	–	*Salicornia fruticosa*	–	[35]
	–	–	*Salicornia procumbens*	–	[35]
	–	–	*Salicornia europaea*	–	[35]
	–	–	*Salicornia perennis*	–	[35]
	–	–	*Sarcocornia fruticosa*	–	[35]
	Drying stalks and inflorescence	Saprobic	*Salicornia* sp.	India	[52]
	Dried inflorescences	Saprobic	*Salicornia virginica*	Bermuda	[35,53]
	–	Saprobic	*Spartina marítima*	Portugal: Alentejo, Lisbon	[54]
	–	–	*Suaeda vermiculata*	–	[35]

Table 1. *Cont.*

Taxon	Host Part	Life Mode	Hosts	Distribution	References
	Drying stalks and inflorescence	Saprobic	*Suaeda* sp.	India	[52]
Mycosphaerella spp.	–	–	*Elymus pungens*	UK	[38]
	Senescent and decaying leaves	Saprobic	*Juncus roemerianus*	USA: Florida, Mississippi	[43,55]
	Decaying leaves, leaf blades	Saprobic	*Spartina alterniflora*	Argentina: Buenos Aires; USA: Alabama, California, Georgia, Mississippi	[25,35,36,55–58]
	–	–	*Spartina* cf. *densiflora*	USA: California	[25,35]
	–	–	*Spartina* cf. *pectinata*	–	[35]
	–	–	*Spartina* sp.	Argentina: Buenos Aires; Canada	[35,36]
	Decaying leaf blades	Saprobic	*Spartina foliosa*	USA: California	[25]
	Leaf sheaths and blades, stem	Saprobic	*Spartina marítima*	Portugal: Alentejo, Lisbon, Centro	[54,59]
Mycosphaerella staticicola (Pat.) Dias	–	–	*Armeria pungens*	–	[35]
Mycosphaerella suaedae-australis Hansf.	–	–	*Suaeda australis*	–	[35]
Rivilata ius Kohlm., Volkm.-Kohlm. and O.E. Erikss.	Tips of senescent, very old, and brittle leaves	Saprobic	*Juncus roemerianus*	USA: North Carolina	[60]
Septoria spp.	Living, senescent, and decaying leaves	Saprobic	*Juncus roemerianus*	USA: Florida	[43]
	Living/decomposing leaf sheaths	Saprobic	*Phragmites australis*	Netherlands: Zeeland	[39]
	Upper leaves, inflorescence, seeds	Saprobic	*Spartina alterniflora*	USA: Rhode Island	[61]
Septoria suaedae-australis Hansf.	Dead stems	Saprobic	*Suaeda australis*	South Australia	[62]
Sphaerulina albispiculata Tubaki	Sheath	Saprobic	*Spartina marítima*	Portugal: Alentejo, Lisbon	[54]
	Stem	Saprobic	*Spartina marítima*	Portugal: Alentejo	[63]
Sphaerulina orae-maris Linder	–	–	*Ammophila arenaria*	–	[35]
	Rhizome and root	Saprobic	*Spartina densiflora*	Argentina: Buenos Aires	[64]

Table 1. *Cont.*

Taxon	Host Part	Life Mode	Hosts	Distribution	References
	Leaf sheaths and blades, stem	Saprobic	*Spartina marítima*	Portugal: Alentejo, Lisbon, Algarve, Centro	[31,54,59,63]
Sphaerulina pedicellata T.W. Johnson	–	Saprobic	*Spartina townsendii*	–	[65]
	Attached culms, stems	Saprobic, parasitic	*Spartina alterniflora*	USA: Rhode Island	[20,61]
Sphaerulina sp.	Senescent and decaying leaves	Saprobic	*Juncus roemerianus*	USA: Florida	[43]
Dothideales					
Saccotheciaceae					
Aureobasidium sp.	Living, senescent, and decaying leaves	Saprobic	*Juncus roemerianus*	USA: Florida	[43]
Pseudoseptoria donacis (Pass.) B. Sutton	Living/decomposing leaf blades and sheaths	Saprobic	*Phragmites australis*	Netherlands: Zeeland	[39,50]
Selenophoma sp.	Senescent and decaying leaves	Saprobic	*Juncus roemerianus*	USA: Florida	[43]
Dothideaceae					
Scirrhia annulata Kohlm., Volkm.-Kohlm. and O.E. Erikss.	Senescent culms and leaves	Saprobic	*Juncus roemerianus*	USA: North Carolina	[66]
Dothideomycetes families *incertae sedis*					
Eriomycetaceae					
Heleiosa barbatula Kohlm., Volkm.-Kohlm. and O.E. Erikss.	Senescent leaves	Saprobic	*Juncus roemerianus*	USA: North Carolina	[66]
Pseudorobillardaceae					
Pseudorobillarda phragmitis (Cunnell) M. Morelet	Decaying stems and leaf sheaths	Saprobic	*Phragmites australis*	China: Hong Kong	[41,67]
Pseudorobillarda sp.	Dead stems	Saprobic	*Spartina alterniflora*	Canada	[36]
Dothideomycetes genera *incertae sedis*					
Bactrodesmium atrum M.B. Ellis	Living/decomposing stems	Saprobic	*Phragmites australis*	Netherlands: Zeeland	[40]
Lautitia danica (Berl.) S. Schatz	–	–	*Elymus pungens*	UK	[38]
	–	–	*Puccinellia maritima*	UK	[38]
Monodictys austrina Tubaki	Senescent leaves	Saprobic	*Juncus roemerianus*	USA: Florida	[43]
Monodictys castaneae (Wallr.) S. Hughes	Leaves	Saprobic	*Spartina* sp.	Canada: Bay of Fundy	[48]
Neottiosporina australiensis B. Sutton and Alcorn	Living/decomposing leaf blades and sheaths, stems	Saprobic	*Phragmites australis*	Netherlands: Zeeland	[39,40,50]
Neottiosporina sp.	Decaying stems and leaf sheaths	Saprobic	*Phragmites australis*	China: Hong Kong	[41]

Table 1. *Cont.*

Taxon	Host Part	Life Mode	Hosts	Distribution	References
Otthia sp.	Senescent leaves	Saprobic	*Juncus roemerianus*	USA: Florida	[43]
Trichometasphaeria setulosa. (Sacc. and Roum.) Apinis and Chesters ined.	–	–	*Elymus pungens*	UK	[38]
Trichometasphaeria sp.	–	–	*Elymus pungens*	UK	[38]
Microthyriales					
Microthyriaceae					
Microthyrium microscopicum Desm.	–	–	*Spartina patens*	–	[68]
Microthyrium gramineum Sacc., E. Bommer and M. Rousseau	–	–	*Elymus pungens*	UK	[38]
Muyocopronales					
Muyocopronaceae					
Ellisiodothis inquinans (Ellis and Everh.) Theiss.	–	Saprobic	*Spartina alterniflora*	Argentina: Buenos Aires	[36]
Mytilinidiales					
Mytilinidiaceae					
Septonema secedens Corda	Living, senescent, and decaying leaves	Saprobic	*Juncus roemerianus*	USA: Florida	[43]
Phaeotrichales					
Phaeotrichaceae					
Trichodelitschia bisporula (P. Crouan and H. Crouan) E. Müll. and Arx	–	–	*Elymus pungens*	UK	[38]
			Spartina townsendii	UK	[38]
Pleosporales					
Amniculicolaceae					
Neomassariosphaeria typhicola (P. Karst.) Y. Zhang ter, J. Fourn. and K.D. Hyde	–	–	*Juncus roemerianus*	–	[35]
	Decaying herbaceous stems	Saprobic	*Spartina densiflora*	Argentina: Buenos Aires	[64]
	–	Saprobic	*Spartina* spp.	Argentina: Buenos Aires	[32,35,36]
	–	Saprobic	Unidentified saltmarsh plants	USA: Mississippi	[58]
Camarosporiaceae					
Camarosporium feurichii Henn.	Living/decomposing leaf sheaths	Saprobic	*Phragmites australis*	Netherlands: Zeeland	[39]

Table 1. *Cont.*

Taxon	Host Part	Life Mode	Hosts	Distribution	References
Camarosporium palliatum Kohlm. and E. Kohlm.	–	–	*Sarcocornia perennis*	–	[35]
	–	–	*Salicornia* sp.	–	[35]
	–	–	*Salicornia virginica*	–	[35]
	–	Saprobic or perthophytic	Salt marsh plants	India: Maharashtra	[52]
	–	–	*Suaeda vermiculata*		[35]
Camarosporium roumeguerei Sacc.	–	–	*Atripex halimus*		[35]
	–	–	*Atripex* sp.		[35]
	–	–	*Distichlis spicata*		[35]
	Twigs	–	*Salicornia europaea*	France	[35,69]
	–	–	*Sarcocornia fruticosa*		[35]
	–	–	*Salicornia* sp.		[35]
	–	Saprobic or perthophytic	Salt marsh plants	India: Gujarat, Maharashtra, Tamil Nadu, Andhara Pradesh, West Bengal	[52]
	Leaf sheaths and blades, stem	Saprobic	*Spartina maritima*	Portugal: Algarve, Centro	[59]
	–	–	*Suaeda maritima*	–	[35]
Camarosporium salicorniae Hansf.	Twigs	–	*Sarcocornia quinqueflora*	South Australia	[62]
Camarosporium spp.	Living/decomposing leaf sheaths and stems	Saprobic	*Phragmites australis*	Netherlands: Zeeland	[39,40]
Camarosporium suaedae-fruticosae S. Ahmad	Dead branches	Saprobic	*Suaeda vermiculata*	Pakistan	[70]
Coniothyriaceae					
Coniothyrium obiones Jaap	–	–	*Atriplex portulacoides*	–	[35]
	–	Saprobic	Salt marsh plants	India: Orissa	[52]
	Leaf sheaths and blades, stem	Saprobic	*Spartina maritima*	Portugal: Algarve	[59]
Coniothyrium spp.	Living, senescent, and decaying leaves	Saprobic	*Juncus roemerianus*	USA: Florida	[43]
Cyclothyriellaceae					

Table 1. *Cont.*

Taxon	Host Part	Life Mode	Hosts	Distribution	References
Massariosphaeria erucacea Kohlm., Volkm.-Kohlm. and O.E. Erikss.	Senescent culms and leaves	Saprobic	*Juncus roemerianus*	USA: North Carolina	[66]
Massariosphaeria scirpina (G. Winter) Leuchtm.	–	Saprobic	*Spartina* sp.	USA: Florida, North Carolina	[71]
Massariosphaeria sp.	Living/decomposing stems	Saprobic	*Phragmites australis*	Netherlands: Zeeland	[40]
Dictyosporiaceae					
Dictyosporium oblongum (Fuckel) S. Hughes	Living/decomposing leaf blades and sheaths, stems	Saprobic	*Phragmites australis*	Netherlands: Zeeland	[39,40,50]
Dictyosporium pelagicum (Linder) G.C. Hughes ex E.B.G. Jones	Decomposing culms	Saprobic	*Spartina alterniflora*	USA: Rhode Island	[35,61]
	–	–	*Spartina* spp.	–	[32]
	Leaf sheaths and blades, stem	Saprobic	*Spartina marítima*	Portugal: Alentejo, Lisbon, Algarve, Centro	[54,59,63]
Jalapriya toruloides (Corda) M.J. D'souza, Hong Y. Su, Z.L. Luo and K.D. Hyde	Stems	Saprobic	*Spartina* sp.	UK	[72]
Didymellaceae					
Ascochyta cf. *arundinariae* Tassi	Living/decomposing leaf blades and sheaths	Saprobic	*Phragmites australis*	Netherlands: Zeeland	[39,50]
Ascochyta leptospora (Trail) Hara	Living/decomposing leaf sheaths	Saprobic	*Phragmites australis*	Netherlands: Zeeland	[39]
Ascochyta salicorniae-patulae (Trotter) Melnik	–	Saprobic, parasitic	*Salicornia* spp.	Canada, Denmark, Germany, India, UK, USA	[52]
Ascochyta spp.	Living/decomposing leaf sheaths	Saprobic	*Phragmites australis*	Netherlands: Zeeland	[39]
	Sheath	Saprobic	*Spartina marítima*	Portugal: Alentejo	[54]
Chaetasbolisia sp.	Decaying stems and leaf sheaths	Saprobic	*Phragmites australis*	China: Hong Kong	[41]
Didymella glacialis Rehm	Living/decomposing leaf blades and sheaths, stems	Saprobic	*Phragmites australis*	Netherlands: Zeeland	[39,40,50]
Didymella glomerata (Corda) Qian Chen and L. Cai	Rhizome and basal area	Saprobic	*Spartina densiflora*	Argentina: Buenos Aires	[64]
Didymella spp.	Living/decomposing leaf blades and sheaths	Saprobic	*Phragmites australis*	Netherlands: Zeeland	[39,50]
	–	Pathogenic	*Spartina cynosuroides*	USA: Louisiana	[44]

Table 1. *Cont.*

Taxon	Host Part	Life Mode	Hosts	Distribution	References
Epicoccum nigrum Link	Leaves	Saprobic	*Distichlis spicata*	Argentina: Buenos Aires	[47]
	Living, senescent, and decaying leaves	Saprobic	*Juncus roemerianus*	USA: Florida	[43]
	Inflorescence, upper leaves, seeds	Saprobic, parasitic	*Spartina alterniflora*	USA: Rhode Island, Connecticut, Virginia, Florida, North Carolina	[36,61,73,74]
Epicoccum sp.	–	–	*Spartina alterniflora*	–	[35]
Microsphaeropsis spp.	Living/decomposing leaf blades and sheaths	Saprobic	*Phragmites australis*	Netherlands: Zeeland	[39,41,50]
Phoma herbarum Westend.	Leaves	Saprobic	*Distichlis spicata*	Argentina: Buenos Aires	[47]
Phoma leveillei Boerema and G.J. Bollen	Leaves	Saprobic	*Distichlis spicata*	Argentina: Buenos Aires	[47]
Phoma suaedae Jaap	Twigs, leaves, stems	Saprobic	*Suaeda maritima, Suaeda* sp.	Germany; India	[75]
	–	–	*Suaeda maritima*	–	[35]
Phoma spp.	–	–	*Crithmum maritimum*	–	[35]
	–	–	*Atriplex portulacoides*	–	[35]
	Living, senescent, and decaying leaves	Saprobic	*Juncus roemerianus*	USA: Florida	[43]
	Living/decomposing leaf blades and sheaths, stems	Saprobic	*Phragmites australis*	China: Hong Kong; Netherlands: Zeeland	[39–41,50]
	–	–	*Salicornia europaea*	–	[35]
	–	–	*Spartina alterniflora*	USA: North Carolina, Rhode Island	[20,35,36,61,73, 74]
	–	Saprobic	*Spartina patens*	USA: Rhode Island	[36]
	–	Saprobic	*Spartina* sp.	Argentina: Buenos Aires; Canada; USA: Maine, South Carolina	[36,71]
	–	–	*Spartina townsendii*	UK: England	[35,49,65]
	Leaf sheaths and blades, stem	Saprobic	*Spartina marítima*	Portugal: Alentejo, Lisbon, Algarve, Centro	[54,59,63]

Table 1. *Cont.*

Taxon	Host Part	Life Mode	Hosts	Distribution	References
Paraboeremia putaminum (Speg.) Qian Chen and L. Cai	Leaves	Saprobic	*Distichlis spicata*	Argentina: Buenos Aires	[47]
Stagonosporopsis salicorniae (Magnus) Died.	–	–	*Salicornia europaea*	–	[35]
	–	–	*Salicornia patula*	–	[35]
Didymosphaeriaceae					
Didymosphaeria lignomaris Strongman and J.D. Mill.	Basal area of the sheath	Saprobic	*Spartina densiflora*	Argentina: Buenos Aires	[64]
	–	–	*Spartina* spp.	–	[32]
Julella herbatilis Kohlm., Volkm.-Kohlm. and O.E. Erikss.	Senescent leaves	Saprobic	*Juncus roemerianus*	USA: North Carolina	[76]
Paraphaeosphaeria apicicola Kohlm., Volkm.-Kohlm. and O.E. Erikss.	Senescent leaves	Saprobic	*Juncus roemerianus*	USA: North Carolina	[51]
Paraphaeosphaeria pilleata Kohlm., Volkm.-Kohlm. and O.E. Erikss.	Senescent culms	Saprobic	*Juncus roemerianus*	USA: North Carolina	[77]
Paraphaeosphaeria michotii (Westend.) O.E. Erikss.	–	–	*Elymus pungens*	UK	[38]
	Living/decomposing leaf sheaths	Saprobic	*Phragmites australis*	Netherlands: Zeeland	[39]
Pseudopithomyces atro-olivaceus (Cooke and Harkn.) G. Guevara, K.C. Cunha and Gené	Senescent and decaying leaves	Saprobic	*Juncus roemerianus*	USA: Florida	[43]
Pseudopithomyces chartarum (Berk. and M.A. Curtis) Jun F. Li, Ariyaw. and K.D. Hyde	Senescent leaves	Saprobic	*Juncus roemerianus*	USA: Florida	[43]
Pseudopithomyces maydicus (Sacc.) Jun F. Li, Ariyaw. and K.D. Hyde	Senescent and decaying leaves	Saprobic	*Juncus roemerianus*	USA: Florida	[43]
	Decaying stems and leaf sheaths	Saprobic	*Phragmites australis*	China: Hong Kong	[41]
Spegazzinia tessarthra (Berk. and M.A. Curtis) Sacc.	Living leaves		*Juncus roemerianus*	USA: Florida	[43]
	Decaying stems and leaf sheaths	Saprobic	*Phragmites australis*	China: Hong Kong	[41]
Tremateia halophila Kohlm., Volkm.-Kohlm. and O.E. Erikss.	Lower and middle parts of senescent culms	Saprobic	*Juncus roemerianus*	USA: North Carolina	[78]
	–	Saprobic	*Spartina marítima*	Portugal: Alentejo, Lisbon	[54]

Table 1. *Cont.*

Taxon	Host Part	Life Mode	Hosts	Distribution	References
Lentitheciaceae					
Halobyssothecium estuariae B. Devadatha, Calabon, K.D. Hyde and E.B.G. Jones	Dead culm	Saprobic	*Phragmites australis*	UK: Pembrokeshire	[79]
Halobyssothecium obiones (P. Crouan and H. Crouan) Dayarathne, E.B.G. Jones and K.D. Hyde	Drift stems, attached and dead culms	Saprobic	*Spartina alterniflora*	India: Maharashtra, Tamil Nadu, Andhara Pradesh; USA: Maine, Rhode Island, Connecticut, Massachusetts, New Jersey, Maryland, Virginia, North Carolina, South Carolina, Florida, Mississippi, Texas	[20,35,52,61,71, 74,80–82]
	–	–	*Spartina cynosuroides*	–	[35]
	Pod and rhizome	Saprobic	*Spartina densiflora*	Argentina: Buenos Aires	[64]
	–	Saprobic	*Spartina patens*	USA: Rhode Island	[36]
	Culms	Saprobic	*Spartina* sp.	UK: England, Hampshire	[79,83]
	Stem	Saprobic	*Spartina townsendii*	UK: Hampshire, Wales	[49,65]
	–	Saprobic	*Spartina* spp.	USA: New Jersey, South Carolina; Mississippi, Argentina: Buenos Aires	[32,35,36,58,84]
	Stem, leaf sheaths, and blades	Saprobic	*Spartina marítima*	Portugal: Alentejo, Lisbon, Algarve, Centro	[31,54,59,63]
	–	Saprobic	Unidentified saltmarsh plants	USA: Mississippi	[55,58]
	–	–	*Elymus pungens*	–	[35]
	–	–	*Atriplex portulacoides*	–	[35]
	–	–	*Spartina townsendii*	–	[35]
Halobyssothecium phragmitis M.S. Calabon, E.B.G. Jones, S. Tibell and K.D. Hyde	Dead culm and stem	Saprobic	*Phragmites* sp.	Sweden: Gotland	[85]
Halobyssothecium versicolor M.S. Calabon, E.B.G. Jones and K.D. Hyde	Dead stem	Saprobic	*Atriplex portulacoides*	UK: Hampshire	[85]

Table 1. *Cont.*

Taxon	Host Part	Life Mode	Hosts	Distribution	References
Keissleriella culmifida (P. Karst.) S.K. Bose	–	–	*Elymus pungens*	UK	[38]
Keissleriella linearis E. Müll. ex Dennis	Living/decomposing stems	Saprobic	*Phragmites australis*	Netherlands: Zeeland	[40]
	Dead culm	Saprobic	*Phragmites* sp.	Sweden: Gotland	[85]
Keissleriella phragmiticola Wanas., E.B.G. Jones and K.D. Hyde	Culms	Saprobic	*Phragmites australis*	UK: Wales	[79]
Keissleriella rara Kohlm., Volkm.-Kohlm. and O.E. Erikss.	Senescent culms	Saprobic	*Juncus roemerianus*	USA: North Carolina	[77]
Keissleriella spp.	Senescent leaves	Saprobic	*Juncus roemerianus*	USA: Florida	[43]
Lentithecium fluviatile (Aptroot and Van Ryck.) K.D. Hyde, J. Fourn. and Ying Zhang	Dead leaf sheaths	Saprobic	*Phragmites australis*	Belgium: East Flanders	[86]
	Living/decomposing leaf blades and sheaths, stems	Saprobic	*Phragmites australis*	Netherlands: Zeeland	[39,40,50]
Setoseptoria arundinacea (Sowerby) Kaz. Tanaka and K. Hiray.	–	–	*Elymus pungens*	UK	[38]
	Living/decomposing leaf blades and sheaths, stems	Saprobic	*Phragmites australis*	Netherlands: Zeeland	[39,40,50]
	–	Saprobic	*Spartina* sp.	USA: North Carolina, Florida	[71]
Setoseptoria phragmitis Quaedvl., Verkley and Crous	Culm	Saprobic	*Phragmites* sp.	Sweden: Södermanland	[87]
Towyspora aestuari Wanas., E.B.G. Jones and K.D. Hyde	–	–	*Phragmites australis*	UK: Wales	[88]
Leptosphaeriaceae					
Leptosphaeria albopunctata (Westend.) Sacc.	–	–	*Juncus maritimus*	–	[35]
	–	–	*Phragmites australis*	–	[35]
	Attached culms	-	*Spartina alterniflora*	USA: Rhode Island	[35,36,61,71,73, 80]
	–	–	*Spartina* spp.	Canada: Bay of Fundy; USA: New Jersey, South Carolina; Argentina: Buenos Aires	[35,36,48,89,90]
	Stem	Saprobic	*Spartina townsendii*	UK: Wales	[35,65]

Table 1. *Cont.*

Taxon	Host Part	Life Mode	Hosts	Distribution	References
Leptosphaeria australiensis (Cribb and J.W. Cribb) G.C. Hughes	Senescent and decaying leaves	Saprobic	*Juncus roemerianus*	USA: Florida	[43]
	Pod	Saprobic	*Spartina densiflora*	Argentina: Buenos Aires	[64]
	–	–	*Spartina* spp.	–	[32]
Leptosphaeria culmifraga (Fr.) Ces. and De Not.	–	–	*Elymus pungens*	UK	[38]
Leptosphaeria littoralis Sacc.	–	–	*Elymus pungens*	UK	[38]
Leptosphaeria marina Ellis and Everh.	–	–	*Juncus roemerianus*		[35]
	–	Saprobic	*Spartina alterniflora*	USA: Maine, Rhode Island, Connecticut, New Jersey, Delaware, Virginia, North Carolina, South Carolina	[35,36,71,73,80]
	–	Saprobic	*Spartina* spp.	Canada; USA: New Jersey	[32,35,36,65,89–91]
	–	–	*Spartina townsendii*	UK	[35,38]
	Leaf sheaths and blades, stem	Saprobic	*Spartina maritima*	Portugal: Algarve	[31,59]
Leptosphaeria orae-maris Linder	–	–	*Arundo donax*	–	[35]
	–	Saprobic	*Lysimachia maritima*	USA: Massachusetts	[35,92]
	–	Saprobic	*Spartina alterniflora*	USA: Massachusetts, Rhode Island, North Carolina, Florida, Texas	[36,71,80,92]
	Rhizome	Saprobic	*Spartina densiflora*	Argentina: Buenos Aires	[64]
	–	–	*Spartina* spp.	–	[32]
	–	Saprobic	*Spartina townsendii*	UK	[35,65,93]

Table 1. *Cont.*

Taxon	Host Part	Life Mode	Hosts	Distribution	References
Leptosphaeria pelagica E.B.G. Jones	–	–	*Elymus pungens*	UK	[35,38]
	–	–	*Puccinellia maritima*	UK	[38]
	Decaying herbaceous stems, dead culms, decaying leaves	Saprobic	*Spartina alterniflora*	USA: Connecticut, Mississippi, Rhode Island; India: Goa, Karanataka	[20,36,52,55,73, 94]
	–	Saprobic	*Spartina densiflora*	Argentina: Buenos Aires	[64]
	–	Saprobic	*Spartina patens*	USA: Rhode Island	[36]
	–	–	*Spartina townsendii*	UK	[38]
	–	–	*Spartina* spp.	UK	[32,65]
	Sheath	Saprobic	*Spartina marítima*	Portugal: Alentejo, Lisbon	[54]
	Stem	Saprobic	*Spartina marítima*	Portugal: Alentejo	[63]
Leptosphaeria peruvianae Speg.	Decaying stems	Saprobic	*Sarcocornia perennis*	Argentina: Buenos Aires; in temperate marine waters	[52]
Leptosphaeria spp.	Decaying leaves	Saprobic	*Juncus roemerianus*	USA: Mississippi	[55]
	Decaying stems and leaf sheaths	Saprobic	*Phragmites australis*	China: Hong Kong	[41]
	–	–	*Spartina alterniflora*	USA: Rhode Island	[74]
	Leaf sheaths and blades, stem	Saprobic	*Spartina maritima*	Portugal: Centro	[59]
Leptosphaeria suaedae Hansf.	Dead twigs	Saprobic	*Suaeda australis*	South Australia	[95]
Lindgomycetaceae					
Arundellina typhae Wanas., E.B.G. Jones and K.D. Hyde	Dead stem	Saprobic	*Typha* sp.	UK: England	[96]
Lophiostomataceae					
Lophiostoma semiliberum (Desm.) Ces. and De Not.	Living/decomposing stems	Saprobic	*Phragmites australis*	Netherlands: Zeeland	[40]
Lophiostoma sp.	–	–	*Elymus pungens*	UK	[38]

Table 1. *Cont.*

Taxon	Host Part	Life Mode	Hosts	Distribution	References
Sigarispora arundinis (Pers.) Thambug., Qing Tian, Kaz. Tanaka and K.D. Hyde	Living/decomposing stems	Saprobic	*Phragmites australis*	Netherlands: Zeeland	[40]
Massarinaceae					
Helminthosporium sp.	Decaying leaf blades	Saprobic	*Spartina alterniflora*	USA: Georgia	[56]
Massarina carolinensis Kohlm., Volkm.-Kohlm. and O.E. Erikss.	Senescent culms	Saprobic	*Juncus roemerianus*	USA: North Carolina	[77]
Massarina igniaria (C. Booth) Aptroot	Decaying leaves	Saprobic	*Juncus roemerianus*	USA: Florida	[43]
Massarina phragmiticola Poon and K.D. Hyde	Decaying stems and leaf sheaths	Saprobic	*Phragmites australis*	China: Hong Kong	[41]
Massarina ricifera Kohlm., Volkm.-Kohlm. and O.E. Erikss.	Lower parts of senescent culms, decaying leaves	Saprobic	*Juncus roemerianus*	USA: Alabama, Mississippi, North Carolina	[55,58,97]
Massarina spp.	Senescent and decaying leaves	Saprobic	*Juncus roemerianus*	USA: Florida	[43]
	Living/decomposing leaf sheaths	Saprobic	*Phragmites australis*	Netherlands: Zeeland	[39]
Stagonospora abundata Kohlm. and Volkm.-Kohlm.	Senescent leaves and bracts	Saprobic	*Juncus roemerianus*	USA: Florida, Georgia, North Carolina	[98]
Stagonospora cylindrica Gunnell	Living/decomposing stems	Saprobic	*Phragmites australis*	Netherlands: Zeeland	[40]
Stagonospora elegans (Berk.) Sacc. and Traverso	Living/decomposing leaf sheaths, stems, culms	Saprobic	*Phragmites australis*	Australis; Netherlands: Zeeland	[39,40,95]
Stagonospora epicalamia (Cooke) Sacc.	–	–	*Phragmites australis*	Australia	[95]
Stagonospora haliclysta Kohlm.	Leaf sheaths and blades, stem	Saprobic	*Spartina maritima*	Portugal: Algarve	[59]
Stagonospora spp.	Living and senescent leaves	Saprobic	*Juncus roemerianus*	USA: Florida	[43]
	Living/decomposing leaf blades and sheaths, stems	Saprobic	*Phragmites australis*	China: Hong Kong; Netherlands: Zeeland	[39–41,50]
	Senescent and dead leaves/inflorescence, living and dead seeds, decaying leaf blades	Saprobic, pathogenic	*Spartina alterniflora*	Canada; USA: Maine, Rhode Island, Georgia, Connecticut, New Jersey, Virginia, Florida, North Carolina; Argentina: Buenos Aires	[35,36,56,73,74]

Table 1. *Cont.*

Taxon	Host Part	Life Mode	Hosts	Distribution	References
	–	Pathogenic	*Spartina cynosuroides*	USA: Maryland	[44]
	–	Saprobic	*Spartina patens*	USA: Rhode Island	[35,36]
	–	Saprobic	*Spartina* spp.	Canada	[35,36]
	Leaf sheaths and blades, stem, limb	Saprobic	*Spartina marítima*	Portugal: Alentejo, Lisbon, Algarve, Centro	[31,54,59]
Stagonospora suaedae Syd. and P. Syd.	Leaves	–	*Suaeda marítima*	Germany	[99]
Melanommataceae					
Aposphaeria spp.	Living/decomposing leaf sheaths, stems	Saprobic	*Phragmites australis*	Netherlands: Zeeland	[39,40]
Bicrouania maritima (P. Crouan and H. Crouan) Kohlm. and Volkm.-Kohlm.	Dead stems	Saprobic	*Atriplex portulacoides*	India	[35,52]
Morosphaeriaceae					
Helicascus kanaloanus Kohlm.	–	–	*Spartina* spp.	–	[32]
Neocamarosporiaceae					
Neocamarosporium artemisiae Dayarathne and E.B.G. Jones	–	Saprobic	*Artemisia maritima*	Sweden: Bohuslän	[100]
Neocamarosporium maritimae Dayarathne and E.B.G. Jones	–	Saprobic	*Artemisia maritima*	Sweden: Bohuslän	[100]
Neocamarosporium obiones (Jaap) Wanas. and K.D. Hyde	–	–	*Atriplex portulacoides*	–	[35]
Neocamarosporium phragmitis D.N. Wanasinghe, E.B.G. Jones and K.D. Hyde	Decaying culms	Saprobic	*Phragmites australis*	UK	[101]
Neocamarosporium salicorniicola Dayar., E.B.G. Jones and K.D. Hyde	Dead stems	Saprobic	*Salicornia* sp.	Thailand	[102]
Periconiaceae					
Periconia cookei E.W. Mason and M.B. Ellis	Senescent and decaying leaves	Saprobic	*Juncus roemerianus*	USA: Florida	[43]
	Living/decomposing leaf blades and sheaths	Saprobic	*Phragmites australis*	Netherlands: Zeeland	[39,50]
Periconia digitata (Cooke) Sacc.	Living, senescent, and decaying leaves	Saprobic	*Juncus roemerianus*	USA: Florida	[43]
Periconia digitata (Cooke) Sacc.	Living/decomposing leaf sheaths	Saprobic	*Phragmites australis*	Netherlands: Zeeland	[39]

Table 1. *Cont.*

Taxon	Host Part	Life Mode	Hosts	Distribution	References
Periconia echinochloae (Bat.) M.B. Ellis	Senescent and decaying leaves	Saprobic	*Juncus roemerianus*	USA: Florida	[43]
Periconia minutissima Corda	Leaves	Saprobic	*Distichlis spicata*	Argentina: Buenos Aires	[47]
	Senescent and decaying leaves	Saprobic	*Juncus roemerianus*	USA: Florida	[43]
	Living/decomposing leaf sheaths	Saprobic	*Phragmites australis*	Netherlands: Zeeland	[39]
Periconia sp.	–	Saprobic	Unidentifed saltmarsh plants	USA: Mississippi	[58]
Phaeosphaeriaceae					
Amarenomyces ammophilae (Lasch) O.E. Erikss.	–	–	*Ammophila arenaria*	–	[35]
	–	–	× *Ammocalamagrostis baltica*	–	[35]
	–	–	*Uniola paniculata*	–	[35]
Amphisphaeria culmicola Sacc.	Stem		*Spartina townsendii*	UK: England	[49]
Camarosporioides phragmitis W.J. Li and K.D. Hyde	Dead stem	Saprobic	*Phragmites australis*	Germany	[96]
Hendersonia culmiseda Sacc.	Living/decomposing leaf blades	Saprobic	*Phragmites australis*	Netherlands: Zeeland	[50]
	Living/decomposing leaf sheaths	Saprobic	*Phragmites australis*	Netherlands: Zeeland	[39]
	–	–	*Spartina townsendii*	UK	[103]
Hendersonia spp.	Living/decomposing leaf blades and sheaths	Saprobic	*Phragmites australis*	Netherlands: Zeeland; USA: Florida	[39,43,50]
Loratospora aestuarii Kohlm. and Volkm.-Kohlm.	Senescent culms	Saprobic	*Juncus roemerianus*	USA: North Carolina	[104]
Loratospora aestuarii Kohlm. and Volkm.-Kohlm.	–	Saprobic	Unidentified saltmarsh plants	USA: Mississippi	[58]
Ophiobolus littoralis (P. Crouan and H. Crouan) Sacc.	–	–	*Elymus pungens*	UK	[38]
Phaeoseptoria sp.	Living/decomposing leaf sheaths	Saprobic	*Phragmites australis*	Netherlands: Zeeland	[39]

Table 1. *Cont.*

Taxon	Host Part	Life Mode	Hosts	Distribution	References
Phaeosphaeria anchiala Kohlm., Volkm.-Kohlm. and C.K.M. Tsui	Senescent leaves	Saprobic	*Juncus roemerianus*	USA: Florida, Georgia, Maryland, North Carolina, Virginia	[105]
Phaeosphaeria caricinella (P. Karst.) O.E. Erikss.	–	–	*Spartina* sp.	USA: Florida, North Carolina	[71]
Phaeosphaeria culmorum (Auersw.) Leuchtm.	Living/decomposing leaf blades and sheaths	Saprobic	*Phragmites australis*	Netherlands: Zeeland	[39,50]
Phaeosphaeria eustoma (Fuckel) L. Holm	Living/decomposing leaf blades and sheaths, stems, culms	Saprobic	*Phragmites australis*	Netherlands: Zeeland	[39,40,50,95]
Phaeosphaeria fuckelii (Niessl) L. Holm	–	–	*Elymus pungens*	UK	[38]
Phaeosphaeria gessneri Shoemaker and C.E. Babc.	–	–	*Spartina* spp.	–	[32]
Phaeosphaeria halima (T.W. Johnson) Shoemaker and C.E. Babc.	Dead culms; Decaying leaves, leaf blades	Saprobic	*Spartina alterniflora*	India: Kerala; USA: California, Georgia, Mississippi, Vancouver, North Carolina	[25,35,52,55–58,71,80]
	Decaying leaf blades	Saprobic	*Spartina densiflora*	USA: California	[25]
			Spartina spp.		[32]
	Decaying leaves	Saprobic	*Spartina foliosa*	USA: California	[25]
	Leaf sheaths and blades, stem	Saprobic	*Spartina maritima*	Portugal: Algarve, Centro	[31]
Phaeosphaeria herpotrichoides (De Not.) L. Holm	–	–	*Spartina patens*	USA: North Carolina, Florida	[71]
Phaeosphaeria juncina (Auersw.) L. Holm	–	Saprobic	*Juncus roemerianus*	USA: Florida	[43]
Phaeosphaeria luctuosa (Niessl ex Sacc.) Y. Otani and Mikawa	Living/decomposing leaf sheaths, stems	Saprobic	*Phragmites australis*	Netherlands: Zeeland	[39,40]
	–	–	*Elymus pungens*	UK	[38]

Table 1. *Cont.*

Taxon	Host Part	Life Mode	Hosts	Distribution	References
Phaeosphaeria macrosporidium (E.B.G. Jones) Shoemaker and C.E. Babc.	Decaying stems	Saprobic	*Spartina sp*	UK: Wales, England	[65]
	Stem	Saprobic	*Spartina marítima*	Portugal: Lisbon	[54,63]
Phaeosphaeria microscopica (P. Karst.) O.E. Erikss.	–	–	*Elymus pungens*	UK	[38]
Phaeosphaeria neomaritima (R.V. Gessner and Kohlm.) Shoemaker and C.E. Babc.	–	–	*Juncus maritimus*	–	[35]
	–	–	*Juncus roemerianus*	–	[35]
	–	Saprobic	*Juncus* sp.	Canada; India: Maharashtra, Karnataka; USA: Virginia, North Carolina	[36,52,71,80]
	–	–	*Spartina alterniflora*	–	[35]
	–	Saprobic	*Spartina* spp.	Canada; USA: North Carolina, Virginia	[32,71,80]
	–	–	*Spartina townsendii*	UK	[35,93]
	Stem	Saprobic	*Spartina marítima*	Portugal: Alentejo	[63]
Phaeosphaeria nigrans (Roberge ex Desm.) L. Holm	–	–	*Elymus pungens*	UK	[38]
Phaeosphaeria olivacea Kohlm., Volkm.-Kohlm. and O.E. Erikss.	Senescent leaves	Saprobic	*Juncus roemerianus*	USA: North Carolina, Mississippi	[58,76]
Phaeosphaeria pontiformis (Fuckel) Leuchtm.	–	–	*Elymus pungens*	UK	[38]
	Living/decomposing leaf blades and sheaths, stems	Saprobic	*Phragmites australis*	Netherlands: Zeeland	[39,40,50]
Phaeosphaeria roemeriani Kohlm., Volkm.-Kohlm. and O.E. Erikss.	Senescent and decaying leaves	Saprobic	*Juncus roemerianus*	USA: Mississippi, North Carolina	[55,58,60]
Phaeosphaeria spartinae (Ellis and Everh.) Shoemaker and C.E. Babc.	–	Saprobic	*Spartina* spp.	India: Kerala	[32,52]
	Decaying herbaceous stems and pod	Saprobic	*Spartina densiflora*	Argentina: Buenos Aires	[64]
	–	Saprobic	*Spartina marítima*	Portugal: Lisbon	[54]

Table 1. *Cont.*

Taxon	Host Part	Life Mode	Hosts	Distribution	References
Phaeosphaeria spartinicola Leuchtm.	–	Saprobic	*Juncus* sp.	India	[52]
	Dead leaves, decaying leaf blades	Saprobic	*Spartina alterniflora*	Mexico; USA: Alabama, California, Georgia, Mississippi; Canada: Nova Scotia, New Brunswick	[25,36,55–58]
	Pod, leaf blades	Saprobic	*Spartina densiflora*	Argentina: Buenos Aires; USA: California	[25,64]
	–	–	*Spartina* spp.	–	[32]
	Leaf blades	Saprobic	*Spartina foliosa*	USA: California	[25]
	Leaf sheaths and blades, stem, limb	Saprobic	*Spartina marítima*	Portugal: Alentejo, Lisbon, Algarve, Centro	[31,54,59,63]
Phaeosphaeria spp.	Living/decomposing leaf blades and sheaths, stems	Saprobic	*Phragmites australis*	Netherlands: Zeeland	[39,40,50]
	–	Saprobic	*Spartina alterniflora*	USA: Rhode Island	[74]
Sclerostagonospora sp.	Decaying stems and leaf sheaths	Saprobic	*Phragmites australis*	China: Hong Kong	[41]
Septoriella phragmitis Oudem.	Living/decomposing leaf sheaths and stems	Saprobic	*Phragmites australis*	Netherlands: Zeeland	[39,40]
Septoriella spp.	Decaying stems and leaf sheaths and blades, stems	Saprobic	*Phragmites australis*	China: Hong Kong; Netherlands: Zeeland	[39–41,50]
Septoriella thalassica (Speg.) Nag Raj	–	–	*Distichlis spicata*	–	[35]
			Distichlis spicata		[35]
Septoriella unigalerita Kohlm. and Volkm.-Kohlm.	Senescent leaves	Saprobic	*Juncus roemerianus*	USA: North Carolina	[98]
Septoriella vagans (Niessl) Y. Marín and Crous	–	–	*Elymus pungens*	UK	[38]
	–	–	*Puccinellia maritima*	UK	[38]
	–	Saprobic	*Spartina alterniflora*	USA: Rhode Island	[74]
Pleomassariaceae					
Splanchnonema sp.	Living, senescent, and decaying leaves	Saprobic	*Juncus roemerianus*	USA: Florida	[43]

Table 1. *Cont.*

Taxon	Host Part	Life Mode	Hosts	Distribution	References
Pleosporaceae					
Alternaria alternata (Fr.) Keissl.	Leaves	Saprobic	*Distichlis spicata*	Argentina: Buenos Aires	[47]
	Living, senescent, and decaying leaves	Saprobic	*Juncus roemerianus*	USA: Florida	[43]
	Living/decomposing leaf blades and sheaths, stems	Saprobic	*Phragmites australis*	Netherlands: Zeeland	[39,41,50]
	–	Saprobic	*Spartina alterniflora*	USA: North Carolina	[74]
	Leaves, stems, and roots	Saprobic	*Spartina* sp.	Canada: Bay of Fundy	[48]
Alternaria infectoria E.G. Simmons	–	–	*Elymus pungens*	UK	[38]
	Living/decomposing leaf sheaths	Saprobic	*Phragmites australis*	Netherlands: Zeeland	[39]
Alternaria longissima Deighton and MacGarvie	Living, senescent, and decaying leaves	Saprobic	*Juncus roemerianus*	USA: Florida	[43]
Alternaria maritima G.K. Sutherl.	Stem	Saprobic, pathogenic	*Spartina townsendii*	UK: England	[49]
Alternaria spp.	–	–	*Atriplex portulacoides*	–	[35]
	–	–	*Juncus roemerianus*	–	[35]
	–	–	*Salsola kali*	–	[35]
	Inflorescence and upper leaves	Saprobic, parasitic	*Spartina alterniflora*	USA: Rhode Island	[35,61]
	Culms	Saprobic	*Spartina* sp.	Thailand	This study
	–	–	*Spartina townsendii*	–	[35]
Bipolaris cynodontis (Marignoni) Shoemaker	Leaves	Saprobic	*Distichlis spicata*	Argentina: Buenos Aires	[47]
Curvularia hawaiiensis (Bugnic. ex M.B. Ellis) Manamgoda, L. Cai and K.D. Hyde	Living and senescent leaves	Saprobic	*Juncus roemerianus*	USA: Florida	[43]
	Decaying stems and leaf sheaths	Saprobic	*Phragmites australis*	China: Hong Kong	[41]
Curvularia protuberata R.R. Nelson and Hodges	Leaves	Saprobic	*Distichlis spicata*	Argentina: Buenos Aires	[47]
	Senescent leaves	Saprobic	*Juncus roemerianus*	USA: Florida	[43]
Curvularia spp.	Living, senescent, and decaying leaves	Saprobic	*Juncus roemerianus*	USA: Florida	[43]
	–	Saprobic	*Spartina altrerniflora*	USA: North Carolina	[74]

Table 1. *Cont.*

Taxon	Host Part	Life Mode	Hosts	Distribution	References
Curvularia tuberculata B.L. Jain	Senescent and decaying leaves	Saprobic	*Juncus roemerianus*	USA: Florida	[43]
Decorospora gaudefroyi (Pat.) Inderb., Kohlm. and Volkm.-Kohlm.	Stems	Saprobic	*Atriplex* sp.	UK: Portsmouth	[106]
	–	–	*Atriplex portulacoides*	–	[35]
	–	–	*Sarcocornia perennis*	–	[35]
	–	–	*Sarcoconia fructicosa*	–	[35]
	–	–	*Salicornia europaea*	–	[35]
	–	–	*Salicornia* sp.	–	[35]
	Leaf sheaths and blades, stem	Saprobic	*Spartina maritima*	Portugal: Algarve	[59]
	–	–	*Suaeda maritima*	–	[35]
Drechslera sp.	Living, senescent, and decaying leaves	Saprobic	*Juncus roemerianus*	USA: Florida	[43]
Exserohilum rostratum (Drechsler) K.J. Leonard and Suggs	–	–	*Distichlis spicata*	–	[35]
	Living, senescent, and decaying leaves	Saprobic	*Juncus roemerianus*	USA: Florida	[43]
	Senescent and dead leaves	Saprobic	*Spartina alterniflora*	USA: Rhode Island, North Carolina, Florida	[35,36,73]
	–	–	*Spartina* spp.	–	[32]
Paradendryphiella arenariae (Nicot) Woudenb. and Crous	Decomposing culms	Saprobic	*Spartina alterniflora*	USA: Rhode Island	[35,61]
	–	–	*Spartina* spp.	–	[32]
Paradendryphiella salina (G.K. Sutherl.) Woudenb. and Crous	–	–	*Atriplex portulacoides*	–	[35]
	Decaying leaves	Saprobic	*Juncus roemerianus*	USA: Florida	[43]
	–	–	*Puccinellia maritima*	–	[35]
	–	–	*Salicornia europaea*	–	[35]
	Decomposing culms	Saprobic	*Spartina alterniflora*	USA: Rhode Island	[35,61]
	–	–	*Spartina* spp.	–	[32]
	–	–	*Spartina townsendii*	–	[35]
	Leaves and stems	Saprobic	*Spartina* sp.	Canada: Bay of Fundy	[48]
	–	–	*Suaeda maritima*	–	[35]

Table 1. *Cont.*

Taxon	Host Part	Life Mode	Hosts	Distribution	References
Pleospora abscondita Sacc. and Roum.	Living/decomposing leaf sheaths	Saprobic	*Phragmites australis*	Netherlands: Zeeland	[39]
Pleospora pelagica T.W. Johnson	Decomposing culms; decaying leaf blades	Saprobic	*Spartina alterniflora*	India: Maharashtra, Kerala; USA: Georgia, Rhode Island, North Carolina, Florida	[35,36,52,56,71, 73,74,80]
	Decaying leaf blades	Saprobic	*Spartina densiflora*	USA: California	[25]
		Saprobic	*Spartina* spp.	USA: South Carolina	[32,36]
			Typha sp.		[35]
Pleospora pelvetiae G.K. Sutherl.	–	Saprobic	Unidentifed saltmarsh plants	USA: Mississippi	[58]
Pleospora spp.	–	–	*Salicornia virginica*	–	[35]
	Dead leaves/culms	Saprobic	*Spartina alterniflora*	USA: Rhode Island	[61]
Pleospora spartinae (J. Webster and M.T. Lucas) Apinis and Chesters	Decaying stems and leaf sheaths	Saprobic	*Phragmites australis*	China: Hong Kong	[41]
	Decaying leaf blades	Saprobic	*Spartina alterniflora*	USA: Georgia	[56]
	Stem	Saprobic	*Spartina* spp.	Canada: Bay of Fundy	[32,48]
	–	–	*Spartina townsendii*	UK	[35,38,107]
Pleospora straminis Sacc. and Speg.	–	–	*Elymus pungens*	UK	[38]
Pleospora vagans Niessl var. vagans	Living/decomposing leaf sheaths	Saprobic	*Phragmites australis*	Netherlands: Zeeland	[39]
	Dead culms	Saprobic	*Spartina alterniflora*	USA: Rhode Island	[73]
Pyrenophora tritici-repentis (Died.) Drechsler	–	–	*Elymus pungens*	UK	[38]
Stemphylium botryosum Wallr.	Leaves	Saprobic	*Distichlis spicata*	Argentina: Buenos Aires	[47]
Stemphylium lycopersici (Enjoji) W. Yamam.	Living leaves	–	*Juncus roemerianus*	USA: Florida	[43]
Stemphylium maritimum T.W. Johnson	–	Saprobic	*Spartina* sp.	UK	[65]
Stemphylium spp.	–	–	*Salsola kali*	–	[35]
	Leaves	Saprobic	*Spartina* spp.	Canada: Bay of Fundy	[35,48]
Stemphylium vesicarium (Wallr.) E.G. Simmons	–	–	*Elymus pungens*	UK	[38]
	Living, senescent and decaying leaves	Saprobic	*Juncus roemerianus*	USA: Florida	[43]
	–	Saprobic	*Lysimachia maritima*	USA: Massachusetts	[92]

Table 1. *Cont.*

Taxon	Host Part	Life Mode	Hosts	Distribution	References
	–	Saprobic	*Spartina alterniflora*	USA: Rhode Island	[61]
	Glumes, rachis	–	*Spartina townsendii*	UK: England	[38,49]
			Spartina sp.	UK	[65]
Stemphylium triglochinicola B. Sutton and Piroz.	–	–	*Triglochin maritima*	Sweden: Västergötland	[35,87]
	Dead leaves and inflorescences	Saprobic	*Triglochin* sp.	India: Kerala; UK	[52,108]
Typhicola typharum (Desm.) Crous	Senescent and dead leaves	Saprobic, pathogenic	*Spartina alterniflora*	Canada; USA: Maine, Rhode Island, Connecticut, New Jersey, Virginia, North Carolina, Florida	[35,36,61,73,74]
	–	Saprobic	*Spartina patens*	USA: Rhode Island	[36]
	–		*Spartina townsendii*	UK	[38]
	–	Saprobic	*Spartina* spp.	Argentina: Buenos Aires; Canada; USA: Maine	[35,36]
	Stems	Saprobic	*Spartina townsendii*	UK: England	[35,49,65]
Pleosporales genera *incertae sedis*					
Phialophorophoma litoralis Linder	Stem and sheath	Saprobic	*Spartina marítima*	Portugal: Alentejo, Lisbon	[54,63]
Phialophorophoma spp.	Living/decomposing leaf sheaths, stems	Saprobic	*Phragmites australis*	Netherlands: Zeeland	[39,40]
Pyrenochaeta sp.	Living leaves	Saprobic	*Juncus roemerianus*	USA: Florida	[43]
Scolecobasidium humicola G.L. Barron and L.V. Busch	Living, senescent, and decaying leaves	Saprobic	*Juncus roemerianus*	USA: Florida	[43]
Roussoellaceae					
Cytoplea sp.	Decaying stems and leaf sheaths	Saprobic	*Phragmites australis*	China: Hong Kong	[41]
Sporormiaceae					
Preussia funiculata (Preuss) Fuckel	–	–	*Spartina townsendii*	UK	[38]
Preussia terricola Cain	–	–	*Elymus pungens*	UK	[38]
Sporormia longipes Massee and E.S. Salmon	–	–	*Elymus pungens*	UK	[38]
Sporormia sp.	Senescent and decaying leaves	Saprobic	*Juncus roemerianus*	USA: Florida	[43]

Table 1. *Cont.*

Taxon	Host Part	Life Mode	Hosts	Distribution	References
Sporormiella intermedia (Auersw.) S.I. Ahmed and Cain ex Kobayasi	–	–	*Elymus pungens*	UK	[38]
Sporormiella lageniformis (Fuckel) S.I. Ahmed and Cain	–	–	*Spartina townsendii*	UK	[38]
Sporormiella minima (Auersw.) S.I. Ahmed and Cain	–	–	*Elymus pungens*	UK	[38]
	–	–	*Spartina townsendii*	UK	[38]
Teichosporaceae					
Teichospora striata (Kohlm. and Volkm.-Kohlm.) Jaklitsch and Voglmayr	Senescent leaves and inflorescences	Saprobic	*Juncus roemerianus*	USA: North Carolina, Virginia	[98]
Teichospora suaedae Speg.	Dead branches	Saprobic	*Suaeda divaricata*	Argentina: Mendoza	[109]
Testudinaceae					
Verruculina enalia (Kohlm.) Kohlm. and Volkm.-Kohlm.	Decaying stems and leaf sheaths	Saprobic	*Phragmites australis*	China: Hong Kong	[41]
Tetraplosphaeriaceae					
Tetraploa aristata Berk. and Broome	Leaves	Saprobic	*Distichlis spicata*	Argentina: Buenos Aires	[47]
	Senescent and decaying leaves	Saprobic	*Juncus roemerianus*	USA: Florida	[43]
	Decaying stems and leaf sheaths	Saprobic	*Phragmites australis*	China: Hong Kong	[41]
Torulaceae					
Torula herbarum (Pers.) Link	Decaying leaves	Saprobic	*Juncus roemerianus*	USA: Florida	[43]
Trematosphaeriaceae					
Halomassarina thalassiae (Kohlm. and Volkm.-Kohlm.) Suetrong, Sakay., E.B.G. Jones, Kohlm., Volkm.-Kohlm. and C.L. Schoch	Decaying stems and leaf sheaths	Saprobic	*Phragmites australis*	China: Hong Kong	[41]
EUROTIOMYCETES					
Chaetothyriales					
Herpotrichiellaceae					
Rhinocladiella spp.	Living, senescent, and decaying leaves	Saprobic	*Juncus roemerianus*	USA: Florida	[43]
	Decaying stems and leaf sheaths	Saprobic	*Phragmites australis*	China: Hong Kong	[41]
Veronaea sp.	Decaying leaves	Saprobic	*Juncus roemerianus*	USA: Florida	[43]
Eurotiales					
Aspergillaceae					

Table 1. *Cont.*

Taxon	Host Part	Life Mode	Hosts	Distribution	References
Aspergillus fumigatus Fresen.	–	–	*Elymus pungens*	UK	[38]
Aspergillus nidulans (Eidam) G. Winter	–	–	*Elymus pungens*	UK	[38]
	–	–	*Spartina townsendii*	UK	[38]
Aspergillus niger Tiegh.	Living, senescent, and decaying leaves	Saprobic	*Juncus roemerianus*	USA: Florida	[43]
Aspergillus spp.	Living, senescent, and decaying leaves	Saprobic	*Juncus roemerianus*	USA: Florida	[43]
	–	–	*Spartina townsendii*	UK: England	[49]
Monascus purpureus Went	–	–	*Elymus pungens*	UK	[38]
Penicillium aurantiogriseum Dierckx	Leaves	Saprobic	*Spartina* sp.	Canada: Bay of Fundy	[48]
Penicillium brevicompactum Dierckx	Roots	Saprobic	*Spartina* sp.	Canada: Bay of Fundy	[48]
Penicillium chrysogenum Thom	Roots	Saprobic	*Spartina* sp.	Canada: Bay of Fundy	[48]
Penicillium lividum Westling	Leaves and stems	Saprobic	*Spartina* sp.	Canada: Bay of Fundy	[48]
Penicillium spp.	Living, senescent, and decaying leaves	Saprobic	*Juncus roemerianus*	USA: Florida	[43]
	Decaying stems and leaf sheaths	Saprobic	*Phragmites australis*	China: Hong Kong	[41]
	–	–	*Spartina townsendii*	UK: England	[49]
Thermoascaceae					
Thermoascus crustaceus (Apinis and Chesters) Stolk	–	–	*Elymus pungens*	UK	[38]
Paecilomyces spp.	Senescent and decaying leaves	Saprobic	*Juncus roemerianus*	USA: Florida	[43]
	Decaying stems and leaf sheaths	Saprobic	*Phragmites australis*	China: Hong Kong	[41]
	–	Saprobic	Salt marsh plants	India: Goa	[52]
Trichocomaceae					
Thermomyces dupontii (Griffon and Maubl.) Houbraken and Samson	–	–	*Elymus pungens*	UK	[38]
Onygenales					
Onygenaceae					
Amauroascus albicans (Apinis) Arx	–	–	*Elymus pungens*	UK	[38]
Amauroascus albicans (Apinis) Arx	–	–	*Spartina townsendii*	UK	[38]

Table 1. *Cont.*

Taxon	Host Part	Life Mode	Hosts	Distribution	References
LECANOROMYCETES					
Ostropales					
Stictidaceae					
Glomerobolus gelineus Kohlm. and Volkm.-Kohlm.	Senescent culms	Saprobic	*Juncus roemerianus*	USA: North Carolina	[110]
Stictis sp.	Living/decomposing leaf sheaths	Saprobic	*Phragmites australis*	Netherlands: Zeeland	[39]
LEOTIOMYCETES					
Helotiales					
Amorphothecaceae					
Amorphotheca resinae Parbery	Roots	Saprobic	*Spartina* sp.	Canada: Bay of Fundy	[48]
Calloriaceae					
Cistella fugiens (W. Phillips) Matheis	Living/decomposing stems	Saprobic	*Phragmites australis*	Netherlands: Zeeland	[40]
Helotiaceae					
Cyathicula culmicola (Desm.) De Not.	–	–	*Elymus pungens*	UK	[38]
Helotium sp.	–	–	*Elymus pungens*	UK	[38]
Lachnaceae					
Brunnipila palearum (Desm.) Baral	–	–	*Elymus pungens*	UK	[38]
	–	–	*Spartina townsendii*	UK	[38]
Lachnum controversum (Cooke) Rehm	–	–	*Elymus pungens*	UK	[38]
Lachnum spartinae S.A. Cantrell	Decaying leaf sheaths	Saprobic	*Spartina alterniflora*	USA: Georgia	[56,111]
	–	–	*Spartina* spp.	–	[32]
Mollisiaceae					
Belonopsis atriella (Cooke) Lindau	–	–	*Spartina cynosuroides*	USA: Louisiana	[68,90,112]
Mollisia hydrophila (P. Karst.) Sacc.	Living/decomposing leaf sheaths	Saprobic	*Phragmites australis*	Netherlands: Zeeland	[39]
Mollisia palustris (P. Karst.) P. Karst.	Living/decomposing leaf sheaths	Saprobic	*Phragmites australis*	Netherlands: Zeeland	[39]
Trichobelonium kneiffii (Wallr.) J. Schröt.	Living/decomposing leaf sheaths, stems	Saprobic	*Phragmites australis*	Netherlands: Zeeland	[39,40]
Ploettnerulaceae					
Cadophora melinii Nannf.	Leaves	Saprobic	*Spartina* sp.	Canada: Bay of Fundy	[48]
Sclerotiniaceae					

Table 1. *Cont.*

Taxon	Host Part	Life Mode	Hosts	Distribution	References
Botrytis cinerea Pers.	Stem		*Spartina townsendii*	UK: England	[49]
	Leaves	Saprobic	*Spartina* sp.	Canada: Bay of Fundy	[48]
Monilia sp.	Decaying leaves	Saprobic	*Juncus roemerianus*	USA: Florida	[43]
Solenopeziaceae					
Halenospora varia (Anastasiou) E.B.G. Jones	Senescent leaves	Saprobic	*Juncus roemerianus*	USA: Florida	[43]
	Basal area of the sheath	Saprobic	*Spartina densiflora*	Argentina: Buenos Aires	[64]
	–	–	*Spartina* spp.	–	[32]
Helotiales genera *incertae sedis*					
Cejpia hystrix (De Not.) Baral	–	–	*Elymus pungens*	UK	[38]
Dactylaria sp.	Decaying stems and leaf sheaths	Saprobic	*Phragmites australis*	China: Hong Kong	[41]
Crocicreas gramineum (Fr.) Fr.	–	–	*Elymus pungens*	UK	[38]
Leotiales					
Leotiales genera *incertae sedis*					
Flagellospora sp.	Living leaves	–	*Juncus roemerianus*	USA: Florida	[43]
Rhytismatales					
Rhytismataceae					
Lophodermium arundinaceum (Schrad.) Chevall.	–	–	*Elymus pungens*	UK	[38]
	Living/decomposing leaf sheaths, stems	Saprobic	*Phragmites australis*	Netherlands: Zeeland	[39,40]
Thelebolales					
Thelebolaceae					
Thelebolus crustaceus (Fuckel) Kimbr.	–	–	*Elymus pungens*	UK	[38]
	–	–	*Puccinellia maritima*	UK	[38]
	–	–	*Spartina townsendii*	UK	[38]
ORBILIOMYCETES					
Orbiliales					
Orbiliaceae					
Arthrobotrys conoides Drechsler	Decaying stems and leaf sheaths	Saprobic	*Phragmites australis*	China: Hong Kong	[41]
Arthrobotrys sp.	Decaying stems and leaf sheaths	Saprobic	*Phragmites australis*	China: Hong Kong	[41]
Orbilia junci Kohlm., Baral and Volkm.-Kohlm.	Tips of senescent leaves	–	*Juncus roemerianus*	USA: North Carolina	[113]

Table 1. *Cont.*

Taxon	Host Part	Life Mode	Hosts	Distribution	References
PEZIZOMYCETES					
Pezizales					
Pezizaceae					
Belonium heteromorphum (Ellis and Everh.) Seaver	–	–	*Spartina cynosuroides*	USA: Louisiana	[68,114]
SACCHAROMYCETES					
Saccharomycetales					
Debaryomycetaceae					
Debaryomyces hansenii (Zopf) Lodder and Kreger-van Rij	Decaying leaf blades	Saprobic	*Spartina alterniflora*	USA: Louisiana	[56]
Scheffersomyces spartinae (Ahearn, Yarrow and Meyers) Kurtzman and M. Suzuki	Decaying leaf blades	Saprobic	*Spartina alterniflora*	USA: Louisiana	[56]
Saccharomycetaceae					
Kluyveromyces lactis (Stell.-Dekk.) Van der Walt	Decaying leaf blades	Saprobic	*Spartina alterniflora*	USA: Louisiana	[56]
SORDARIOMYCETES					
Amphisphaeriales					
Amphisphaeriaceae					
Massariella sp.	–	–	*Spartina townsendii*	UK	[38]
Ommatomyces coronatus Kohlm., Volkm.-Kohlm. and O.E. Erikss.	Lower parts of senescent culms	Saprobic	*Juncus roemerianus*	USA: North Carolina	[97]
Pestalotia sp.	Living, senescent and decaying leaves	Saprobic	*Juncus roemerianus*	USA: Florida	[43]
Apiosporaceae					
Arthrinium arundinis (Corda) Dyko and B. Sutton	Living/decomposing leaf sheaths	Saprobic	*Phragmites australis*	Netherlands: Zeeland	[39]
	Dead culms	Saprobic	*Phragmites* sp.	South Australia	[62]
Arthrinium phaeospermum (Corda) M.B. Ellis	Living/decomposing leaf blades and sheaths, stems	Saprobic	*Phragmites australis*	Netherlands: Zeeland	[39,40,50]
	–	Saprobic	*Spartina patens*	USA: Rhode Island	[61]
	Inflorescence and upper leaves	Saprobic	*Spartina alterniflora*	USA: Rhode Island	[36]

Table 1. *Cont.*

Taxon	Host Part	Life Mode	Hosts	Distribution	References
Arthrinium spp.	Living leaves	Saprobic	*Juncus roemerianus*	USA: Florida	[43]
	Decaying stems and leaf sheaths	Saprobic	*Phragmites australis*	China: Hong Kong	[41]
Nigrospora oryzae (Berk. and Broome) Petch	Leaves	Saprobic	*Distichlis spicata*	Argentina: Buenos Aires	[47]
	Living, senescent, and decaying leaves	Saprobic	*Juncus roemerianus*	USA: Florida	[43]
Beltraniaceae					
Beltrania querna Harkn.	Decaying leaves	Saprobic	*Juncus roemerianus*	USA: Florida	[43]
Hyponectriaceae					
Phragmitensis ellipsoidea M.K.M. Wong, Goh and K.D. Hyde	Intertidal to aerial culms	Saprobic	*Phragmites australis*	China: Hong Kong	[115]
Phragmitensis marina M.K.M. Wong, Poon and K.D. Hyde	Decaying stems and leaf sheaths	Saprobic	*Phragmites australis*	China: Hong Kong	[41]
Physalospora citogerminans Kohlm., Volkm.-Kohlm. and O.E. Erikss.	Lower and upper parts of senescent culms	Saprobic	*Juncus roemerianus*	USA: North Carolina	[116]
Sporocadaceae					
Discostroma sp.	Living/decomposing leaf sheaths	Saprobic	*Phragmites australis*	Netherlands: Zeeland	[39]
Pestalotiopsis juncestris Kohlm. and Volkm.-Kohlm.	Senescent involucral leaves and culms	Saprobic	*Juncus roemerianus*	USA: North Carolina	[117]
Pestalotiopsis planimi (Vize) Steyaert	–	–	*Spartina alterniflora*	USA: Rhode Island	[61]
Pestalotiopsis sp.	Decaying stems and leaf sheaths	Saprobic	*Phragmites australis*	China: Hong Kong	[41]
Coronophorales					
Ceratostomataceae					
Melanospora sp.	Decaying leaves	Saprobic	*Juncus roemerianus*	USA: Florida	[43]
Microthecium fimicola (E.C. Hansen) Y. Marín, Stchigel, Guarro and Cano	–	–	*Elymus pungens*	UK	[38]
Microthecium levitum Udagawa and Cain	Dead leaves/culms	Saprobic	*Spartina alterniflora*	USA: Rhode Island	[61]
Coronophorales genera *incertae sedis*					
Papulaspora halima Anastasiou	Living and decaying leaves	Saprobic	*Juncus roemerianus*	USA: Florida	[43]
Papulosa amerospora Kohlm. and Volkm.-Kohlm.	Senescent culms	Saprobic	*Juncus roemerianus*	USA: North Carolina	[118]
Diaporthales					
Diaporthaceae					

Table 1. *Cont.*

Taxon	Host Part	Life Mode	Hosts	Distribution	References
Phomopsis spp.	Senescent and decaying leaves	Saprobic	*Juncus roemerianus*	USA: Florida	[43]
	Decaying stems and leaf sheaths	Saprobic	*Phragmites australis*	China: Hong Kong	[41]
	–	–	*Spartina* sp.	–	[71]
Gnomoniaceae					
Gnomonia salina E.B.G. Jones (*probably a nomen dubiumand possibly a Halosarpheia species)*	–	Saprobic	*Spartina alterniflora*	USA: Connecticut	[36]
			Spartina spp.		[32,35]
	–	–	*Spartina townsendii*	UK	[35,65]
Diaporthales *incertae sedis*					
Botryodiplodia sp.	Senescent and decaying leaves	Saprobic	*Juncus roemerianus*	USA: Florida	[43]
Glomerellales					
Glomerellaceae					
Colletotrichum sp.	Decaying stems and leaf sheaths	Saprobic	*Phragmites australis*	China: Hong Kong	[41]
Plectosphaerellaceae					
Stachylidium bicolor Link	Senescent leaves	Saprobic	*Juncus roemerianus*	USA: Florida	[43]
Hypocreales					
Bionectriaceae					
Acremonium spp.	Leaves	Saprobic	*Distichlis spicata*	Argentina: Buenos Aires	[47]
	Living, senescent, and decaying leaves	Saprobic	*Juncus roemerianus*	USA: Florida	[43]
	Decaying stems and leaf sheaths	Saprobic	*Phragmites australis*	China: Hong Kong	[41]
Clonostachys rosea (Link) Schroers, Samuels, Seifert and W. Gams	Leaves	Saprobic	*Spartina* sp.	Canada: Bay of Fundy	[48]
Fusariella obstipa (Pollack) S. Hughes	Decaying leaves	Saprobic	*Juncus roemerianus*	USA: Florida	[43]
Gliomastix spp.	Senescent and decaying leaves	Saprobic	*Juncus roemerianus*	USA: Florida	[43]
midrule *Hydropisphaera arenula* (Berk. and Broome) Rossman and Samuels	Decaying stems and leaf sheaths	Saprobic	*Phragmites australis*	China: Hong Kong	[41]
	Living/decomposing leaf sheaths	Saprobic	*Phragmites australis*	Netherlands: Zeeland	[39]
Hydropisphaera erubescens (Roberge ex Desm.) Rossman and Samuels	Decaying leaf blades	Saprobic	*Spartina alterniflora*	USA: Georgia	[56]
	–	–	*Spartina* spp.	–	[32]

Table 1. *Cont.*

Taxon	Host Part	Life Mode	Hosts	Distribution	References
Clavicipitaceae					
Atkinsonella hypoxylon (Peck) Diehl	–	–	*Spartina cynosuroides*	–	[68]
	–	Saprobic	*Phragmites australis*	UK: England (Southampton Hampshire, Sussex, Oxon)	[119,120]
	Replaced seeds in the inflorescence, ovaries of the flowers	Saprobic, parasitic	*Spartina alterniflora*	USA: Rhode Island; Argentina	[36,61,68,73,121,122]
Claviceps purpurea (Fr.) Tul.	–	Pathogenic	*Spartina anglica*	UK	[123]
	–	Saprobic, parasitic	*Spartina cynosuroides*	USA: New York, Florida, Mississippi	[44,68,121,124]
			Spartina patens	USA: Maryland, Mississippi	[44,68,124,125]
	–	–	*Spartina townsendii*	UK: England	[120,126]
	–	–	*Spartina* sp.	Argentina	[122]
Claviceps sp.	–	–	*Spartina foliosa*	USA: California	[127]
Metarhizium anisopliae (Metschn.) Sorokīn	Leaves	Saprobic	*Distichlis spicata*	Argentina: Buenos Aires	[47]
Hypocreaceae					
Cladobotryum sp.	Decaying leaves	Saprobic	*Juncus roemerianus*	USA: Florida	[43]
Gliocladium sp.	Senescent leaves	Saprobic	*Juncus roemerianus*	USA: Florida	[43]
Trichoderma citrinum (Pers.) Jaklitsch, W. Gams and Voglmayr	Leaves	Saprobic	*Spartina* sp.	Canada: Bay of Fundy	[48]
Trichoderma sp.	Decaying stems and leaf sheaths	Saprobic	*Phragmites australis*	China: Hong Kong	[41]
Trichoderma viride Pers.	Living, senescent, and decaying leaves	Saprobic	*Juncus roemerianus*	USA: Florida	[43]
Nectriaceae					
Calonectria sp.	–	–	*Elymus pungens*	UK	[38]
Fusarium fujikuroi Nirenberg	–	Saprobic	*Suaeda australis*	South Australia	[62]
Fusarium graminearum Schwabe	Living/decomposing leaf sheaths, stems	Saprobic	*Phragmites australis*	Netherlands: Zeeland	[39,40]
Fusarium heterosporum Nees and T. Nees	–	–	*Spartina maritima*	–	[128]

Table 1. *Cont.*

Taxon	Host Part	Life Mode	Hosts	Distribution	References
Fusarium incarnatum (Desm.) Sacc.	Leaves	Saprobic	*Distichlis spicata*	Argentina: Buenos Aires	[47]
Fusarium oxysporum Schltdl.	Leaves	Saprobic	*Distichlis spicata*	Argentina: Buenos Aires	[47]
	Leaves and roots	Saprobic	*Spartina* sp.	Canada: Bay of Fundy	[48]
Fusarium poae (Peck) Wollenw.	Leaves	Saprobic	*Distichlis spicata*	Argentina: Buenos Aires	[47]
Fusarium solani (Mart.) Sacc.	Decaying stems and leaf sheaths	Saprobic	*Phragmites australis*	China: Hong Kong	[41]
Fusarium spp.	Leaves	Saprobic	*Distichlis spicata*	Argentina: Buenos Aires	[47]
	Living, senescent, and decaying leaves	Saprobic	*Juncus roemerianus*	USA: Florida	[43]
	Living/decomposing leaf sheaths, stems	Saprobic	*Phragmites australis*	China: Hong Kong; Netherlands: Zeeland	[39–41]
	Leaf sheaths and blades, stem	Saprobic	*Spartina maritima*	Portugal: Algarve	[59]
Gibberella sp.	–	Saprobic	*Spartina alterniflora*	Argentina: Buenos Aires	[36]
Nectria sp.	Senescent and decaying leaves	Saprobic	*Juncus roemerianus*	USA: Florida	[43]
Tubercularia pulverulenta Speg.	–	–	*Sarcocornia perennis*	–	[35]
	–	–	*Salicornia europaea*	–	[35]
	–	Saprobic	Unidentified saltmarsh plants	USA: Mississippi	[58]
	–	–	*Sarcocornia fruticosa*	–	[35]
Tubercularia sp.	Decaying leaf blades	Saprobic	*Spartina alterniflora*	USA: Georgia	[56]
Volutella ciliata (Alb. and Schwein.) Fr.	Leaves	Saprobic	*Distichlis spicata*	Argentina: Buenos Aires	[47]
Sarocladiaceae					
Sarocladium implicatum (J.C. Gilman and E.V. Abbott) A. Giraldo, Gené and Guarro	Leaves	Saprobic	*Distichlis spicata*	Argentina: Buenos Aires	[47]
Sarocladium sp.	Decaying stems and leaf sheaths	Saprobic	*Phragmites australis*	China: Hong Kong	[41]
Stachybotryaceae					
Albifimbria verrucaria (Alb. and Schwein.) L. Lombard and Crous	Leaves	Saprobic	*Distichlis spicata*	Argentina: Buenos Aires	[47]
Paramyrothecium roridum (Tode) L. Lombard and Crous	Living, senescent, and decaying leaves	Saprobic	*Juncus roemerianus*	USA: Florida	[43]

Table 1. *Cont.*

Taxon	Host Part	Life Mode	Hosts	Distribution	References
Stachybotrys chartarum (Ehrenb.) S. Hughes	Senescent and decaying leaves	Saprobic	*Juncus roemerianus*	USA: Florida	[43]
Stachybotrys cylindrosporus C.N. Jensen	Decaying leaves	Saprobic	*Juncus roemerianus*	USA: Florida	[43]
Stachybotrys echinatus (Rivolta) G. Sm.	Senescent and decaying leaves	Saprobic	*Juncus roemerianus*	USA: Florida	[43]
Stachybotrys kampalensis Hansf.	Senescent leaves	Saprobic	*Juncus roemerianus*	USA: Florida	[43]
Stachybotrys nephrosporus Hansf.	Senescent and decaying leaves	Saprobic	*Juncus roemerianus*	USA: Florida	[43]
Stachybotrys spp.	Senescent and decaying leaves	Saprobic	*Juncus roemerianus*	USA: Florida	[43]
	Decaying stems and leaf sheaths	Saprobic	*Phragmites australis*	China: Hong Kong	[41]
	Decaying leaf blades	Saprobic	*Spartina alterniflora*	USA: Georgia	[56]
Striaticonidium cinctum (Corda) L. Lombard and Crous	Living/decomposing leaf sheaths	Saprobic	*Phragmites australis*	Netherlands: Zeeland	[39]
Xepicula jollymannii (N.C. Preston) L. Lombard and Crous	Senescent and decaying leaves	Saprobic	*Juncus roemerianus*	USA: Florida	[43]
Hypocreales genera *incertae sedis*					
Cephalosporium spp.	Dead leaves/culms	Saprobic	*Spartina alterniflora*	USA: Rhode Island	[61]
Lulworthiales					
Lulworthiaceae					
Cumulospora marina I. Schmidt	Dead culm	Saprobic	*Phragmites australis*	Iraq, Egypt, Germany, Thailand	[129]
	–	–	*Spartina* spp.	–	[32]
Halazoon fuscus (I. Schmidt) Abdel-Wahab, K.L. Pang, Nagah., Abdel-Aziz and E.B.G. Jones	Decaying rhizomes	Saprobic	*Phragmites australis*	France, Germany, Japan	[35,130]
	Rhizomes and culms	Saprobic	*Phragmites* sp.	Sweden	[87]
Halazoon melhae Abdel-Aziz, Abdel-Wahab and Nagahama	Decaying stem	Saprobic	*Phragmites australis*	Egypt: Port Said	[130]
Lulworthia floridana Meyers	–	Saprobic	*Spartina alterniflora*	USA: North Carolina, Rhode Island	[20,131]

Table 1. *Cont.*

Taxon	Host Part	Life Mode	Hosts	Distribution	References
Lulworthia medusa (Ellis and Everh.) Cribb and J.W. Cribb	–	–	*Elymus pungens*	UK	[38]
	–	Saprobic	*Spartina cynosuroides*	USA: New Jersey	[89,132]
	–		*Spartina* spp.	USA: New Jersey	[32,89]
	Stems	Saprobic	*Spartina townsendii*	UK: England (Wales); USA: Virginia, North Carolina, South Carolina, Florida, Texas	[38,49,71,72,89, 132–134]
Lulworthia spp.	–	–	*Elymus pungens*	–	[35]
	–	–	*Juncus roemerianus*	–	[35,36]
	Dead culms	Saprobic	*Spartina alterniflora*	Argentina: Buenos Aires; USA: Rhode Island, North Carolina	[35,36,61,73,74]
	–	–	*Spartina cynosuroides*	–	[35]
	–	Saprobic	*Spartina* sp.	Argentina: Buenos Aires; Canada; USA: Maine, North Carolina	[36]
	–	–	*Spartina townsendii*	–	[35]
	Leaf sheaths and blades, stem	Saprobic	*Spartina maritima*	Portugal: Alentejo, Lisbon, Algarve, Centro	[31,54,59,63]
Moleospora maritima Abdel-Wahab, Abdel-Aziz and Nagah.	Decayed stems	Saprobic	*Phragmites australis*	Egypt: Port Said	[130]
Magnaporthales					
Ceratosphaeriaceae					
Ceratosphaeria sp.	Senescent leaves	Saprobic	*Juncus roemerianus*	USA: Florida	[43]
Magnaporthaceae					
Buergenerula spartinae Kohlm. and R.V. Gessner	Lower stem and leaf sheath during the growth phase of the plant/living and dead; decaying leaf blades	Saprobic, parasitic	*Spartina alterniflora*	USA: Alabama, Rhode Island, Maine, New Hampshire, Connecticut, Mississippi, New Jersey, Virginia, North Carolina, Florida, Georgia	[20,35,36,55,56, 58,61,73,74,82,92]

Table 1. *Cont.*

Taxon	Host Part	Life Mode	Hosts	Distribution	References
	Leaves	Saprobic	*Spartina* spp.	Canada: Bay of Fundy; USA: South Carolina; UK	[32,35,36,48,65] this study
	Leaf sheaths and blades, stem	Saprobic	*Spartina maritima*	Portugal: Alentejo, Lisbon, Algarve, Centro	[31,54,59]
Gaeumannomyces sp.	Decaying stems and leaf sheaths	Saprobic	*Phragmites australis*	China: Hong Kong	[41]
Kohlmeyeriopsis medullaris (Kohlm., Volkm.-Kohlm. and O.E. Erikss.) Klaubauf, M.-H. Lebrun and Crous	Lower parts of senescent culms	Saprobic	*Juncus roemerianus*	USA: North Carolina	[97,135]
Utrechtiana roumeguerei (Cavara) Videira and Crous	Living/decomposing leaf blades and sheaths	Saprobic	*Phragmites australis*	Netherlands: Zeeland	[39,50]
Pseudohalonectriaceae					
Pseudohalonectria falcata Shearer	Decaying stems and leaf sheaths	Saprobic	*Phragmites australis*	China: Hong Kong	[41]
Pseudohalonectria halophila Kohlm. and Volkm.-Kohlm.	Fragments of leaves and culms in the wrack	Saprobic	*Juncus roemerianus*	USA: North Carolina	[105]
Meliolales					
Meliolaceae					
Meliola arundinis Pat.	–	–	*Phragmites australis*	Australia: Queensland	[62]
Microascales					
Halosphaeriaceae					
Aniptodera chesapeakensis Shearer and M.A. Mill.	Dead leaves	Saprobic	*Juncus roemerianus*	USA: North Carolina	[35]
	Decaying stems and leaf sheaths	Saprobic	*Phragmites australis*	China: Hong Kong	[41]
	–	–	*Spartina alterniflora*	–	[35]
	–	–	*Spartina* spp.	–	[32]
	Leaf sheaths and blades, stem	Saprobic	*Spartina maritima*	Portugal: Alentejo, Algarve, Centro	[59,63]
Aniptodera juncicola Volkm.-Kohlm. and Kohlm.	Dead standing culms of	Saprobic	*Juncus roemerianus*	India: Kerala, West Bengal, Tamil Nadu; USA: North Carolina	[52,136]
Aniptodera phragmiticola O. K. Poon et K. D. Hyde	Decaying stems and leaf sheaths	Saprobic	*Phragmites australis*	China: Hong Kong	[41]

Table 1. *Cont.*

Taxon	Host Part	Life Mode	Hosts	Distribution	References
Ceriosporopsis halima Linde	–	–	*Arundo donax*	–	[35]
	Submerged seeds	Saprobic	*Spartina alterniflora*	USA	[137]
	–	–	*Spartina* spp.	–	[32]
	–	–	*Spartina townsendii*	UK	[35,38]
	Stem	Saprobic	*Spartina maritima*	Portugal: Alentejo	[63]
Cirrenalia macrocephala (Kohlm.) Meyers and R.T. Moore	–	–	*Ammophila arenaria*	–	[35]
	Decaying culms	Saprobic	*Juncus roemerianus*	USA: Florida	[43]
	Decomposing culms, submerged seeds	Saprobic	*Spartina alterniflora*	USA: Rhode Island	[35,61,137]
	–	–	*Spartina* spp.	–	[32]
	Stem	Saprobic	*Spartina maritima*	Portugal: Alentejo	[63]
Cirrenalia pseudomacrocephala Kohlm.	Senescent leaves	Saprobic	*Juncus roemerianus*	USA: Florida	[43]
Corollospora maritima Werderm.	Submerged seeds, decomposing culms	Saprobic	*Spartina alterniflora*	USA: Rhode Island	[20,35,61,137]
	–	–	*Spartina* spp.	–	[32]
	Stem	Saprobic	*Spartina maritima*	Portugal: Alentejo	[63]
	–	Saprobic	Unidentified saltmarsh plants	USA: Mississippi	[58]
Corollospora ramulosa (Meyers and Kohlm.) E.B.G. Jones and Abdel-Wahab	–	Saprobic	Unidentified saltmarsh plants	USA: Mississippi	[58]
	–	Saprobic	*Zostera marina*	USA: North Carolina	[74]
Haligena elaterophora Kohlm.	–	–	*Spartina alterniflora*	–	[35]
	–	–	*Spartina tonwsendii*	UK	[38]
	–	–	*Spartina* spp.	–	[32]
Halosarpheia culmiperda Kohlm., Volkm.-Kohlm. and O.E. Erikss.	Lower parts of senescent culms	Saprobic	*Juncus roemerianus*	USA: North Carolina	[97]
Halosarpheia sp.	Stem	Saprobic	*Spartina maritima*	Portugal: Alentejo	[63]
Halosarpheia viscosa I. Schmidt ex Shearer and J.L. Crane	Decaying leaf blades	Saprobic	*Spartina alterniflora*	USA: Georgia	[56]
	–	Saprobic	*Spartina maritima*	Portugal: Lisbon	[54]
Halosphaeria appendiculata Linder	–	–	*Arundo donax*	–	[35]
Halosphaeria sp.	Submerged seeds	Saprobic	*Spartina alterniflora*	USA	[137]
Lautisporopsis circumvestita (Kohlm.) E.B.G. Jones, Yusoff and S.T. Moss	–	–	*Arundo donax*	–	[35]

Table 1. *Cont.*

Taxon	Host Part	Life Mode	Hosts	Distribution	References
Lignincola laevis Höhnk	–	–	*Elymus pungens*	–	[35]
	–	Saprobic	*Spartina* spp.	USA: North Carolina	[32,138]
	–	–	*Spartina townsendii*	–	[35]
	Stem	Saprobic	*Spartina maritima*	Portugal: Alentejo	[63]
Magnisphaera spartinae (E.B.G. Jones) J. Campb., J.L. Anderson and Shearer	–	–	*Elymus farctus*	–	[35]
	–	–	*Elymus pungens*	–	[35]
	Living/decomposing stems	Saprobic	*Phragmites australis*	Netherlands: Zeeland	[40]
	–	Saprobic	*Spartina alterniflora*	USA: Rhode Island	[20,35,61]
	–		*Spartina* spp.		[32]
	–	Saprobic	*Spartina patens*	USA: Rhode Island	[36]
	Stem	Saprobic	*Spartina townsendii*	UK: Wales	[35,139]
	–		*Typha* sp.		[35]
Nais inornata Kohlm.	Decomposing culms	Saprobic	*Spartina alterniflora*	USA: Rhode Island	[20,35,61]
			Spartina spp.		[32]
Natantispora unipolaris K.L. Pang, S.Y. Guo and E.B.G. Jones	Dead stem	Saprobic	*Phragmites australis*	Taiwan: Nankunshen	[140]
Natantispora retorquens (Shearer and J.L. Crane) J. Campb., J.L. Anderson and Shearer	Leaf sheaths and blades, stem	Saprobic	*Spartina maritima*	Portugal: Alentejo, Lisbon, Algarve, Centro	[31,54,59,63]
Oceanitis unicaudata (E.B.G. Jones and Camp.-Als.) J. Dupont and E.B.G. Jones	Decaying stems and leaf sheaths	Saprobic	*Phragmites australis*	China: Hong Kong	[41]
	Stem	Saprobic	*Spartina maritima*	Portugal: Alentejo	[63]
Panorbis viscosus (I. Schmidt) J. Campb., J.L. Anderson and Shearer	Leaf sheaths and blades, stem	Saprobic	*Spartina maritima*	Portugal: Alentejo, Algarve	[59,63]
Remispora hamata (Höhnk) Kohlm.	–	–	*Elymus pungens*	UK	[35,38]
	Senescent and decaying leaves	Saprobic	*Juncus roemerianus*	USA: Florida	[43]
	Living/decomposing leaf blades and sheaths, stems	Saprobic	*Phragmites australis*	Netherlands: Zeeland	[39,40,50]

Table 1. *Cont.*

Taxon	Host Part	Life Mode	Hosts	Distribution	References
	–	Saprobic	*Phragmites* sp.	Sweden	[87]
	Dead leaves	Saprobic	*Spartina alterniflora*	USA: Rhode Island, Maine, Florida	[20,35,36,61,73]
	–	Saprobic	*Spartina patens*	USA: Rhode Island	[36]
	–	Saprobic	*Spartina* sp.	USA: North Carolina; Argentina: Buenos Aires	[36,138]
	–	–	*Spartina townsendii*	–	[35]
	–	–	*Typha* sp.	–	[35]
Remispora trullifera Kohlm.	Leaf sheaths and blades, stem	Saprobic	*Spartina maritima*	Portugal: Centro	[59]
Tirispora unicaudata E.B.G. Jones and Vrijmoed	Stem	Saprobic	*Spartina maritima*	Portugal: Alentejo	[63]
Microascaceae					
Scopulariopsis spp.	Living, senescent, and decaying leaves	Saprobic	*Juncus roemerianus*	USA: Florida	[43]
Myrmecridiales					
Myrmecridiaceae					
Myrmecridium schulzeri (Sacc.) Arzanlou, W. Gams and Crous	Leaves	Saprobic	*Distichlis spicata*	Argentina: Buenos Aires	[47]
Ophiostomatales					
Ophiostomataceae					
Sporothrix sp.	Senescent leaves	Saprobic	*Juncus roemerianus*	USA: Florida	[43]
Phomatosporales					
Phomatosporaceae					
Phomatospora bellaminuta Kohlm., Volkm.-Kohlm. and O.E. Erikss.	Lower parts of senescent culms	Saprobic	*Juncus roemerianus*	USA: North Carolina	[116]
Phomatospora berkeleyi Sacc.	Living/decomposing leaf blades and sheaths, stems	Saprobic	*Phragmites australis*	Netherlands: Zeeland	[39,40,50]
Phomatospora dinemasporium J. Webster	Decaying stems and leaf sheaths, stems	Saprobic	*Phragmites australis*	China: Hong Kong; Netherlands: Zeeland	[40,41]
	Dead leaves	Saprobic	*Phragmites* sp.	South Australia	[62]
	–	–	*Spartina townsendii*	UK	[38]
Phomatospora phragmiticola Poon and K.D. Hyde	Decaying stems and leaf sheaths	Saprobic	*Phragmites australis*	China: Hong Kong	[41]

Table 1. *Cont.*

Taxon	Host Part	Life Mode	Hosts	Distribution	References
Phomatospora spp.	Senescent and decaying leaves	Saprobic	*Juncus roemerianus*	USA: Florida	[43]
	Living/decomposing leaf sheaths, stems	Saprobic	*Phragmites australis*	Netherlands: Zeeland	[39,40]
Phyllachorales					
Phyllachoraceae					
Phyllachora graminis (Pers.) Fuckel	–	–	*Elymus pungens*	UK	[38]
	–	Saprobic, pathogenic	*Spartina alterniflora*	USA: Massachusetts	[44]
	–	–	*Spartina cynosuroides*	–	[68]
Phyllachora cynodontis Niessl.	–	Saprobic, pathogenic	*Spartina alterniflora*	USA	[68]
	–	Saprobic, pathogenic	*Spartina foliosa*	USA: California	[44,112,141]
Phyllachora paludicola Kohlm. and Volkm.-Kohlm.	Dead leaves (lower half of standing culms)	Saprobic	*Spartina alterniflora*	USA: Florida, Georgia, North Carolina, Maryland, Delaware	[142]
Phyllachora sylvatica Sacc. and Speg.	–	Saprobic	*Spartina patens*	USA: South Carolina	[141]
Savoryellales					
Savoryellaceae					
Savoryella paucispora (Cribb and J.W. Cribb) J. Koch	–	–	*Elymus pungens*	–	[35]
	–	–	*Juncus roemerianus*	–	[35]
	–	–	*Spartina alterniflora*	–	[35]
	–	–	*Spartina* sp.	–	[35]
	–	–	*Spartina townsendii*	–	[35]
Sordariales					
Chaetomiaceae					
Achaetomium sp.	Decaying leaves	Saprobic	*Juncus roemerianus*	USA: Florida	[43]
Chaetomium elatum Kunze	–	–	*Elymus pungens*	UK	[38]
	–	–	*Puccinellia maritima*	UK	[38]
	–	–	*Spartina townsendii*	UK	[38]
	–	–			
Chaetomium globosum Kunze	–	–	*Elymus pungens*	UK	[38]
	Decaying stems and leaf sheaths	Saprobic	*Phragmites australis*	China: Hong Kong	[41]

Table 1. *Cont.*

Taxon	Host Part	Life Mode	Hosts	Distribution	References
	–	–	*Puccinellia maritima*	UK	[38]
	–	–	*Spartina townsendii*	UK	[38]
Chaetomium spirale Zopf	–	–	*Elymus pungens*	UK	[38]
Chaetomium thermophilum La Touche	–	–	*Elymus pungens*	UK	[38]
	–	–	*Puccinellia maritima*	UK	[38]
	–	–	*Spartina townsendii*	UK	[38]
Chaetomium sp.	Stem	Saprobic	*Typha* sp.	UK	This study
Corynascus sepedonium (C.W. Emmons) Arx	–	–	*Elymus pungens*	UK	[38]
	–	–	*Puccinellia maritima*	UK	[38]
	–	–	*Spartina townsendii*	UK	[38]
Dichotomopilus funicola (Cooke) X.Wei Wang and Samson	–	–	*Elymus pungens*	UK	[38]
	–	–	*Spartina alterniflora*	USA: Rhode Island	[61]
	–	–	*Spartina townsendii*	UK	[38]
Dichotomopilus indicus (Corda) X.Wei Wang and Samson	–	–	*Elymus pungens*	UK	[38]
Humicola sp.	Living, senescent, and decaying leaves	Saprobic	*Juncus roemerianus*	USA: Florida	[43]
	Decaying stems and leaf sheaths	Saprobic	*Phragmites australis*	China: Hong Kong	[41]
Thermothielavioides terrestris (Apinis) X. Wei Wang and Houbraken	–	–	*Elymus pungens*	UK	[38]
	–	–	*Puccinellia maritima*	UK	[38]
Trichocladium constrictum I. Schmidt	Stem	Saprobic	*Spartina maritima*	Portugal: Alentejo	[63]
Trichocladium crispatum (Fuckel) X. Wei Wang and Houbraken	–	–	*Elymus pungens*	UK	[38]
	–	–	*Spartina townsendii*	UK	[38]
Lasiosphaeriaceae	**–**	**–**			
Schizothecium hispidulum (Speg.) N. Lundq.	Living/decomposing leaf sheaths	Saprobic	*Phragmites australis*	Netherlands: Zeeland	[39]
Zopfiella latipes (N. Lundq.) Malloch and Cain	Decaying stems and leaf sheaths	Saprobic	*Phragmites australis*	China: Hong Kong	[41]
Sordariaceae					
Neurospora calospora (Mouton) Dania García, Stchigel and Guarro	–	–	*Elymus pungens*	UK	[38]

Table 1. *Cont.*

Taxon	Host Part	Life Mode	Hosts	Distribution	References
Sordaria fimicola (Roberge ex Desm.) Ces. and De Not.	–	–	*Elymus pungens*	UK	[38]
	Leaves	Saprobic	*Distichlis spicata*	Argentina: Buenos Aires	[47]
	–	–	*Puccinellia maritima*	UK	[38]
	–	–	*Spartina townsendii*	UK	[38]
Sordariomycetes families *incertae sedis*					
Koorchaloma galateae Kohlm. and Volkm.-Kohlm.	Senescent culms	Saprobic	*Juncus roemerianus*	USA: North Carolina	[117]
Koorchaloma spartinicola V.V. Sarma, S.Y. Newell and K.D. Hyde	Decaying leaf blades	Saprobic	*Spartina alterniflora*	USA: Georgia	[56]
Koorchaloma sp.	Decaying leaf blades	Saprobic	*Spartina alterniflora*	USA: Georgia	[56]
Lautospora simillima Kohlm., Volkm.-Kohlm. and O.E. Erikss.	Lower parts of senescent, soft culms	Saprobic	*Juncus roemerianus*	USA: North Carolina	[78]
Sordariomycetes genera *incertae sedis*					
Aquamarina speciosa Kohlm., Volkm.-Kohlm. and O.E. Erikss.	Senescent culms		*Juncus roemerianus*	USA: Georgia, North Carolina, Virginia	[77]
Aropsiclus junci (Kohlm. and Volkm.-Kohlm.) Kohlm. and Volkm.-Kohlm.	Senescent culms	Saprobic	*Juncus roemerianus*	USA: North Carolina	[143]
Zalerion maritima (Linder) Anastasiou	Basal area of the sheath	Saprobic	*Spartina densiflora*	Argentina: Buenos Aires	[64]
	–	–	*Spartina* spp.	–	[32]
Ellisembia sp.	Decaying stems and leaf sheaths	Saprobic	*Phragmites australis*	China: Hong Kong	[41]
Torpedosporales					
Juncigenaceae					
Juncigena adarca Kohlm., Volkm.-Kohlm. and O.E. Erikss.	Senescent leaves	Saprobic	*Juncus roemerianus*	USA: North Carolina	[76]
Moheitospora adarca (Kohlm., Volkm.-Kohlm. and O.E. Erikss.) Abdel-Wahab, Abdel-Aziz and Nagah	Stems	Saprobic	*Juncus roemerianus*	USA	[130]
Moheitospora fruticosae Abdel-Wahab, Abdel-Aziz and Nagah.	Decayed stems	Saprobic	*Suaeda vermiculata*	Egypt: Alexandria	[130]
Torpedospora radiata Meyers	–	Saprobic	Unidentified saltmarsh plants	USA: Mississippi	[58]
Tracyllalales					

Table 1. *Cont.*

Taxon	Host Part	Life Mode	Hosts	Distribution	References
Tracyllaceae					
Tracylla spartinae (Peck) Tassi	–	Saprobic, pathogenic	*Spartina patens*	USA: Mississippi	[44,68]
Xylariales					
Diatrypaceae					
Cryptovalsa suaedicola Spooner	Dead twigs	Saprobic	*Suaeda vermiculata*	UK: Great Britain	[144]
Halocryptovalsa salicorniae Dayar. and K.D. Hyde	Dead stem	Saprobic	*Salicornia* sp.	Thailand: Prachuap Khiri Khan	[145]
Xylariaceae					
Anthostomella atroalba Kohlm., Volkm.-Kohlm. and O.E. Erikss.	Senescent leaves	Saprobic	*Juncus roemerianus*	USA: North Carolina	[60]
Anthostomella lugubris (Roberge ex Desm.) Sacc.	–	–	*Elymus pungens*	UK	[38]
Anthostomella phaeosticta (Berk.) Sacc.	–	–	*Elymus pungens*	UK	[38]
Anthostomella poecila Kohlm., Volkm.-Kohlm. and O.E. Erikss.	Lower and upper parts of senescent culms, decaying leaves	Saprobic	*Juncus roemerianus*	USA: Alabama, Mississippi, North Carolina	[55,58,116]
Anthostomella punctulata (Roberge ex Desm.) Sacc.	Living/decomposing leaf blades and sheaths	Saprobic	*Phragmites australis*	Netherlands: Zeeland	[39,50]
Anthostomella semitecta Kohlm., Volkm.-Kohlm. and O.E. Erikss.	Senescent culms	–	*Juncus roemerianus*	USA: North Carolina	[116]
Anthostomella spissitecta Kohlm. and Volkm.-Kohlm.	Leaf sheaths of senescent culms	Saprobic	*Spartina alterniflora, S. densiflora.*	USA: Connecticut, Florida, North Carolina, Rhode Island; Argentina: Buenos Aires	[32]
	–	–	*Spartina* sp.	–	[32]
	Leaf sheaths and blades, stem	Saprobic	*Spartina maritima*	Portugal: Algarve	[59]
Anthostomella spp.	–	–	*Elymus pungens*	UK	[38]
	–	Saprobic	*Spartina alterniflora*	USA: Connecticut, Florida, North Carolina, Rhode Island; Argentina	[36,61]
	–	–	*Spartina townsendii*	UK	[38]
Anthostomella torosa Kohlm. and Volkm.-Kohlm.	Senescent culms (restricted to short culms)	Saprobic	*Juncus roemerianus*	USA: North Carolina	[32]
Geniculosporium sp.	Living, senescent, and decaying leaves	Saprobic	*Juncus roemerianus*	USA: Florida	[43]

Table 1. *Cont.*

Taxon	Host Part	Life Mode	Hosts	Distribution	References
Rosellinia sp.	Dead leaves/culms	Saprobic	*Spartina alterniflora*	USA: Rhode Island	[61]
Virgaria nigra (Link) Nees	Senescent leaves	Saprobic	*Juncus roemerianus*	USA: Florida	[43]
Zygosporiaceae					
Zygosporium gibbum (Sacc., M. Rousseau and E. Bommer) S. Hughes	Decaying leaves	Saprobic	*Juncus roemerianus*	USA: Florida	[43]
Zygosporium masonii S. Hughes	Decaying leaves	Saprobic	*Juncus roemerianus*	USA: Florida	[43]
Zygosporium sp.	Decaying leaves	Saprobic	*Juncus roemerianus*	USA: Florida	[43]
Xylariales genera *incertae sedis*					
Circinotrichum maculiforme Nees	Decaying leaves	Saprobic	*Juncus roemerianus*	USA: Florida	[43]
Xylariomycetidae family *incertae sedis*					
Cainiaceae					
Atrotorquata lineata Kohlm. and Volkm.-Kohlm.	Senescent culms	Saprobic	*Juncus roemerianus*	USA: North Carolina	[104]
		Saprobic	Unidentified saltmarsh plant	USA: Mississippi	[58]
Ascomycota genera *incertae sedis*					
Asteromyces cruciatus C. Moreau and Moreau ex Hennebert	–	–	*Agropyron* sp.	–	[35]
	–	–	*Ammophila arenaria*	–	[35]
	–	–	*Spartina* spp.	–	[32,35]
	–	Saprobic	*Zostera* sp.	USA: California	[74]
Cremasteria cymatilis Meyers and R.T. Moore Nomen dubium	Senescent leaves	Saprobic	*Juncus roemerianus*	USA: Florida	[43]
Cytoplacosphaeria phragmiticola Poon and K.D. Hyde	Decaying stems and leaf sheaths	Saprobic	*Phragmites australis*	China: Hong Kong	[41]
Cytoplacosphaeria rimosa (Oudem.) Petr.	Living/decomposing leaf sheaths, stems	Saprobic	*Phragmites australis*	Netherlands: Zeeland	[39,40]
Cytosporina sp.	Living, senescent, and decaying leaves	Saprobic	*Juncus roemerianus*	USA: Florida	[43]
Didymosamarospora euryhalina T.W. Johnson and H.S. Gold	Culms	Saprobic	*Juncus roemerianus*	USA: North Carolina	[146]
Haplobasidion lelebae Sawada ex M.B. Ellis	Living, senescent, and decaying leaves	Saprobic	*Juncus roemerianus*	USA: Florida	[43]
Hymenopsis chlorothrix Kohlm. and Volkm.-Kohlm.	Senescent culms	Saprobic	*Juncus roemerianus*	USA: North Carolina	[147]
Hyphopolynema juncatile Kohlm. and Volkm.-Kohlm.	Senescent leaves	Saprobic	*Juncus roemerianus*	USA: North Carolina	[148]

Table 1. *Cont.*

Taxon	Host Part	Life Mode	Hosts	Distribution	References
Kolletes undulatus Kohlm. and Volkm.-Kohlm.	Senescent leaves and culms	Saprobic	*Juncus roemerianus*	USA: North Carolina	[105]
Minimidochium parvum Cabello, Aramb. and Cazau	Leaves	Saprobic	*Distichlis spicata*	Argentina: Buenos Aires	[47]
Monodictys pelagica (T. Johnson) E.B.G. Jones	–	–	*Juncus* sp.	–	[35]
	Decomposing culms	Saprobic	*Spartina alterniflora*	USA: Rhode Island	[20,35,61,73]
	–	–	*Spartina* spp.	–	[32]
Neottiospora sp.	Living, senescent, and decaying leaves	Saprobic	*Juncus roemerianus*	USA: Florida	[43]
Octopodotus stupendus Kohlm. and Volkm.-Kohlm.	Dead leaves (lower half of standing culms)	Saprobic	*Spartina alterniflora*	USA: North Carolina	[142]
Pycnodallia dupla Kohlm. and Volkm.-Kohlm.	Senescent inflorescences (involucral leaves and branchlets)	Saprobic	*Juncus roemerianus*	USA: North Carolina	[147]
Sphaeronaema sp.	Senescent leaves	Saprobic	*Juncus roemerianus*	USA: Florida	[43]
Stauronema sp.	Decaying stems and leaf sheaths	Saprobic	*Phragmites australis*	China: Hong Kong	[41]
Tetranacriella papillata Kohlm. and Volkm.-Kohlm.	Senescent leaves	Saprobic	*Juncus roemerianus*	USA: North Carolina	[117]
Tetranacrium sp.	Decaying stems and leaf sheaths	Saprobic	*Phragmites australis*	China: Hong Kong	[41]
Zythia spp.	Living, senescent, and decaying leaves	Saprobic	*Juncus roemerianus*	USA: Florida	[43]
Psammina sp.	Senescent leaves	Saprobic	*Juncus roemerianus*	USA: Florida	[43]
BASIDIOMYCOTA					
AGARICOMYCETES					
Agaricales					
Niaceae					
Merismodes bresadolae (Grelet) Singer	Living/decomposing stems	Saprobic	*Phragmites australis*	Netherlands: Zeeland	[40]
Nia globispora Barata and Basilio	Stem	Saprobic	*Spartina maritima*	Portugal: Alentejo	[63]
Nia vibrissa R.T. Moore and Meyers	Old stem	Saprobic	*Spartina alterniflora*	USA: North Carolina	[35,149]
	–	Saprobic	*Spartina* spp.	USA: North Carolina	[32,150]
	Stem	Saprobic	*Spartina maritima*	Portugal: Alentejo	[63]

Table 1. *Cont.*

Taxon	Host Part	Life Mode	Hosts	Distribution	References
AGARICOSTILBOMYCETES					
Agaricostilbales					
Chionosphaeraceae					
Stilbum sp.	Decaying leaves	Saprobic	*Juncus roemerianus*	USA: Florida	[43]
BARTHELETIOMYCETES					
Sebacinales					
Sebacinaceae					
Chaetospermum camelliae Agnihothr.	Decaying stems and leaf sheaths	Saprobic	*Phragmites australis*	China: Hong Kong	[41]
MICROBOTRYOMYCETES					
Sporidiobolales					
Sporidiobolaceae					
Sporobolomyces roseus Kluyver and C.B. Niel	Leaves	Saprobic	*Spartina* sp.	Canada: Bay of Fundy	[48]
Sporobolomyces spp.	Living/decomposing leaf blades and sheaths	Saprobic	*Phragmites australis*	Netherlands: Zeeland	[39,50]
PUCCINIOMYCETES					
Pucciniales					
Pucciniaceae					
Puccinia distichlidis Ellis and Everh.	–	–	*Distichlis spicata*	USA	[151]
Puccinia magnusiana Körn.	Living/decomposing leaf blades and sheaths	Saprobic	*Phragmites australis*	Netherlands: Zeeland	[39,50]
Puccinia phragmitis (Schumach.) Tul.	Living/decomposing leaf blades and sheaths	Saprobic	*Phragmites australis*	Netherlands: Zeeland	[39,50]
Puccinia sparganioidis Ellis and Barthol.	–	Saprobic, parasitic	*Spartina alterniflora*	USA: Maine, New Hampshire, Massachusetts, Rhode Island, Delaware, Virginia, North Carolina, Florida, Mississippi	[36,44,68,73,152]
	–	–	*Spartina cynosuroides*	USA: New Jersey, Delaware, Maryland, South Carolina, Florida, Louisiana	[44,68,153]

Table 1. *Cont.*

Taxon	Host Part	Life Mode	Hosts	Distribution	References
	–	Saprobic, pathogenic	*Spartina patens*	USA: Connecticut, Maryland, New Jersey, New York	[44,68,153]
Uromyces acuminatus Arthur	–	Saprobic, pathogenic	*Spartina alterniflora*	USA: Maine, New Hampshire, Massachusetts, Connecticut, New York, New Jersey, Delaware, Maryland, Florida	[44,68,152]
	–	Saprobic, pathogenic	*Spartina cynosuroides*	USA: Florida	[44,68,153]
	–	Saprobic	*Spartina patens*	USA: Connecticut, Delaware, Florida, Maine, Maryland, Massachusetts, New Hampshire, New Jersey,	[44,68]
Uromyces argutus F. Kern	–	Saprobic, pathogenic	*Spartina alterniflora*	France; USA: Florida	[44,68,152]
Uromyces salicorniae (DC.) de Bary	–	–	*Salicornia* sp.	South Australia	[95]
Pucciniales genera *incertae sedis*					
Aecidium suaedae Thüm.	Leaves	–	*Suaeda verae*	Egypt	[154]
TREMELLOMYCETES					
Tremellales					
Tremellaceae					
Tremella spicifera Van Ryck., Van de Put and P. Roberts	Living/decomposing leaf sheaths and stems	Saprobic	*Phragmites australis*	Netherlands: Zeeland	[39,40]
USTILAGINOMYCETES					
Ustilaginales					
Ustilaginaceae					
Tranzscheliella distichlidis (McAlpine) Vánky	–	Pathogenic	*Distichlis spicata*	Australia: Victoria	[155]
Ustilaginales genera *incertae sedis*					
Parvulago marina (Durieu) R. Bauer, M. Lutz, Piątek, Vánky and Oberw.	–	–	*Eleocharis parvula*	Finland, France, Germany, UK, Norway, Sweden	[156]
Urocystidales					
Urocystidaceae					

Table 1. *Cont.*

Taxon	Host Part	Life Mode	Hosts	Distribution	References
Flamingomyces ruppiae (Feldmann) R. Bauer, M. Lutz, Piątek, Vánky and Oberw.	–	Parasitic	*Ruppia maritima*	France	[156]
MUCOROMYCOTA					
MUCOROMYCETES					
Mucorales					
Choanephoraceae					
Blakeslea trispora Thaxt.	Senescent and decaying leaves	Saprobic	*Juncus roemerianus*	USA: Florida	[43]
Mucoraceae					
Mucor sp.	Senescent leaves	Saprobic	*Juncus roemerianus*	USA: Florida	[43]
	Roots	Saprobic	*Spartina* sp.	Canada: Bay of Fundy	[48]
Rhizopodaceae					
Rhizopus stolonifer (Ehrenb.) Vuill.	Stems	Saprobic	*Spartina townsendii*	UK: England	[49]
Syncephalastraceae					
Syncephalastrum racemosum Cohn ex J. Schröt.	Living and senescent leaves	Saprobic	*Juncus roemerianus*	USA: Florida	[43]

2. Taxonomic Classification of Salt Marsh Fungi

2.1. Phyla

Calado and Barata [34] documented 332 taxa associated with *Juncus roemerianus*, *Phragmites australis*, and *Spartina* spp. In this review, we list 486 taxa that belong to three phyla (Ascomycota, Basidiomycota, Mucoromoycota) (Table 1, Figure 3) and selected species are illustrated in Figure 4. Ascomycota dominates the taxa from salt marsh ecosystems, accounting for 95.27% (463 taxa). Nineteen species in twelve genera (*Aecidium*, *Chaetospermum*, *Falmingomyces*, *Merismodes*, *Nia*, *Parvulago*, *Puccinia*, *Sporobolomyces*, *Stilbum*, *Tranzscheliella*, *Tremella*, *Uromyces*) belong to Basidiomycota (3.91%), while Mucoromycota account for 0.82% (four species) of the salt marsh fungi.

2.2. Class

Salt marsh fungi are distributed into 17 classes (Table 1, Figure 5). Dothideomycetes has the highest number of taxa, which comprises 47.12% (229 taxa), followed by Sordariomycetes with 167 taxa (34.36%). Twenty-one species (in 20 genera) can be referred to as Ascomycota genera *incertae sedis*. The Ascomycetes with the least number of species include Leotiomycetes (21 species, 4.32%), Eurotiomycetes (16 species, 3.29%), Orbiliomycetes (3 species, 0.62%), Saccharomycetes (3 species, 0.62%), Lecanoromycetes (2 species, 0.41%), and Pezizomycetes (1 species, 0.21%).

Seven classes represent the Basidiomycota (Figure 5). Puccinomycetes has the highest number of taxa documented (eight species, three genera) followed by Agaricomycetes (three species, two genera), Ustilaginomycetes (three species, three genera), and Microbotryomycetes (two taxa, one genus). Agaricostilbomycetes, Bartheletiomycetes, and Tremellomycetes have one representative taxon each.

The Mucoromoycota account for the taxa *Blakeslea trispora*, *Mucor* sp., *Rhizopus stolonifera*, and *Syncephalastrum racemosum* [43,48,49].

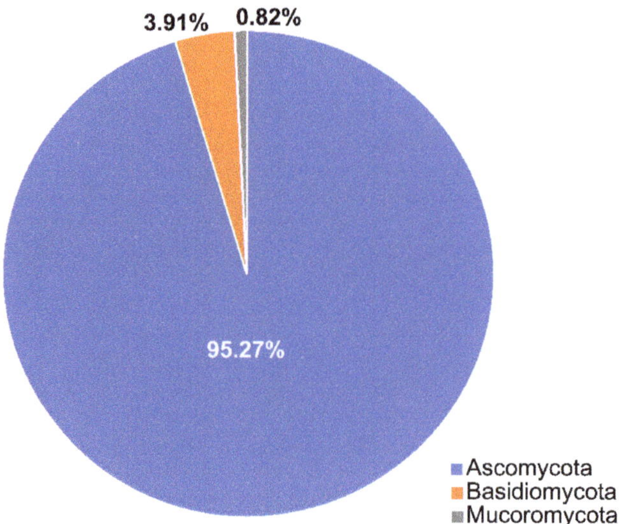

Figure 3. The distribution of salt marsh fungi among three fungal phyla.

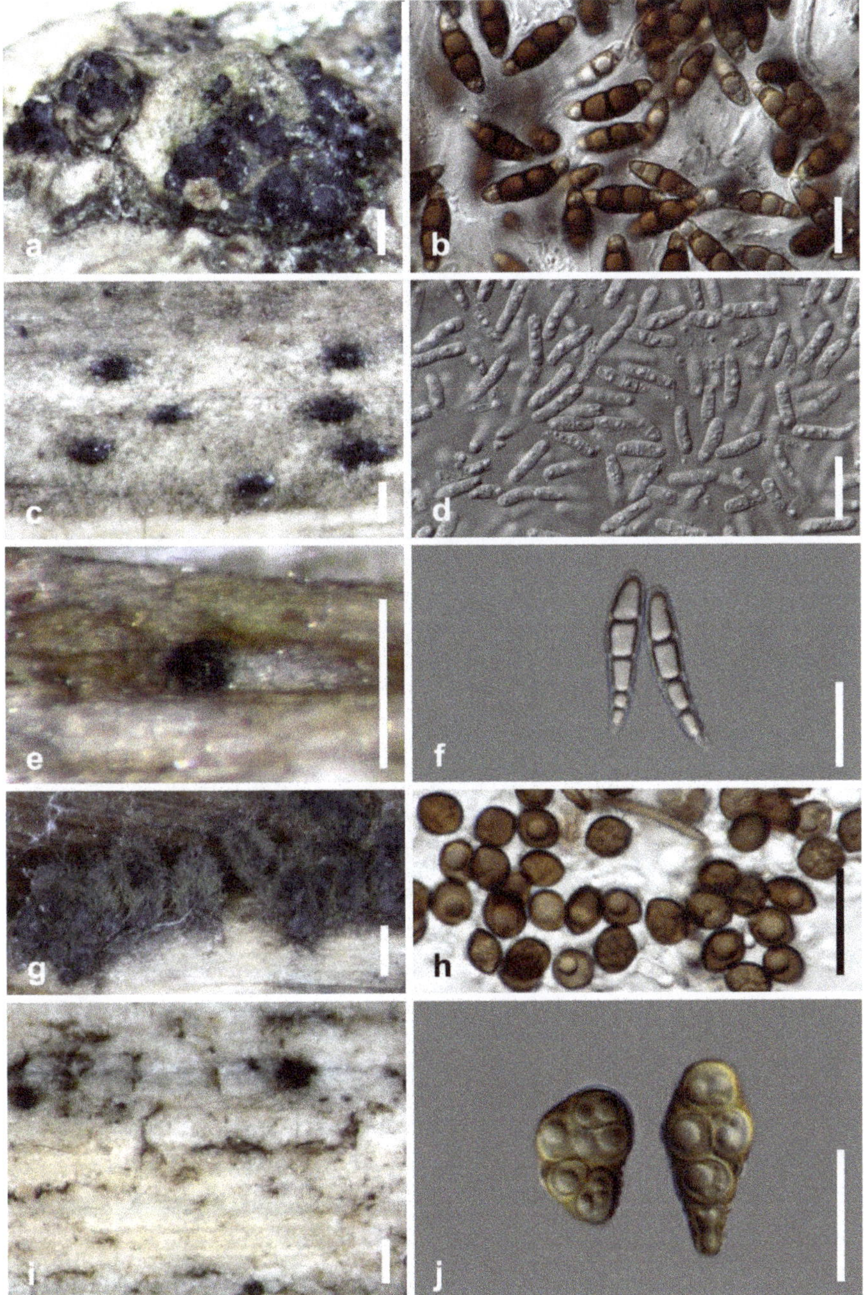

Figure 4. Salt marsh fungi. (**a**,**b**) *Halobyssothecium obiones* from *Atriplex portulacoides*; (**c**,**d**) *Halobyssothecium phragmites* from culms of *Phragmites* sp.; (**e**,**f**) *Buergenerula spartinae* from culms of *Spartina* sp.; (**g**,**h**) *Chaetomium* sp. from stem of *Typha* sp.; (**i**,**j**) *Alternaria* sp. from culms of *Spartina* sp. Scale bars: (**a**,**g**) = 500 μm; (**b**,**d**,**f**,**h**,**j**) = 20 μm; (**c**,**i**) = 200 μm; (**e**) = 100 μm.

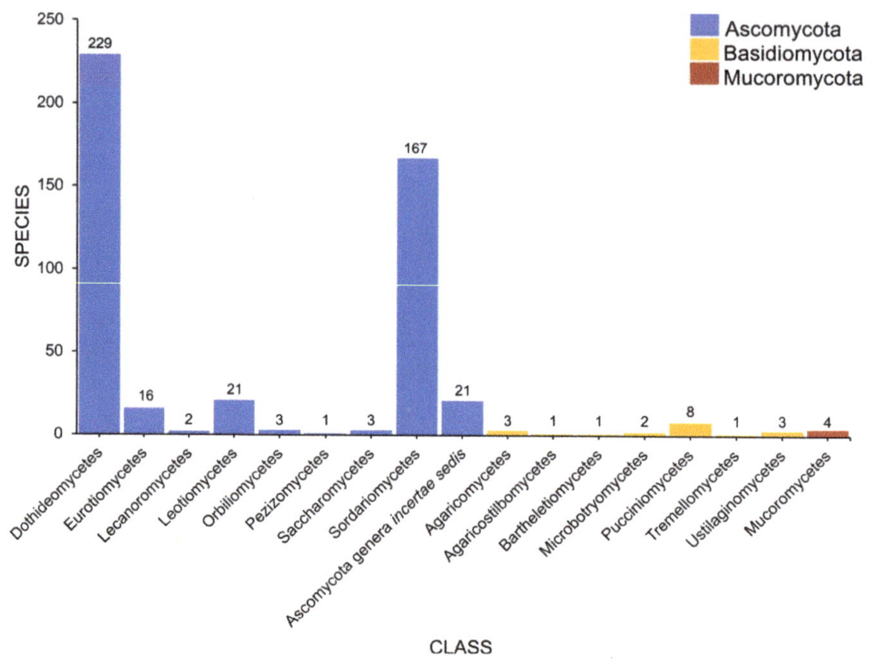

Figure 5. The distribution of salt marsh fungi in different fungal classes.

2.3. Orders

Salt marsh fungi recorded from different halophytes were distributed among 48 orders (Table 1, Figure 6). The Pleosporales is the largest order, with 178 taxa recorded followed by Hypocreales (41), Microascales (26), Capnodiales (22), Helotiales (18), Xylariales (17), Sordariales (16), Amphisphaeriales (15), and Eurotiales (13). The remaining 41 orders have less than 10 species (Table 1, Figure 5). Forty-two taxa belong to *incertae sedis* (Ascomycota genera *incertae sedis*: 21; Dothideomycetes families *incertae* sedis: 11; Sordariomycetes families *incertae sedis*: 9; Xylariomycetidae family *incertae sedis:* 1).

2.4. Families

A total of 108 families and 12 *incertae sedis* were recorded to be associated with salt marsh fungi (Table 1, Figure 7). Phaeosphaeriaceae and Pleosporaceae account for the largest families with 34 and 31 taxa recorded, respectively. Thirteen families have ten or more than taxa and include Nectriaceae (25), Halosphaeriaceae (25), Didymellaceae (17), Mycosphaerellaceae (14), Lentitheciaceae (13), Massarinaceae (13), Chaetomiaceae (12), Xylariaceae (11), Didymosphaeriaceae (10), Leptosphaeriaceae (10), and Aspergillaceae (10). The remaining 95 families have less than ten species recorded. Forty-four taxa are placed as *incertae sedis*, wherein 21 of these belong to Ascomycota genera *incertae sedis*.

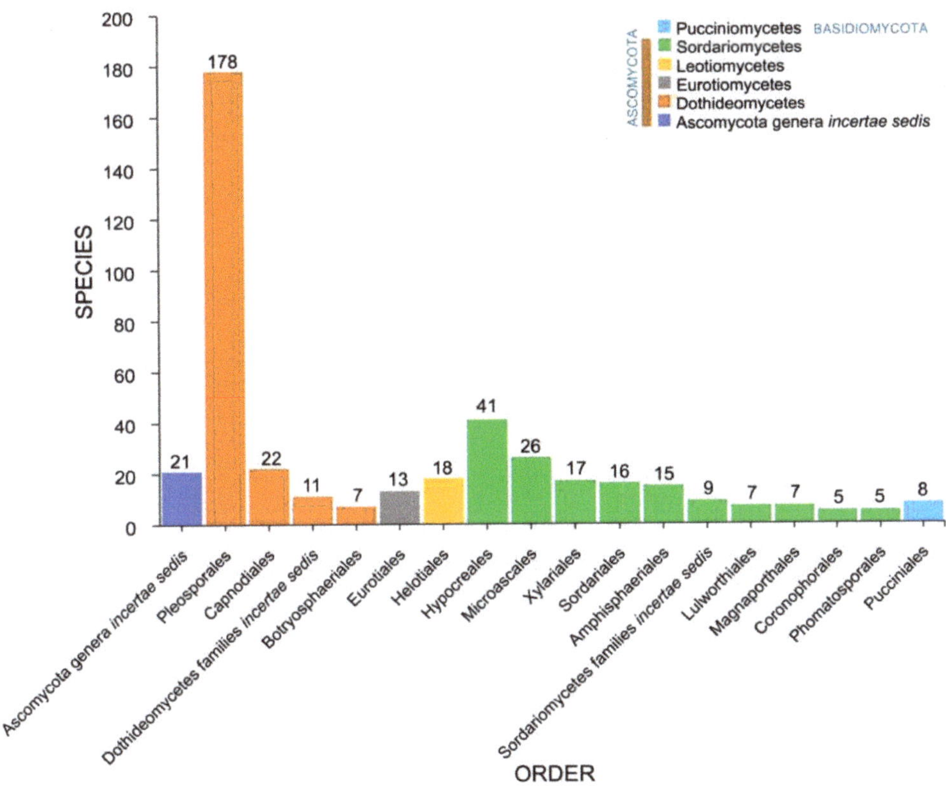

Figure 6. The distribution of salt marsh fungi in major fungal orders.

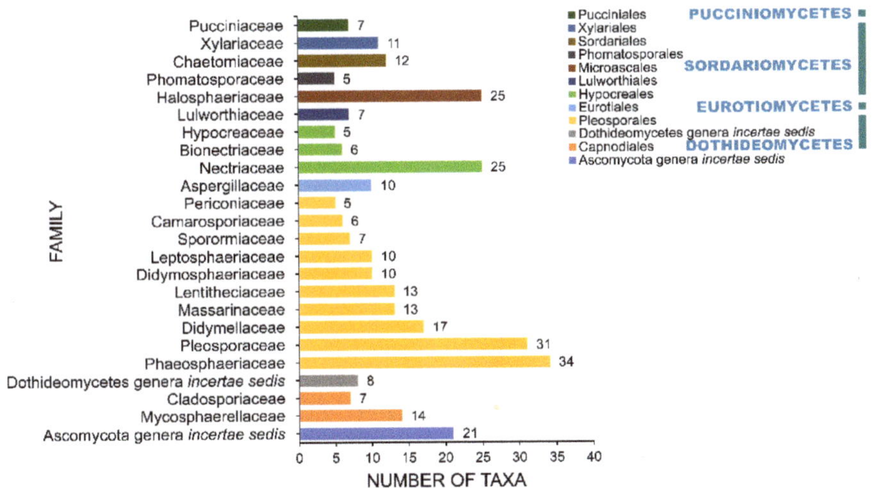

Figure 7. The distribution of salt marsh fungi among major fungal families.

3. Diversity of Fungi in Halophytes

Twenty-seven genera under 11 families (Amaranthaceae, Apiaceae, Caryophyllaceae, Compositae, Juncaceae, Juncaginaceae, Plumbaginaceae, Poaceae, Poaceae, Primulaceae, Ruppiaceae, Typhaceae, Zosteraceae) of halophytes were reviewed for its fungal associates (Table 1, Figure 8). Halophytic species are represented in Figures 1 and 2.

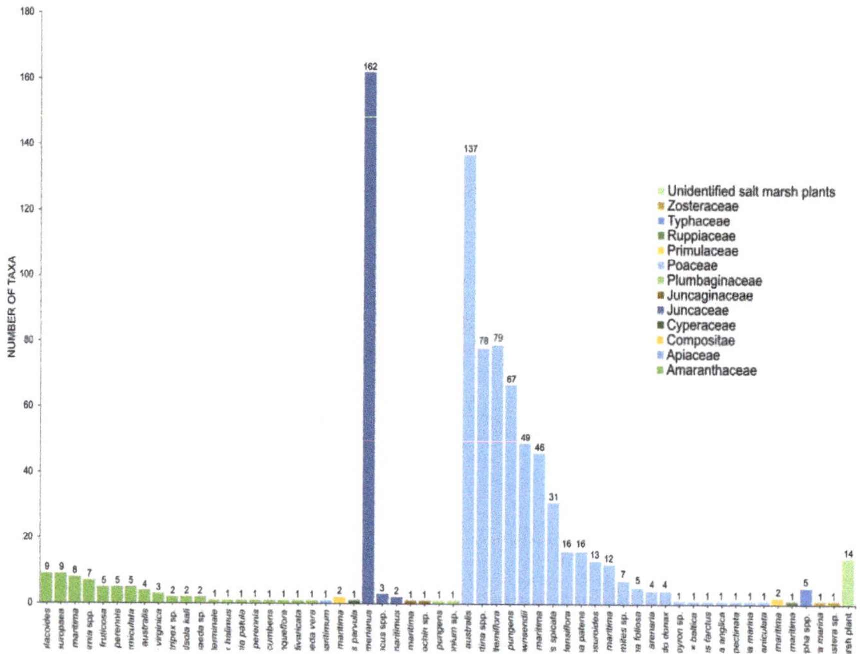

Figure 8. The number of taxa observed from different hosts in salt marsh ecosystems.

3.1. *Amaranthaceae*

Six genera (*Arthrocnemum, Atriplex, Salicornia, Salsola, Sarcocornia, Suaeda*) represent the Amaranthaceae. *Suaeda* and *Salicornia* are the most studied hosts in Amaranthaceae. Ascomycota account for 96.30% of the 52 taxa recorded in Amaranthaceae (Figure 9, Table 1). Two Pucciniomycetes species, *Aecidium suaedae* [154] and *Uromyces salicorniae* [95], represent Basidiomycota. The taxa in Amaranthaceae represent three classes wherein *Dothideomycetes* accounts for 85.19% (46 taxa), followed by *Sordariomycetes* with six taxa reported.

Fungi associated with *Suaeda* total 18 taxa. *Dothideomycetes* was represented by 14 taxa (77.78%), while three taxa were Sordariomycetes (*Cryptovalsa suaedicola* [144], *Fusarium fujikuroi* [62], *Moheitospora fruticosae* [130]) and one taxon of *Pucciniomycetes* (*Aecidium suaedae* [154]).

A total of 14 taxa were documented in *Salicornia*. Eleven of these belong to Dothideomycetes (Pleosporales: 10; Capnodiales: 1), followed by Sordariomycetes (two taxa: *Halocryptovalsa salicorniae* [145], *Tubercularia pulverulenta* [35]), and Pucciniomycetes (one taxon: *Uromyces salicorniae* [95]).

Fungi from *Atriplex* total 11 taxa (10 genera) and all of these belong to Pleosporales (Dothideomycetes). *Sarcocornia* harbors seven taxa (six Dothideomycetes, one Sordariomycetes). Only two taxa (*Alternaria* spp., *Stemphylium* spp.) and a single taxon (*Mycosphaerella salicorniae*) were reported from *Salsola* [35] and *Arthrocnemum* [35], respectively.

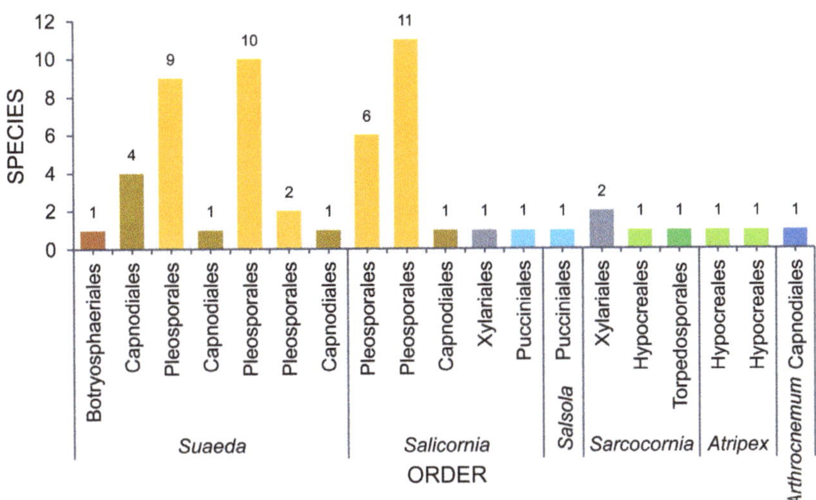

Figure 9. The number of taxa observed from Amaranthaceae.

3.2. *Poaceae*

The association of fungi with grasses have been documented and most of the host plants are members of Poaceae. Ten genera of salt marsh grasses under Poaceae are included in this review wherein *Spartina* is the most studied of halophytic hosts for direct observation of marine fungi. In addition to *Spartina*, salt marsh grasses such as *Phragmites* and *Distichlis* were well studied also for their fungal associates.

Salt marsh fungi are not well-documented from grasses such as *Spartina anglica*, *S. pectinata*, *Spergularia marina*, *Uniola paniculata*, *Elymus farctus*, × *Ammocalamagrostis baltica*, and *Agropyron* sp. with one taxon recorded for each host [35]. Furthermore, there are few studies on the fungal composition of *Arundo donax* (4 taxa) [35] and *Ammophila arenaria* (four taxa). Marram grass (*Ammophila arenaria*) is more common in sand dunes and supports quite a diverse fungal community [157,158], while arbuscular mycorrhizal fungi (AMF) play a key role in the establishment, growth, and survival of plants [159].

3.2.1. *Distichlis spicata*

Ascomycota dominates the taxa associated with *Distichlis spicata* (93.55%) wherein 16 and 13 species are members of Dothideomycetes and Sordariomycetes, respectively. Pleosporalean taxa constitute the majority of fungi associated with *D. spicata* (14 species), followed by Hypocreales with nine species recorded. *Puccinia distichlidis* and *Tranzscheliella distichlidis* represent the Basidiomycota. A total of 26 genera were recorded as associates of *D. spicata* and were mostly observed on senescent and decaying leaves.

3.2.2. *Elymus pungens*

Sixty-seven taxa were recorded in *Elymus pungens* and belong to Ascomycota. Most of the taxa belong to Dothideomycetes (32 taxa), followed by Sordariomycetes (21 taxa), Leotiomycetes, and Eurotiomycetes (6 taxa) (Table 1, Figure 10).

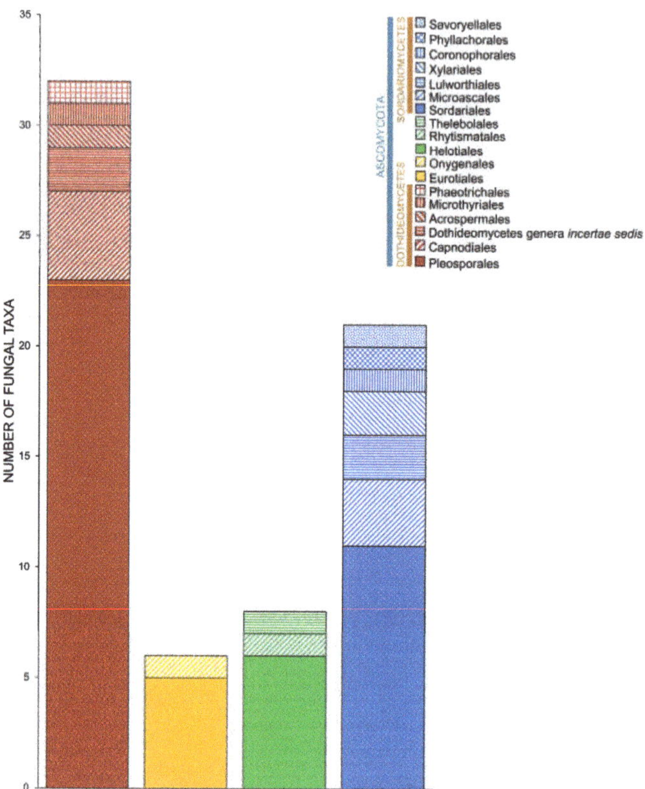

Figure 10. The distribution of fungal taxa associated with *Elymus pungens*.

3.2.3. *Puccinellia maritima*

A total of 12 taxa (six Sordariomycetes; the following five Dothideomycetes: *Micronectriella agropyri*, *Lautitia danica*, *Leptosphaeria pelagica*, *Septoriella vagans*, *Paradendryphiella salina*; one Leotiomycetes: *Thelebolus crustaceus*) were recorded in *Puccinellia maritima* [38]. All the taxa from Sordariomycetes belong to Sordariales (*Chaetomium elatum*, *C. globosum*, *C. thermophilum*, *Corynascus sepedonium*, *Thermothielavioides terrestris*, *Sordaria fimicola*) [38].

3.2.4. *Spartina*

A total of 149 taxa (141 Ascomycota, 6 Basidiomycota, 2 Mucoromycota) were recorded in *Spartina*. The majority of the taxa belong to Dothideomycetes (70 taxa), followed by Sordariomycetes (59 taxa). Pleosporaceae and Halosphaeriaceae dominate the fungi documented in *Spartina* with 19 and 17 taxa recorded, respectively. *Spartina alterniflora*, *S. maritima*, and *Spartina* × *townsendii* harbor 79, 46, and 49 taxa, respectively (Figure 11, Table 1). A total of 78 taxa were recorded in the unidentified *Spartina* species. The identification of the *Spartina* species can be challenging, wherein species are morphologically similar.

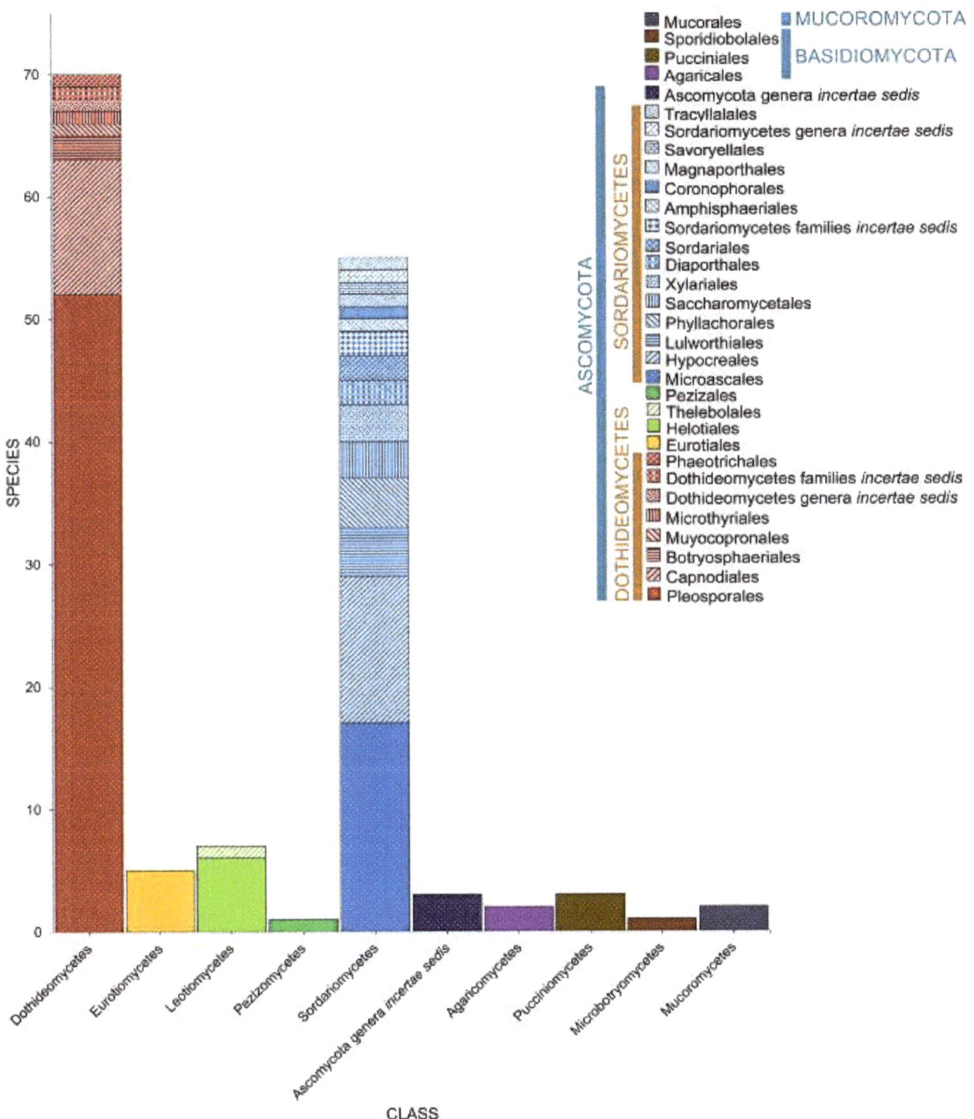

Figure 11. The distribution of fungal taxa associated with *Spartina*.

Halobyssothecium obiones was recorded from six species of *Spartina* (*S. alterniflora* [20,35, 52,61,71,74,80–82], *S. cynosuroides* [35], *S. densiflora* [64], *S. maritima* [31,54,59,63], *S. patens* [36], *S. townsendii* [49,65], and the unidentified *Spartina* sp. [32,35,36,58,84]), while six *Spartina* spp. harbors unidentified *Mycosphaerella* species. Six species (*Leptosphaeria pelagica*, *Lulworthia* spp., *Phaeosphaeria halima*, *Phaeosphaeria spartinicola*, *Phoma* spp., *Stagonospora* spp.) were recorded in five different hosts. The unidentified *Spartina* species harbors 28 unique species. Amongst the taxa found in *Spartina*, 32 species can only be found in *S. alterniflora*, while *S. maritima* harbors 21 unique species, the most intensively surveyed species.

3.2.5. *Phragmites*

A total of 138 taxa have been documented in *Phragmites* (Figure 12, Table 1). Most of the taxa belong to Ascomycota (131 taxa), while six taxa represent the Basidiomycota. Dothideomycetes dominates half of the taxa in *Phragmites* (71 taxa, 51.45%) followed by Sordariomycetes (44 taxa, 31.88%), Leotiomycetes (6 taxa, 4.35%), Ascomycota genera *incertae sedis* (5 taxa, 3.62%), Eurotiomycetes (3 taxa, 2.17%), Orbiliomycetes (2 taxon, 1.45%), and Pucciniomycetes (1 taxa, 1.45%). One taxon each were recorded to Agaricomycetes [40], Bartheletiomycetes [41], Lecanoromycetes [39], Microbotryomycetes [39,50], and Tremellomycetes [39,40]. Pleosporalean taxa accounts for the highest number of fungi associated with *Phragmites* (42.75%, 59 taxa).

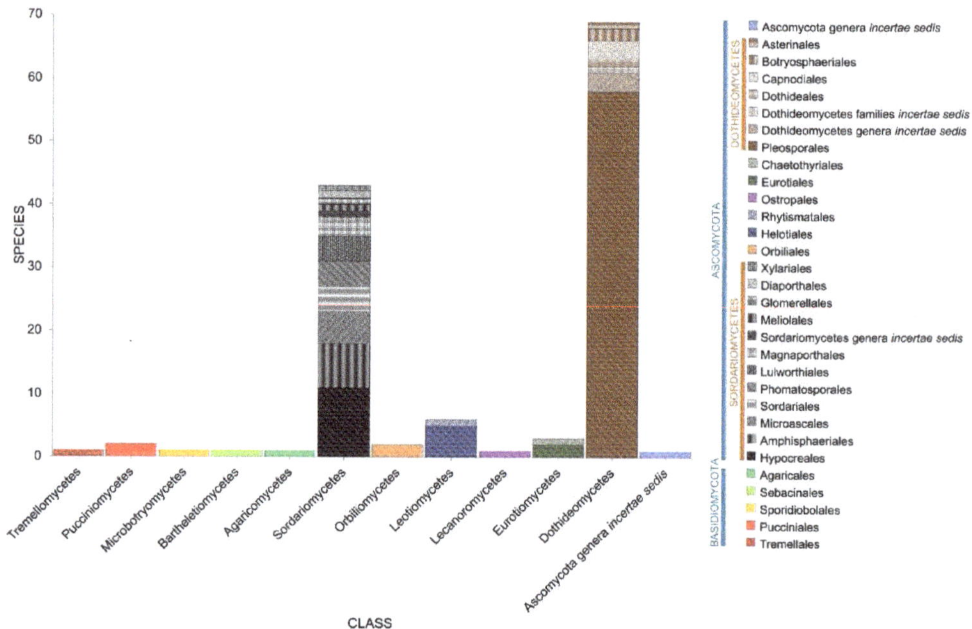

Figure 12. The distribution of fungal taxa associated with *Phragmites*.

Phragmites australis harbors diverse fungi that totals to 137 taxa (101 genera) [39–41,50,79,115]. Seven species (*Arthrinium arundinis* [62], *Halazoon fuscus* [87], *Halobyssothecium phragmitis* [85], *Keissleriella linearis* [85], *Phomatospora dinemasporium* [62], *Remispora hamata* [87], *Setoseptoria phragmitis* [87]) were recorded in unidentified *Phragmites* species.

3.3. *Juncaceae*

Juncus roemerianus, *J. maritimus*, and an unidentified *Juncus* species represent Juncaceae. Salt marsh fungi are diverse in *Juncus* and dominated by Ascomycota, which constitutes 97.58% of the 165 reported taxa (Figure 13, Table 1). *Stilbum* sp. represented the Basidiomycota, while three taxa (*Blakeslea trispora*, *Mucor* sp., *Syncephalastrum racemosum*) of Mucoromycota were recorded. Dothideomycetes and Sordariomycetes account for the highest number of *Juncus*-associated fungi with 72 (43.64%) and 64 (38.79%) taxa documented.

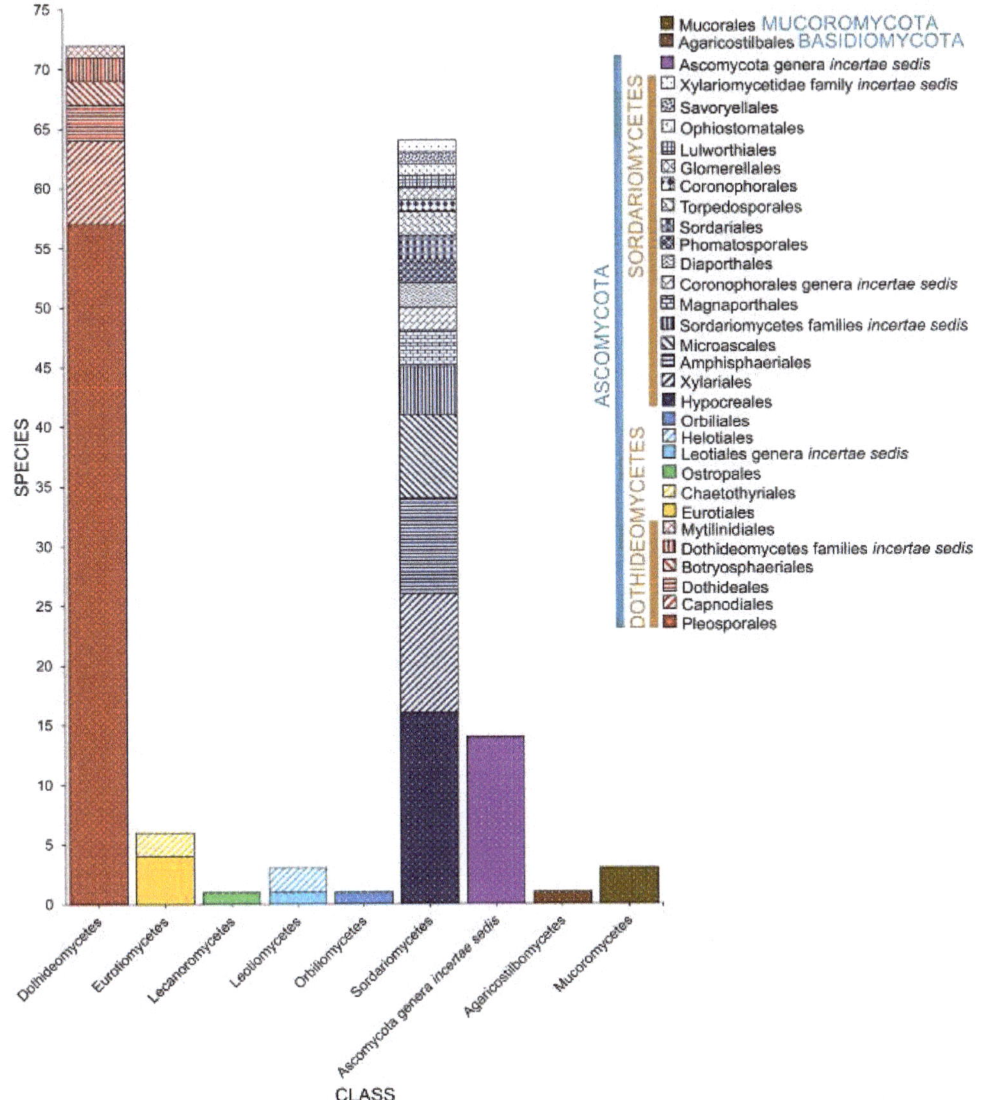

Figure 13. The distribution of fungal taxa associated with *Juncus*.

Juncus roemerianus has been extensively studied for its associates with 162 documented taxa [32,42,43,60,66,76–78,97,98,104,105,110,116–118,135,147,148]. Few species were reported to *Juncus maritimus* that harbor only two taxa (*Leptosphaeria albopunctata*, *Phaeosphaeria neomaritima*) [35]. *Phaeosphaeria neomaritima* [36,52,71,80], *P. spartinicola* [52], and *Monodictys pelagica* [35] were observed in an unidentified species of *Juncus*.

Phragmites australis harbors diverse fungi that totals to 137 taxa (101 genera) [39–41,50,79,115]. Seven species (*Arthrinium arundinis* [62], *Halazoon fuscus* [87], *Halobyssothecium phragmitis* [85], *Keissleriella linearis* [85], *Phomatospora dinemasporium* [62], *Remispora hamata* [87], *Setoseptoria phragmitis* [87]) were recorded in unidentified *Phragmites* species.

3.4. Other Families

Few reports on salt marsh fungi are from the following hosts: Apiaceae: *Crithmum maritimum* (one taxon: *Phoma* sp.), Typhaceae: *Typha* spp. (five taxa: *Arundellina typhae*, *Chaetomium* sp., *Magnisphaera spartinae*, *Pleospora pelagica*, *Remispora hamata*); Compositae: *Artemisia maritima* (two taxon: *Neocamarosporium artemisiae*, *N. maritimae*); Caryophyllaceae: *Spergularia marina* (one taxon: *Cladosporium algarum*); Plumbaginaceae: *Limonium* sp. (one taxon: *Mycosphaerella salicorniae*); *Armeria pungens* (one taxon: *Mycosphaerella staticicola*); Juncaginaceae: *Triglochin* sp. and *T. maritima* (one taxon: *Stemphylium triglochinicola*); Primulaceae: *Lysimachia maritima* (two taxa: *Leptosphaeria orae-maris*, *Stemphylium vesicarium*); Ruppiaceae: *Ruppia maritima* (one taxon: *Flamingomyces ruppiae*); and Zosteraceae: *Zostera marina* (one taxon: *Corollospora ramulosa*) and *Zostera* sp. (*Asteromyces cruciatus*). Alva et al. [160] report *Penicillium chrysogenum* as an endophyte from *Zostera japonica*.

Fourteen taxa were documented from unidentified salt marsh plants. All of the taxa belong to Ascomycota (seven Dothideomycetes, five Sordariomycetes, one Eurotiomycetes). Pleosporalean taxa from six families account for half of the taxa (the following seven species: *Camarosporium palliatum*, *C. roumeguerei*, *Coniothyrium obiones*, *Halobyssothecium obiones*, *Periconia* sp., *Loratospora aestuarii*, *Pleospora pelvetiae*).

4. Geographical Distribution of Salt Marsh Fungi

The salt marsh fungi reported are from countries of three major oceans, as documented in Figure 14. The Atlantic Ocean consists of 12 countries, wherein the USA had the highest number of species recorded (232 taxa) followed by the UK (101 taxa), the Netherlands (74 taxa), and Argentina (51 taxa). China had the highest number of salt marsh fungi in the Pacific Ocean with 165 taxa reported, while in the Indian Ocean, India reported the highest taxa (16 taxa). Most of the biodiversity studies documenting salt marsh fungi in the Atlantic Ocean are mostly from the USA and the UK and this reflects the high number of taxa [32,36,38,49,61]. China ranked second with the most number of salt marsh fungal taxa, mainly due to the biodiversity study in *Phragmites australis* conducted by Poon et al. [41].

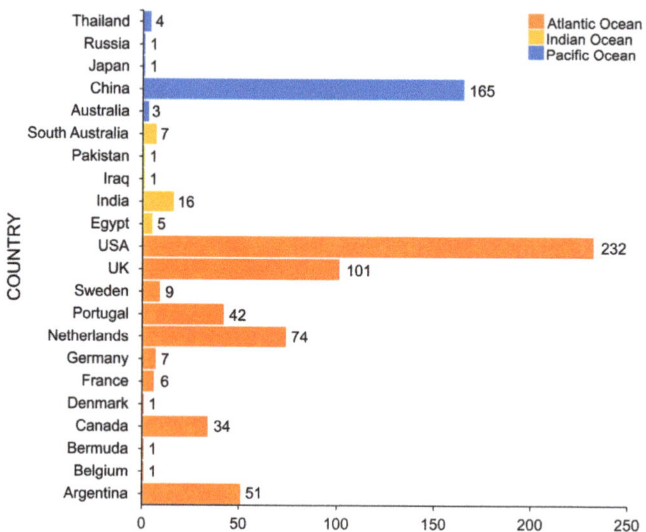

Figure 14. The number of salt marsh fungi reported in the Pacific, Atlantic, and Indian Oceans.

The geographical distribution of salt marsh fungi and the different halophytes are presented in Figure 15. The fungi associated with salt marsh grass *Phragmites australis* have been studied in different countries (Australia, Belgium, Egypt, France, Germany, China, Iraq, Japan, the Netherlands, South Australia, Thailand). *Spartina alterniflora* was recorded in countries along the Atlantic (Argentina, Canada, France, USA) and the Indian Ocean (India), but lacks data from countries in the Pacific Ocean.

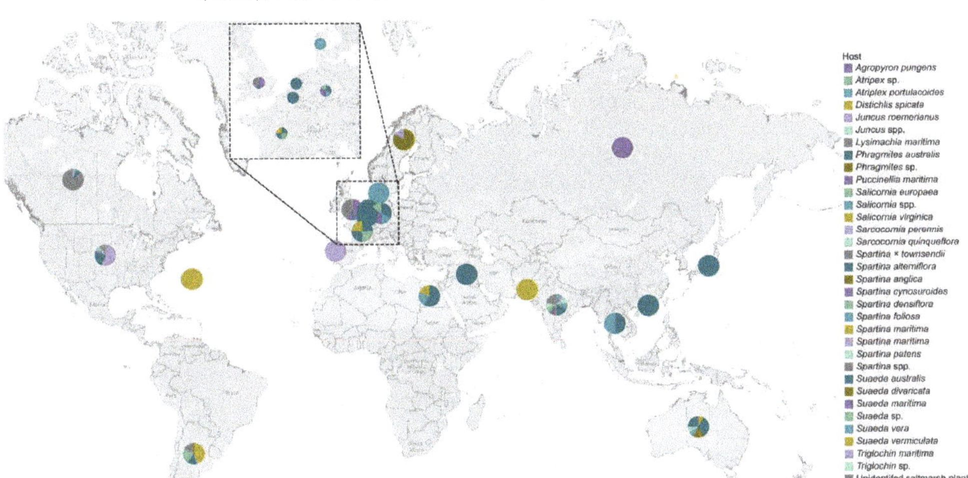

Figure 15. Map of countries showing the global distribution of fungal diversity studies in halophytes. The different color of each pie chart represents the hosts, and the angle measured the number of their fungal associates.

United States of America

Most of the studies of halophytes-associated fungi were concentrated on the United States of America (USA) (Figure 16). Table 1 lists the salt marsh fungi in 20 states. Florida has been the frequently studied, wherein seven hosts (*Juncus roemerianus*: 108 taxa; *Spartina* × *townsendii:* 1; *Spartina alterniflora*: 16; *Spartina cynosuroides*: 3; *Spartina densiflora*: 1; *Spartina patens*: 2; *Spartina* spp.: 3) were observed for salt marsh fungi. Six hosts were studied in North Carolina, wherein *Juncus roemerianus* harbored the highest number of fungi (48 taxa). In Rhode Island, *Spartina alterniflora* accounts for the highest number of fungi, with 41 taxa recorded.

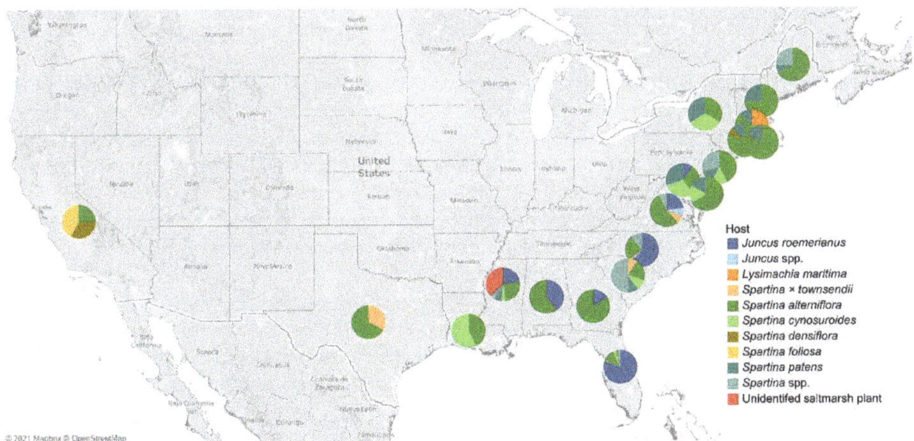

Figure 16. Map of the United States of America (USA) showing the distribution of fungal diversity studies of halophytes in different states. The different color of each pie chart represents the hosts, and the angle measured the number of their fungal associates.

5. Conclusions and Future Perspectives

Most studies of fungi on salt marsh plants are from *Spartina*, *Juncus*, and *Phragmites*, probably due to the huge biomass generated by these taxa. The mycota of less bulky halophytes (e.g., *Limonium*, *Triglochin*, *Uniola*) and litter from the surrounding sea grass beds washed off to marsh areas (e.g., *Zostera japonica*, *Z. marina*, *Z. noltii*) are also less represented, or these hosts are yet to be explored. The checklist presented in the current study updates the list of Calado and Barata [34] and the inclusion of fungi associated with rarely studied halophytes record 486 taxa worldwide. Ascomycota dominate the taxa (463 taxa) and are comprised mostly of Dothideomycetes with their ability to eject their ascospores forcibly and widely, spore type, the formation of ascomata or ascostromata under a clypeus or just immersed in thin leaves, and an ability to decompose lignocellulose substrates [57,161]. Meyers et al. [162] showed that salt marsh yeasts and the ascomycete, *Buergenerula spartinae*, produce degradative enzymes and utilize simple carbon and nitrogen compounds. The yeast, *Pichia spartinae*, produces β-glucosidase and other degradative enzymes. Gessner [74] demonstrated that a number of salt marsh fungi isolated from *Spartina alterniflora*, *Zostera* sp., and *Z. marina* produced enzymes capable of degrading cellulose, cellobiose, lipids, pectin, starch, tannic acid, and xylan and, thus, play a key role in the degradation of storage and structural compounds. Salt marsh fungi might possess high biotransformation and metabolic abilities, which could be related to their ecology.

Basidiomycota (19 taxa) and Mucoromycota (4 taxa) are poorly represented in salt marsh ecosystems as they are in other marine habitats [163]. There are no records of Chytridiomycota listed in the present work and only a few authors detected this group, and other basal fungal lineages, in salt marsh ecosystems using molecular analysis [164–167]. These groups are worth exploring to determine the overall fungal communities in the salt marsh ecosystems. Many chytrids and other basal fungi are more challenging to cultivate and require different isolation methods (e.g., baiting techniques in liquid culture) than the saprobes, methods that have rarely been applied in the study of saltmarsh plants. When appropriate techniques are used, chytrids and other zoosporic organisms have been reported. For example, the fungal-like organism *Phytophthora inundata* has been recovered from the halophilic plants *Aster tripolium* and *Salicornia europaea*, while *P. gemini* and *P. chesapeakensis* occur on *Zostera marina*, and *Salisapilia nakagirii* on the decaying litter of *Spartina alterniflora* (www.marinefungi.org; accessed on 10 May 2021, [163]). Marine chytrids have been isolated from substrates such as seaweeds and mangrove leaves [163].

The taxa listed are mostly saprobes and these can be attributed to the inclusion of salt marsh fungi observed directly from the different host parts, which are mostly submerged decaying substrates. When compared to saprobic fungi in halophytes, few studies have been carried out on the diversity of endophytes and pathogens and their interaction in the salt marsh ecosystems. Surveys on endophytic fungi from halophytes using cultivation-dependent methods coupled with molecular approaches, showed that endophytes were dominated by Ascomycota and a few belonged to Basidiomycota and Zygomycota [168–175]. Pathogenic fungi from salt marsh ecosystems are poorly documented but play a significant role in the dynamics of the ecosystem [176–178]. For example, Govers et al. [179] reported that the fungal-like organisms *Phytophthora gemini* and *P. inundata* caused widespread infection of the common seagrass species, *Zostera marina* (eelgrass), across the northern Atlantic and Mediterranean that threatened the conservation and restoration of vegetated marine coastal systems. Likewise, *Claviceps purpurea* affects the viability of *Spartina townsedii* in south coast UK salt marshes. Fisher et al. [180] noted that *Cl. purpea* in the Alabama and Mississippi coastlines rendered the seeds of one of the primary salt marsh grasses sterile. Raybold et al. [181] recorded epidemics of *C. purpurea* on *Spartina anglica* in Poole Harbor (UK) and that ergot growth was detrimental to seed production. These underexplored fungal groups are worthy to be explored for their ecological and biotechnological importance.

This shows how salt marsh fungal studies were concentrated in countries in the Atlantic Ocean specifically the USA (232 taxa) and the UK (101 taxa). Many salt marsh areas remain unexplored, especially those in the Indian and Pacific Oceans, and these areas are hotspots of biodiversity and novel fungal taxa based on the exploration of various habitats [85,100,163,182–187]. Recently, novel species were isolated in halophytes [85,100,145] and further taxa remain to be discovered, isolated, and sequenced, while vast areas worldwide have yet to be surveyed. For example, salt marsh plants are immensely numerous, diverse, and common along the south-east coast of Australia, yet little is known of their fungal associates [188].

The salt marsh vegetation and its fungal associates are adapted to salt stress and inundation and are subjected to extreme environmental conditions such as being periodically wet to different lengths of time leading to drying out at low tides and exposure to high temperatures and drying out at midday. Many are well adapted to prevailing conditions by their fleshy leaves (*Suaeda australis*), others can tolerate high flooding.

Few data are currently available on the specificity of fungi on their salt marsh hosts. Figure 17 shows the number of fungal taxa recorded from the three commonly studied hosts, *Juncus*, *Phragmites*, and *Spartina*, wherein there is little overlap in the species composition. One of the common species on *Spartina* plants is undoubtedly *Halobyssothecium obiones*, while *Leptosphaeria pelagica* is common. A common ascomycete on *Atriplex portulacoides* and *Suaeda maritima* is *Decorospora gaudefroyi*. Host plants that have been little surveyed for fungi are *Limonium vulgare* (sea lavender) and *Atriplex portulacoides* (sea purslane), yet they do support a number of taxa, e.g., *Neocamarosporium obiones* and *Amarenomyces ammophilae*. The fungal community reported on *Juncus roemerianus* in the salt marsh at North Carolina is significantly different from those on *Spartina* and *Phragmites*. It remains to be seen if this is due to the host plant or its geographical location.

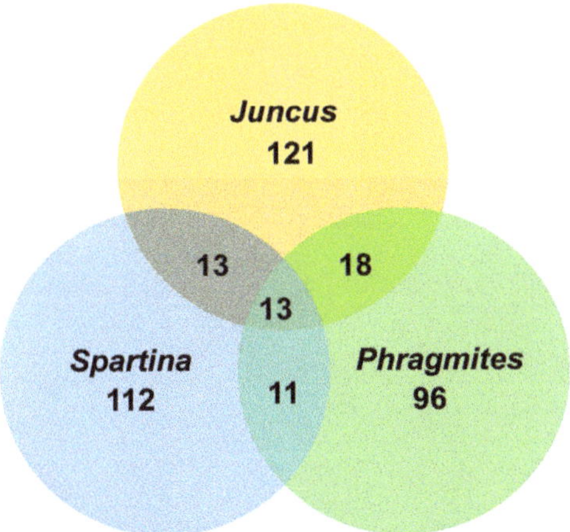

Figure 17. Venn diagram showing the association of salt marsh fungi from commonly studied halophytes.

Another groups of fungi that have not been fully studied in the salt marsh habitat are yeasts, as these also require specific techniques for their isolation from the water column or from plant tissue. Spencer et al. [189] recovered a number of yeasts from the vicinity of *Spartina townsendii*, as follows: very numerous *Cryptococcus* spp.; *Trichosporon cutaneum*; *Trichosporon pullulans*; the relatively rare species, *Metschnikowia bicuspidata and Cryptococcus flavus*; and *Saturnospora ahearnii* [190]. Although marine yeasts are common in sea water and deep seawater vents [163], their large-scale sampling in salt marshes remains a challenge for the future.

Currently, the salt marsh ecosystem has been threatened both by global warming and human activity. Sea-level rises brought about by climate change alter the location and character of the land–sea interface wherein salt marsh vegetation moves upward and inland. The increase in the sea level may not lead to the loss of coastal marshes, but the resiliency will depend on the ability of halophytes to migrate upland. Susceptible areas are organogenic marshes and areas where sediment is limited, potentially leading to catastrophic shifts and marsh loss. In this paper, a total of 57 plant taxa under 27 genera were reviewed for their fungal associates. The halophytes included here are only approximately 11% of the total number of species of salt marsh plants worldwide. Thus, many salt marsh fungi await discovery with wider host plant sampling and the use of a wider range techniques for their isolation. For this reason, it is imperative to study the halophytic fungi to document not just biodiversity but also to discover novel taxa restricted only to this kind of habitat.

Author Contributions: Conceptualization: M.S.C., K.D.H. and E.B.G.J.; methodology: M.S.C., K.D.H. and E.B.G.J.; formal analysis and investigation: M.S.C., K.D.H. and E.B.G.J.; resources: K.D.H. and E.B.G.J.; writing—original draft preparation, M.S.C.; writing—review and editing, K.D.H., E.B.G.J. and I.P.; supervision, K.D.H. and E.B.G.J.; funding acquisition, K.D.H. and E.B.G.J. All authors have read and agreed to the published version of the manuscript.

Funding: K.D.H. thanks the Thailand Research Fund for the grant entitled "Impact of climate change on fungal diversity and biogeography in the Greater Mekong Subregion" (Grant No. RDG6130001). E.B.G.J. is supported under the Distinguished Scientist Fellowship Program (DSFP), King Saud University, Kingdom of Saudi Arabia.

Institutional Review Board Statement: Not applicable.

Informed Consent Statement: Not applicable.

Data Availability Statement: Not applicable.

Acknowledgments: M.S.C. is grateful to the Mushroom Research Foundation and the Department of Science and Technology—Science Education Institute (Philippines). K.D.H. thanks Chiang Mai University.

Conflicts of Interest: The authors declare no conflict of interest.

References

1. Bertness, M.D. *Atlantic Shorelines: Natural History and Ecology*; Princeton University Press: Princeton, NJ, USA, 2008.
2. Öztürk, M.; Altay, V.; Altundağ, E.; Gücel, S. Halophytic plant diversity of unique habitats in Turkey: Salt mine caves of Çankırı and Iğdır. In *Halophytes for Food Security in Dry Lands*; Khan, M.A., Ozturk, M., Gul, B., Ahmed, M.Z., Eds.; Academic Press: New York, NY, USA, 2016; pp. 291–315.
3. Adam, P. The saltmarsh biota. In *Saltmarsh Ecology*; Adam, P., Ed.; Cambridge University Press: Cambridge, UK, 2011; pp. 72–145.
4. Macdonald, K.B. Plant and animal communities of Pacific North American salt marshes. In *Wet Coastal Formations*; Chapman, V.J., Ed.; Elsevier: Amsterdam, The Netherlands, 1976; pp. 167–191.
5. Saenger, P.; Specht, M.M.; Specht, R.L.; Chapman, V.J. Mangal and coastal salt-marsh communities in Australasia. In *Wet Coastal Ecosystems*; Chapman, V.J., Ed.; Elsevier: Amsterdam, The Netherlands, 1977; pp. 293–345.
6. Macdonald, K.B. Coastal salt marsh. In *Terrestrial Vegetation of California*; Barbour, M.G., Major, J., Eds.; Wiley: New York, NY, USA, 1988; pp. 263–294.
7. Costanza, R.; D'Arge, R.; De Groot, R.; Farber, S.; Grasso, M.; Hannon, B.; Limburg, K.; Naeem, S.; O'Neill, R.V.; Paruelo, J.; et al. The value of the world's ecosystem services and natural capital. *Nature* **1997**, *387*, 253–260. [CrossRef]
8. Mcowen, C.J.; Weatherdon, L.V.; Van Bochove, J.W.; Sullivan, E.; Blyth, S.; Zockler, C.; Stanwell-Smith, D.; Kingston, N.; Martin, C.S.; Spalding, M.; et al. A global map of saltmarshes. *Biodivers. Data J.* **2017**, *5*, 11764. [CrossRef]
9. Silliman, B.R. Salt marshes. *Curr. Biol.* **2014**, *24*, R348–R350. [CrossRef] [PubMed]
10. Teal, J.M. Salt marshes and mud flats. In *Encyclopedia of Ocean Sciences*; Thorpe, S.A., Turekian, K.K., Eds.; Academic Press: Cambridge, MA, USA, 2008; pp. 43–48.
11. Roman, C.T. Salt marsh vegetation. In *Encyclopedia of Ocean Sciences*; Thorpe, S.A., Turekian, K.K., Eds.; Academic Press: Cambridge, MA, USA, 2001; pp. 2487–2490.
12. Garbutt, A.; de Groot, A.; Smit, C.; Pétillon, J. European salt marshes: Ecology and conservation in a changing world. *J. Coast. Conserv.* **2017**, *21*, 405–408. [CrossRef]
13. Davy, A.J. Development and structure of salt marshes: Community patterns in time and space. In *Concepts and Controversies in Tidal Marsh Ecology*; Weinstein, M.P., Kreeger, D.A., Eds.; Springer: Dordrecht, The Netherlands, 2005; pp. 137–156.
14. Pennings, S.C.; Callaway, R.M. Salt marsh plant zonation: The relative importance of competition and physical factors. *Ecology* **1992**, *73*, 681–690. [CrossRef]
15. Pomeroy, L.R.; Darley, W.M.; Dunn, E.L.; Gallagher, J.L.; Haines, E.B.; Whitney, D.M. Primary production. In *The Ecology of Salt Marsh*; Pomeroy, L.R., Wiegert, R.G., Eds.; Springer-Verlag: New York, NY, USA, 1981; pp. 39–67.

16. Howarth, R.W.; Hobbie, J.E. *The Regulation of Decomposition and Heterotrophic Microbial Activity in Salt Marsh Soils: A Review*; Kennedy, V.S., Ed.; Academic Press: Cambridge, MA, USA, 1982.
17. Long, S.P.; Mason, C.F. *Saltmarsh Ecology*; Blackie: Glasgow, UK, 1983.
18. Valiela, I.; Teal, J.M.; Allen, S.D.; Van Etten, R.; Goehringer, D.; Volkmann, S. Decomposition in salt marsh ecosystems: The phases and major factors affecting disappearance of above-ground organic matter. *J. Exp. Mar. Biol. Ecol.* **1985**, *89*, 29–54. [CrossRef]
19. Pomeroy, L.R.; Imberger, J. The physical and chemical environment. In *The Ecology of Salt Marsh*; Pomeroy, L.R., Weigert, R.G., Eds.; Springer-Verlag: New York, NY, USA, 1981; pp. 21–36.
20. Gessner, R.V.; Goos, R.D. Fungi from decomposing *Spartina alterniflora*. *Can. J. Bot.* **1973**, *51*, 51–55. [CrossRef]
21. Gessner, R.V.; Goos, R.D.; Sieburth, J.M.N. The fungal microcosm of the internodes of *Spartina alterniflora*. *Mar. Biol.* **1972**, *16*, 269–273. [CrossRef]
22. Benner, R.; Newell, S.Y.; Maccubbin, A.E.; Hodson, R.E. Relative contributions of bacteria and fungi to rates of degradation of lignocellulosic detritus in salt-marsh sediments. *Appl. Environ. Microbiol.* **1984**, *48*, 36–40. [CrossRef]
23. Bergbauer, M.; Newell, S.Y. Contribution to lignocellulose degradation and DOC formation from a salt marsh macrophyte by the ascomycete *Phaeosphaeria spartinicola*. *FEMS Microbiol. Ecol.* **1992**, *86*, 34–348. [CrossRef]
24. Benner, R.; Maccubbin, A.E.; Hodson, R.E. Preparation, characterization, and microbial degradation of specifically radiolabeled [C] lignocelluloses from marine and freshwater macrophytes. *Appl. Environ. Microbiol.* **1984**, *47*, 381–389. [CrossRef]
25. Lyons, J.I.; Alber, M.; Hollibaugh, J.T. Ascomycete fungal communities associated with early decaying leaves of *Spartina* spp. from central California estuaries. *Oecologia* **2010**, *162*, 435–442. [CrossRef]
26. Maccubbin, A.E.; Hodson, R.E. Mineralization of detrital lignocelluloses by salt marsh sediment microflora. *Appl. Environ. Microbiol.* **1980**, *40*, 735–740. [CrossRef]
27. Newell, S.Y.; Porter, D. Microbial secondary production from salt marsh-grass shoots, and its known and potential fates. In *Concepts and Controversies in Tidal Marsh Ecology*; Weinstein, M.P., Kreeger, D.A., Eds.; Springer: Dordrecht, The Netherlands, 2005; pp. 159–185.
28. Newell, S.Y.; Fallon, R.D.; Miller, J.D. Decomposition and microbial dynamics for standing, naturally positioned leaves of the salt-marsh grass Spartina alterniflora. *Mar. Biol.* **1989**, *101*, 471–481. [CrossRef]
29. Hernández, E.G.; Baraza, E.; Smit, C.; Berg, M.P.; Salles, J.F. Salt marsh elevation drives root microbial composition of the native invasive grass *Elytrigia atherica*. *Microorganisms* **2020**, *8*, 1619. [CrossRef] [PubMed]
30. Wang, M.; Li, E.; Liu, C.; Jousset, A.; Salles, J.F. Functionality of root-associated bacteria along a salt marsh primary succession. *Front. Microbiol.* **2017**, *8*, 2102. [CrossRef] [PubMed]
31. Calado, M.d.L.; Carvalho, L.; Barata, M.; Pang, K.L. Potential roles of marine fungi in the decomposition process of standing stems and leaves of *Spartina maritima*. *Mycologia* **2019**, *111*, 371–383. [CrossRef] [PubMed]
32. Kohlmeyer, J.; Volkmann-Kohlmeyer, B. Fungi on Juncus and Spartina: New marine species of *Anthostomella*, with a list of marine fungi known on *Spartina*. *Mycol. Res.* **2002**, *106*, 365–374. [CrossRef]
33. Wijayawardene, N.N.; Hyde, K.D.; Al-ani, L.K.T.; Tedersoo, L.; Haelewaters, D.; Rajeshkumar, K.C.; Zhao, R.; Aptroot, A.; Leontyev, D.V.; Ramesh, K.; et al. Outline of Fungi and fungus-like taxa. *Mycosphere* **2020**, *11*, 1–367. [CrossRef]
34. Calado, M.d.L.; Barata, M. Salt marsh fungi. In *Marine Fungi and Fungal-like Organisms*; Jones, E.B.G., Pang, K.L., Eds.; Walter de Gruyter GmbH & Co. KG: Berlin, Germany, 2012; pp. 345–381.
35. Kohlmeyer, J.; Kohlmeyer, E. *Marine Mycology: The Higher Fungi*; Academic Press: New York, NY, USA, 1979.
36. Gessner, R.V.; Kohlmeyer, J. Geographical distribution and taxonomy of fungi from salt marsh *Spartina*. *Can. J. Bot.* **1976**, *54*, 2023–2037. [CrossRef]
37. Quattrocchi, U. *CRC World Dictionary of Grasses*; CRC Press: Boca Raton, FL, USA, 2006.
38. Apinis, A.E.; Chesters, C.G.C. Ascomycetes of some salt marshes and sand dunes. *Trans. Br. Mycol. Soc.* **1964**, *47*, 419–435. [CrossRef]
39. Van Ryckegem, G.; Verbeken, A. Fungal ecology and succession on *Phragmites australis* in a brackish tidal marsh. I. Leaf sheaths. *Fungal Divers.* **2005**, *19*, 157–187.
40. Van Ryckegem, G.; Verbeken, A. Fungal ecology and succession on *Phragmites australis* in a brackish tidal marsh. II. Stems. *Fungal Divers.* **2005**, *20*, 209–233.
41. Poon, M.O.K.; Hyde, K.D. Biodiversity of intertidal estuarine fungi on *Phragmites* at Mai Po Marshes, Hong Kong. *Bot. Mar.* **1998**, *41*, 141–155. [CrossRef]
42. Kohlmeyer, J.; Volkmann-Kohlmeyer, B. Fungi on *Juncus roemerianus*. *7. Tiarosporella halmyra* sp. nov. *Mycotaxon* **1996**, *59*, 79–83.
43. Fell, J.W.; Hunter, I.L. Fungi associated with the decomposition of the Black Rush, *Juncus roemerianus*, in South Florida. *Mycologia* **1979**, *71*, 322–342. [CrossRef]
44. United State Department of Agriculture Crops Research Division. *Index of Plant Diseases in the United States: Agriculture Handbook*; USDA: Washington, DC, USA, 1960.
45. Brunaud, P. Champignons nouvellement obsrevés aux environs de Saintes, Charente-inférieure. *J. Hist. Nat. Bord Sud. Oust. Bord.* **1888**, *7*, 4.
46. Lobik, A.I. Materialen zur Mykoflora des Terskikreises. *Morbi Plant. Leningr.* **1928**, *17*, 157–199.
47. Elíades, L.A.; Voget, C.E.; Arambarri, A.M.; Cabello, M.N. Fungal communities on decaying saltgrass (*Distichlis spicata*) in Buenos Aires province (Argentina). *Sydowia* **2007**, *59*, 227–234.

48. Miller, J.D.; Whitney, N.J. Fungi of the Bay of Fundy V: Fungi from living species of *Spartina* Schreber. *Proc. Nov. Scotian Inst. Sci.* **1983**, *33*, 75–83.
49. Goodman, P.J. The possible role of pathogenic fungi in die-back of *Spartina townsendii* agg. *Trans. Br. Mycol. Soc.* **1959**, *42*, 409–415. [CrossRef]
50. Van Ryckegem, G.; Gessner, M.O.; Verbeken, A. Fungi on leaf blades of *Phragmites australis* in a brackish tidal marsh: Diversity, succession, and leaf decomposition. *Microb. Ecol.* **2007**, *53*, 600–611. [CrossRef]
51. Kohlmeyer, J.; Volkmann-Kohlmeyer, B.; Eriksson, O.E. Fungi on *Juncus roemerianus* 12. Two new species of *Mycosphaerella* and *Paraphaeosphaeria* (Ascomycotina). *Bot. Mar.* **1999**, *42*, 505–511. [CrossRef]
52. Borse, B.D.; Bhat, D.; Borse, K.; Tuwar, A.; Pawar, N. *Marine fungi of India (Monograph)*; Broadway Book Centre: Panaji, India, 2012.
53. Kohlmeyer, J.; Kohlmeyer, E. Bermuda marine fungi. *Trans. Br. Mycol. Soc.* **1977**, *68*, 207–219. [CrossRef]
54. Barata, M. Fungi on the halophyte *Spartina maritima* in salt marshes. In *Fungi in Marine Environments*; Hyde, K.D., Ed.; Fungal Diversity Press: Hong Kong, China, 2002; pp. 179–193.
55. Walker, A.K.; Campbell, J. Marine fungal diversity: A comparison of natural and created salt marshes of the north-central Gulf of Mexico. *Mycologia* **2010**, *102*, 513–521. [CrossRef]
56. Buchan, A.; Newell, S.Y.; Moreta, J.I.L.; Moran, M.A. Analysis of internal transcribed spacer (ITS) regions of rRNA genes in fungal communities in a southeastern U.S. salt marsh. *Microb. Ecol.* **2002**, *43*, 329–340. [CrossRef] [PubMed]
57. Buchan, A.; Newell, S.Y.; Butler, M.; Biers, E.J.; Hollibaugh, J.T.; Moran, M.A. Dynamics of bacterial and fungal communities on decaying salt marsh grass. *Appl. Environ. Microbiol.* **2003**, *69*, 6676–6687. [CrossRef] [PubMed]
58. Walker, A.K. *Marine Fungi of U.S. GULF of Mexico Barrier Island Beaches: Biodiversity and Sampling Strategy*; The University of Southern Mississippi: Hattiesburg, MS, USA, 2012.
59. Calado, M.d.L.; Carvalho, L.; Pang, K.L.; Barata, M. Diversity and ecological characterization of sporulating higher filamentous marine fungi associated with *Spartina maritima* (Curtis) Fernald in two Portuguese salt marshes. *Microb. Ecol.* **2015**, *70*, 612–633. [CrossRef]
60. Kohlmeyer, J.; Volkmann-Kohlmeyer, B.; Eriksson, O.E. Fungi on *Juncus roemerianus*. 11. More new ascomycetes. *Can. J. Bot.* **1998**, *76*, 467–477. [CrossRef]
61. Gessner, R.V.; Goos, R.D. Fungi from *Spartina alterniflora* in Rhode Island. *Mycologia* **1973**, *65*, 1296–1301. [CrossRef]
62. Hansford, C.G. Australian Fungi. II. New species and revisions. *Proc. Linn. Soc. N. S. W.* **1954**, *79*, 97–141.
63. Barata, M. Marine fungi from Mira River salt marsh in Portugal. *Rev. Iberoam. Micol.* **2006**, *23*, 179–184. [CrossRef]
64. Peña, N.I.; Arambarri, A.M. Hongos marinos lignícolas de la laguna costera de Mar Chiquita (provincia de Buenos Aires, Argentina). L. Ascomycotina y Deuteromycotina sobre *Spartina densiflora*. *Darwiniana* **1998**, *35*, 61–67.
65. Jones, E.B.G. Marine fungi. I. *Trans. Br. Mycol. Soc.* **1962**, *45*, 93–114. [CrossRef]
66. Kohlmeyer, J.; Volkmann-Kohlmeyer, B.; Eriksson, O.E. Fungi on *Juncus roemerianus*. 8. New bitunicate ascomycetes. *Can. J. Bot.* **1996**, *74*, 1830–1840. [CrossRef]
67. Johnson, T.W.; Hughes, G.C. *Robillarda phragmitis* Cunnell in estuarine waters. *Trans. Br. Mycol. Soc.* **1960**, *43*, 523–524. [CrossRef]
68. Seymour, A.B. *Host Index of the Fungi of North America*; Harvard University Press: Cambridge, MA, USA, 1929.
69. Saccardo, P.A. Fungi Gallici lecti a cl. viris P. Brunaud, Abb. Letendre, A. Malbranche, J. Therry, vel editi in Mycotheca Gallica C. Roumeguèri. Series II. *Michelia* **1880**, *2*, 39–135.
70. Ahmad, S. Contributions to the fungi of West Pakistan. VI. *Biol. Lahore* **1967**, *13*, 15–42.
71. Johnson, T.W.; Sparrow, F.K. *Fungi in Oceans and Estuaries*; Cramer: Weinheim, Germany, 1961.
72. Jones, E.B.G. Marine fungi: II. Ascomycetes and deuteromycetes from submerged wood and drift *Spartina*. *Trans. Br. Mycol. Soc.* **1963**, *46*, 135–144. [CrossRef]
73. Gessner, R.V. Seasonal occurrence and distribution of fungi associated with *Spartina alterniflora* from a Rhode Island estuary. *Mycologia* **1977**, *69*, 477–491. [CrossRef]
74. Gessner, R.V. Degradative enzyme production by salt-marsh fungi. *Bot. Mar.* **1980**, *23*, 133–139. [CrossRef]
75. Jaap, O. Weitere Beiträge zur Pilzflora der nordfriesischen Inseln. *Schr. Des Nat. Ver. Für Schleswig-Holst.* **1907**, *14*, 15–33.
76. Kohlmeyer, J.; Volkmann-Kohlmeyer, B.; Eriksson, O.E. Fungi on *Juncus roemerianus* 9. New obligate and facultative marine ascomycotina. *Bot. Mar.* **1997**, *40*, 291–300. [CrossRef]
77. Kohlmeyer, J.; Volkmann-Kohlmeyer, B.; Eriksson, O.E. Fungi on *Juncus roemerianus*. New marine and terrestrial ascomycetes. *Mycol. Res.* **1996**, *100*, 393–404. [CrossRef]
78. Kohlmeyer, J.; Volkmann-Kohlmeyer, B.; Eriksson, O.E. Fungi on *Juncus roemerianus* 2. New dictyosporous Ascomycetes. *Bot. Mar.* **1995**, *38*, 165–174. [CrossRef]
79. Devadatha, B.; Calabon, M.S.; Abeywickrama, P.D.; Hyde, K.D.; Jones, E.B.G. Molecular data reveals a new holomorphic marine fungus, *Halobyssothecium estuariae*, and the asexual morph of *Keissleriella phragmiticola*. *Mycology* **2020**, *11*, 167–183. [CrossRef] [PubMed]
80. Johnson, T.W. Marine Fungi. I. *Leptosphaeria* and *Pleospora*. *Mycologia* **1956**, *48*, 495–505. [CrossRef]
81. Wagner, D. Ecological studies on *Leptospheria discors*, a graminicolous fungus of salt marshes. *Nov. Hedwig.* **1969**, *18*, 383–396.
82. Webber, E.E. Marine ascomycetes from New England. *Bull. Torrey Bot. Club* **1970**, *97*, 119–120. [CrossRef]
83. Dayarathne, M.C.; Wanasinghe, D.N.; Jones, E.B.G.; Chomnunti, P.; Hyde, K.D. A novel marine genus, *Halobyssothecium* (Lentitheciaceae) and epitypification of *Halobyssothecium obiones* comb. nov. *Mycol. Prog.* **2018**, *17*, 1161–1171. [CrossRef]

84. Saccardo, P.A. *Sylloge fungorum Omnium Hucusque Cognitorum*; R. Friedländer & Sohn.: Berlin, Germany, 1883; Volume 2.
85. Calabon, M.S.; Jones, E.B.G.; Hyde, K.D.; Boonmee, S.; Tibell, S.; Tibell, L.; Pang, K.L.; Phookamsak, R. Phylogenetic assessment and taxonomic revision of *Halobyssothecium* and *Lentithecium* (Lentitheciaceae, Pleosporales). *Mycol. Prog.* **2021**, *20*, 701–720. [CrossRef]
86. Van Ryckegem, G.; Aptroot, A. A new *Massarina* and a new *Wettsteinina* (Ascomycota) from freshwater and tidal reeds. *Nov. Hedwig.* **2001**, *73*, 161–166. [CrossRef]
87. Tibell, S.; Tibell, L.; Pang, K.L.; Calabon, M.S.; Jones, E.B.G. Marine fungi of the Baltic Sea. *Mycology* **2020**, *11*, 195–213. [CrossRef] [PubMed]
88. Li, G.J.; Hyde, K.D.; Zhao, R.L.; Hongsanan, S.; Abdel-Aziz, F.A.; Abdel-Wahab, M.A.; Alvarado, P.; Alves-Silva, G.; Ammirati, J.F.; Ariyawansa, H.A.; et al. Fungal diversity notes 253–366: Taxonomic and phylogenetic contributions to fungal taxa. *Fungal Divers.* **2016**, *78*, 1–237. [CrossRef]
89. Ellis, J.B.; Everhart, B.M. New Fungi. *J. Mycol.* **1885**, *1*, 42–44. [CrossRef]
90. Ellis, J.B.; Everhart, B.M. *The North American Pyrenomycetes. A Contribution to Mycologic Botany*; Ellis & Everhart: Newfield, NJ, USA, 1892.
91. Lucas, M.T. Culture studies on portuguese species of *Leptosphaeria* I. *Trans. Br. Mycol. Soc.* **1963**, *46*, 361–367. [CrossRef]
92. Webber, E.E. Fungi from a Massachusetts salt marsh. *Trans. Am. Microsc. Soc.* **1966**, *85*, 556–558. [CrossRef]
93. Dennis, R.W.G. *British Ascomycetes*; J. Cramer: Lehre, Germany, 1968.
94. Poli, A.; Vizzini, A.; Prigione, V.; Varese, G.C. Basidiomycota isolated from the Mediterranean Sea—Phylogeny and putative ecological roles. *Fungal Ecol.* **2018**. [CrossRef]
95. Hansford, C.G. Australian fungi IV. New species and revisions (cont'd). *Proc. Linn. Soc. N. S. W.* **1957**, *82*, 209–229.
96. Hyde, K.D.; Hongsanan, S.; Jeewon, R.; Bhat, D.J.; McKenzie, E.H.C.; Jones, E.B.G.; Phookamsak, R.; Ariyawansa, H.A.; Boonmee, S.; Zhao, Q.; et al. Fungal diversity notes 367–490: Taxonomic and phylogenetic contributions to fungal taxa. *Fungal Divers.* **2016**, *80*, 1–270. [CrossRef]
97. Kohlmeyer, J.; Volkmann-Kohlmeyer, B.; Eriksson, O.E. Fungi on *Juncus roemerianus*. 4. New marine ascomycetes. *Mycologia* **1995**, *87*, 532–542. [CrossRef]
98. Kohlmeyer, J.; Volkmann-Kohlmeyer, B. Fungi on *Juncus roemerianus*. 14. Three new coelomycetes, including *Floricola*, anam.-gen. nov. *Bot. Mar.* **2000**, *43*, 385–392. [CrossRef]
99. Sydow, H. Mycotheca germanica. Fasc. XX–XXI. *Ann. Mycol.* **1911**, *9*, 554–558.
100. Dayarathne, M.; Jones, E.; Maharachchikumbura, S.; Devadatha, B.; Sarma, V.; Khongphinitbunjong, K.; Chomnunti, P.; Hyde, K. Morpho-molecular characterization of microfungi associated with marine based habitats. *Mycosphere* **2020**, *11*, 1–188. [CrossRef]
101. Hyde, K.D.; Chaiwan, N.; Norphanphoun, C.; Boonmee, S.; Camporesi, E.; Chethana, K.W.T.; Dayarathne, M.C.; de Silva, N.I.; Dissanayake, A.J.; Ekanayaka, A.H.; et al. Mycosphere notes 169–224. *Mycosphere* **2018**, *9*, 271–430. [CrossRef]
102. Wanasinghe, D.N.; Hyde, K.D.; Jeewon, R.; Crous, P.W.; Wijayawardene, N.N.; Jones, E.B.G.; Bhat, D.J.; Phillips, A.J.L.; Groenewald, J.Z.; Dayarathne, M.C.; et al. Phylogenetic revision of *Camarosporium* (Pleosporineae, Dothideomycetes) and allied genera. *Stud. Mycol.* **2017**, *87*, 207–256. [CrossRef]
103. Grove, W.B. *British Stem- and Leaf-Fungi Vol. II*; Cambridge University Press: Cambridge, UK, 1937.
104. Kohlmeyer, J.; Volkmann-Kohlmeyer, B. Atrotorquata and Loratospora: New ascomycete genera on *Juncus roemerianus*. *Syst. Ascomycetum* **1993**, *12*, 7–22.
105. Kohlmeyer, J.; Volkmann-Kohlmeyer, B.; Tsui, C.K.M. Fungi on *Juncus roemerianus*. 17. New ascomycetes and the hyphomycete genus *Kolletes* gen. nov. *Bot. Mar.* **2005**, *48*, 306–317. [CrossRef]
106. Yusoff, M.; Moss, S.T.; Jones, E.B.G. Ascospore ultrastructure of *Pleospora gaudefroyi* (Pleosporaceae, Loculoascomycetes, Ascomycotina). *Can. J. Bot.* **1994**, *72*, 1–6. [CrossRef]
107. Webster, J.; Lucas, M.T. Observations on British species of Pleospora. II. *Trans. Br. Mycol. Soc.* **1961**, *44*, 417–436. [CrossRef]
108. Sutton, B.C.; Pirozynski, K.A. Notes on British microfungi. I. *Trans. Br. Mycol. Soc.* **1963**, *46*, 505–522. [CrossRef]
109. Spegazzini, C. Mycetes Argentinenses. Series IV. *An. del Mus. Nac. Hist. Nat. Buenos Aires. Ser. 3* **1909**, *12*, 257–458.
110. Kohlmeyer, J.; Volkmann-Kohlmeyer, B. Fungi on *Juncus roemerianus*. 6. *Glomerobolus* gen. nov., the first ballistic member of Agonomycetales. *Mycologia* **1996**, *88*, 328–337. [CrossRef]
111. Cantrell, S.A.; Hanlin, R.T.; Newell, S.Y. A new species of *Lachnum* on *Spartina alterniflora*. *Mycotaxon* **1996**, *57*, 479–485.
112. Ellis, J.B.; Everhart, B.M. New species of fungi from various localities. *J. Mycol.* **1888**, *4*, 121–124. [CrossRef]
113. Kohlmeyer, J.; Baral, H.-O.; Volkmann-Kohlmeyer, B. Fungi on *Juncus roemerianus*. 10. A new *Orbilia* with ingoldian anamorph. *Mycologia* **1998**, *90*, 303–309. [CrossRef]
114. Seaver, F.J. *The North American Cup-Fungi (Inoperculates)*; Lancaster Press Inc.: Lancaster, PA, USA, 1951.
115. Wong, M.K.M.; Goh, T.K.; Hyde, K.D. A new species of *Phragmitensis* (ascomycetes) from senescent culms of *Phragmites australis*. *Fungal Divers.* **1999**, *2*, 175–180.
116. Kohlmeyer, J.; Volkmann Kohlmeyer, B.; Eriksson, O.E. Fungi on *Juncus roemerianus*. 3. New ascomycetes. *Bot. Mar.* **1995**, *38*, 175–186. [CrossRef]
117. Kohlmeyer, J.; Volkmann-Kohlmeyer, B. Fungi on *Juncus roemerianus*. 16. More new coelomycetes, including *Tetranacriella*, gen. nov. *Bot. Mar.* **2001**, *44*, 147–156. [CrossRef]

118. Kohlmeyer, J.; Volkmann-Kohlmeyer, B. Two new genera of Ascomycotina from saltmarsh *Juncus*. *Syst. Ascomycetum* **1993**, *11*, 95–106.
119. Loveless, A.R.; Peach, J.M. Evidence from ascospores for host restriction in *Claviceps purpurea*. *Trans. Br. Mycol. Soc.* **1986**, *86*, 603–610. [CrossRef]
120. Loveless, A.R. Conidial evidence for host restriction in *Claviceps purpurea*. *Trans. Br. Mycol. Soc.* **1971**, *56*, 419–434. [CrossRef]
121. Sprague, R. *Septoria* disease of Gramineae in western United States. *Oregon State Monogr. Stud. Bot.* **1944**, *6*, 1–151.
122. Spegazzini, C. Fungi Argentini additis nonnullis brasiliensibus montevidensibusque. *An. la Soc. Cient. Argent.* **1882**, *13*, 60–64.
123. Peach, J.M.; Loveless, A.R. A comparison of two methods of inoculating *Triticum aestivum* with spore suspensions of *Claviceps purpurea*. *Trans. Br. Mycol. Soc.* **1975**, *64*, 328–331. [CrossRef]
124. Eleuterius, L.N.; Meyers, S.P. *Claviceps purpurea* on *Spartina* in coastal marshes. *Mycologia* **1974**, *66*, 978–986. [CrossRef]
125. Sprague, R. *Diseases of Cereals and Grasses in North America: Fungi Except Smuts and Rusts*; Ronald Press Company: New York, NY, USA, 1950.
126. Mantle, P.G. Development of alkaloid production in vitro by a strain of *Claviceps purpurea* from *Spartina townsendii*. *Trans. Br. Mycol. Soc.* **1969**, *52*, 381–392. [CrossRef]
127. Moberley, D.G. Taxonomy and distribution of the genus Spartina. *Iowa State J. Sci.* **1956**, *30*, 471–574.
128. Saccardo, P.A. *Sylloge Fungorum Omnium Hucusque Cognitorum*; R. Friedländer & Sohn.: Berlin, Germany, 1886; Volume 4.
129. Abdullah, S.K.; Abdulkadder, M.A.; Goos, R.D. *Basramyces marinus* nom.nov. (hyphomycete) from southern marshes of Iraq. *Int. J. Mycol. Lichenol.* **1989**, *4*, 181–186.
130. Abdel-Wahab, M.A.; Pang, K.L.; Nagahama, T.; Abdel-Aziz, F.A.; Jones, E.B.G. Phylogenetic evaluation of anamorphic species of *Cirrenalia* and *Cumulospora* with the description of eight new genera and four new species. *Mycol. Prog.* **2010**, *9*, 537–558. [CrossRef]
131. Johnson, T.W. Marine Fungi. IV. *Lulworthia* and *Ceriosporopsis*. *Mycologia* **1958**, *50*, 151–163. [CrossRef]
132. Mounce, I.; Diehl, W.W. A new *Ophiobolus* on eelgrass. *Can. J. Res.* **1934**, *11*, 242–246. [CrossRef]
133. Johnson, T.W. Marine fungi. II. Ascomycetes and Deuteromycetes from submerged wood. *Mycologia* **1956**, *48*, 841–851. [CrossRef]
134. Lloyd, L.S.; Wilson, I.M. Development of the perithecium in *Lulworthia medusa* (Ell. & Ev.) Cribb & Cribb, a saprophyte on *Spartina townsendii*. *Trans. Br. Mycol. Soc.* **1962**, *45*, 359–372. [CrossRef]
135. Kohlmeyer, J.; Volkmann-Kohlmeyer, B. Fungi on *Juncus roemerianus*. 1. *Trichocladium medullare* sp. nov. *Mycotaxon* **1995**, *53*, 349–353.
136. Volkmann-Kohlmeyer, B.; Kohlmeyer, J. A new *Aniptodera* (Ascomycotina) from saltmarsh *Juncus*. *Bot. Mar.* **1994**, *37*, 109–114. [CrossRef]
137. Gessner, R.V. *Spartina alterniflora* seed fungi. *Can. J. Bot.* **1978**, *56*, 2942–2947. [CrossRef]
138. Kohlmeyer, J.; Kohlmeyer, E. *Icones Fungorum Maris. Fasc. 1–9*; J. Cramer: Weinheim/Lehre, Germany, 1967.
139. Jones, E.B.G. *Haligena spartinae* sp. nov., a pyrenomycete on *Spartina townsendii*. *Trans. Br. Mycol. Soc.* **1962**, *45*, 245–248. [CrossRef]
140. Liu, J.K.; Hyde, K.D.; Jones, E.B.G.; Ariyawansa, H.A.; Bhat, D.J.; Boonmee, S.; Maharachchikumbura, S.S.N.; McKenzie, E.H.C.; Phookamsak, R.; Phukhamsakda, C.; et al. Fungal diversity notes 1–110: Taxonomic and phylogenetic contributions to fungal species. *Fungal Divers.* **2015**, *72*, 1–197. [CrossRef]
141. Orton, C.R. Graminicolous species of Phyllachora in North America. *Mycologia* **1944**, *36*, 18–53. [CrossRef]
142. Kohlmeyer, J.; Volkmann-Kohlmeyer, B. *Octopodotus stupendus* gen. & sp. nov. and *Phyllachora paludicola* sp. nov., two marine fungi from *Spartina alterniflora*. *Mycologia* **2003**, *95*, 117–123. [CrossRef]
143. Kohlmeyer, J.; Volkmann-Kohlmeyer, B. *Aropsiclus* nom. nov. (Ascomycotina) to replace Sulcospora Kohlm. & Volk.-Kohlm. *Syst. Ascomycetum* **1994**, *13*, 24.
144. Spooner, B.M. New records and species of British microfungi. *Trans. Br. Mycol. Soc.* **1981**, *76*, 265–301. [CrossRef]
145. Dayarathne, M.C.; Wanasinghe, D.N.; Devadatha, B.; Abeywickrama, P.; Gareth Jones, E.B.; Chomnunti, P.; Sarma, V.V.; Hyde, K.D.; Lumyong, S.; Mckenzie, E.H.C. Modern taxonomic approaches to identifying diatrypaceous fungi from marine habitats, with a novel genus *Halocryptovalsa* Dayarathne & K.D.Hyde, gen. Nov. *Cryptogam. Mycol.* **2020**, *41*, 21–67. [CrossRef]
146. Johnson, T.W.J.; Gold, H.S. *Didymosamarospora*, a new genus of fungi from fresh and marine waters. *J. Elisha Mitchell Sci. Soc.* **1957**, *73*, 103–108.
147. Kohlmeyer, J.; Volkmann-Kohlmeyer, B. Fungi on *Juncus roemerianus*: New coelomycetes with notes on *Dwayaangam junci*. *Mycol. Res.* **2001**, *105*, 500–505. [CrossRef]
148. Kohlmeyer, J.; Volkmann-Kohlmeyer, B. Fungi on *Juncus roemerianus*. 13. *Hyphopolynema juncatile* sp. nov. *Mycotaxon* **1999**, *70*, 489–495.
149. Kohlmeyer, J.; Kohlmeyer, E. Marine fungi from tropical America and Africa. *Mycologia* **1971**, *63*, 831–861. [CrossRef]
150. Kohlmeyer, J.; Kohlmeyer, E. New marine fungi from mangroves and trees along eroding shorelines. *Nov. Hedwig.* **1965**, *9*, 89–104.
151. Ellis, J.B.; Everhart, B.M. New species of North American fungi from various localities. *Proc. Acad. Nat. Sci. Phila.* **1893**, *45*, 128–172. [CrossRef]
152. Cummins, G.B. *The Rust Fungi of Cereals, Grasses and Bamboos*; Springer: New York, NY, USA; Berlin/Heidelberg, Germany, 1971.
153. Orton, C.R. *Manual of the Rusts in United States and Canada*; Purdue Research Foundation: Lafayette, IN, USA, 1934.
154. von Thümen, F. Fungi Egyptiaci, Ser. III. *Flora (Regensbg.)* **1880**, *63*, 477–479.

155. McAlpine, D. *The Smuts of Australia, Their Structure, Life History, Treatment, and Classification*; Department of Agriculture of Victoria, Melbourn, Australia: Melbourne, VIC, Australia, 1910.
156. Bauer, R.; Lutz, M.; Piatek, M.; Vánky, K.; Oberwinkler, F. *Flamingomyces* and *Parvulago*, new genera of marine smut fungi (Ustilaginomycotina). *Mycol. Res.* **2007**, *111*, 1199–1206. [CrossRef] [PubMed]
157. Dennis, R.W.G. Fungi of *Ammophila arenaria* in Europe. *Rev. Biol.* **1983**, *12*, 15–48.
158. Treigienė, A. Fungi associated with *Ammophila arenaria* in Lithuania and taxonomical notes on some species. *Bot. Lith.* **2011**, *17*, 47–53.
159. Rodríguez-Echeverría, S.; Hol, W.H.G.; Freitas, H.; Eason, W.R.; Cook, R. Arbuscular mycorrhizal fungi of *Ammophila arenaria* (L.) Link: Spore abundance and root colonisation in six locations of the European coast. *Eur. J. Soil Biol.* **2008**, *44*, 30–36. [CrossRef]
160. Alva, P.; McKenzie, E.H.C.; Pointing, S.B.; Pena-Murala, R.; Hyde, K.D. Do sea grasses harbour endophytes? In *Fungi in Marine Environments*; Hyde, K.D., Ed.; Fungal Diversity Press: Hong Kong, China, 2002; pp. 167–178.
161. Wagner, D.T. Developmental morphology of *Leptosphaeria* discors (Saccardo and Ellis) Saccardo and Ellis. *Nov. Hedwig.* **1965**, *9*, 45–61.
162. Meyers, S.P.; Ahearn, D.G.; Alexander, S.K.; Cook, W.L. *Pichia spartinae*, a dominant yeast of the *Spartina* salt marsh. *Dev. Ind. Microbiol.* **1975**, *16*, 261–267.
163. Jones, E.B.G.; Pang, K.L.; Abdel-Wahab, M.A.; Scholz, B.; Hyde, K.D.; Boekhout, T.; Ebel, R.; Rateb, M.E.; Henderson, L.; Sakayaroj, J.; et al. An online resource for marine fungi. *Fungal Divers.* **2019**, *96*, 347–433. [CrossRef]
164. Mohamed, D.J.; Martiny, J.B.H. Patterns of fungal diversity and composition along a salinity gradient. *ISME J.* **2011**, *5*, 379–388. [CrossRef]
165. Stoeck, T.; Epstein, S. Novel eukaryotic lineages inferred from small-subunit rRNA analyses of oxygen-depleted marine environments. *Appl. Environ. Microbiol.* **2003**, *69*, 2657–2663. [CrossRef]
166. Dini-Andreote, F.; Pylro, V.S.; Baldrian, P.; Van Elsas, J.D.; Salles, J.F. Ecological succession reveals potential signatures of marine-terrestrial transition in salt marsh fungal communities. *ISME J.* **2016**, *10*, 1984–1997. [CrossRef]
167. D'entremont, T.W. Saltmarsh Sediment Fungal Communities and Arbuscular Mycorrhizal Fungi in Sporobolus Pumilus (Roth) (Poaceae) (Spartina Patens) of the Minas Basin, Nova Scotia; Identification, Abundance and Role in Restoration. Ph.D. Thesis, Acadia University, Wolfville, NS, Canada, 2019.
168. Khalmuratova, I.; Kim, H.; Nam, Y.J.; Oh, Y.; Jeong, M.J.; Choi, H.R.; You, Y.H.; Choo, Y.S.; Lee, I.J.; Shin, J.H.; et al. Diversity and plant growth promoting capacity of endophytic fungi associated with halophytic plants from the west coast of Korea. *Mycobiology* **2015**, *43*, 373–383. [CrossRef] [PubMed]
169. You, Y.H.; Yoon, H.; Kang, S.M.; Shin, J.H.; Choo, Y.S.; Lee, I.J.; Lee, J.M.; Kim, J.G. Fungal diversity and plant growth promotion of endophytic fungi from six halophytes in Suncheon Bay. *J. Microbiol. Biotechnol.* **2012**, *22*, 1549–1556. [CrossRef] [PubMed]
170. Khalmuratova, I.; Choi, D.H.; Woo, J.R.; Jeong, M.J.; Oh, Y.; Kim, Y.G.; Lee, I.J.; Choo, Y.S.; Kim, J.G. Diversity and plant growth-promoting effects of fungal endophytes isolated from salt-tolerant plants. *J. Microbiol. Biotechnol.* **2020**, *30*, 1680–1687. [CrossRef] [PubMed]
171. Kandalepas, D.; Blum, M.J.; Van Bael, S.A. Shifts in symbiotic endophyte communities of a foundational salt marsh grass following oil exposure from the deepwater horizon oil spill. *PLoS ONE* **2015**, *10*, e0122378. [CrossRef] [PubMed]
172. Maciá-Vicente, J.G.; Jansson, H.B.; Abdullah, S.K.; Descals, E.; Salinas, J.; Lopez-Llorca, L.V. Fungal root endophytes from natural vegetation in Mediterranean environments with special reference to *Fusarium* spp. *FEMS Microbiol. Ecol.* **2008**, *64*, 90–105. [CrossRef]
173. You, Y.-H.; Yoon, H.-J.; Woo, J.-R.; Seo, Y.-G.; Kim, M.-A.; Lee, G.-M.; Kim, J.-G. Diversity of endophytic fungi from the roots of halophytes growing in Go-chang salt marsh. *Korean J. Mycol.* **2012**, *40*, 86–92. [CrossRef]
174. Khalmuratova, I.; Choi, D.H.; Yoon, H.J.; Yoon, T.M.; Kim, J.G. Diversity and plant growth promotion of fungal endophytes in five halophytes from the Buan salt marsh. *J. Microbiol. Biotechnol.* **2021**, *31*, 408–418. [CrossRef]
175. Kalyanasundaram, I.; Nagamuthu, J.; Muthukumaraswamy, S. Antimicrobial activity of endophytic fungi isolated and identified from salt marsh plant in Vellar Estuary. *J. Microbiol. Biotechnol.* **2015**, *7*, 13–20. [CrossRef]
176. Elmer, W.H.; Marra, R.E. New species of *Fusarium* associated with dieback of *Spartina alterniflora* in Atlantic salt marshes. *Mycologia* **2011**, *103*, 806–819. [CrossRef]
177. Alber, M.; Swenson, E.M.; Adamowicz, S.C.; Mendelssohn, I.A. Salt marsh dieback: An overview of recent events in the US. *Estuar. Coast. Shelf Sci.* **2008**, *80*, 1–11. [CrossRef]
178. Elmer, W.H. Pathogenic microfungi associated with *Spartina* in salt marshes. In *Biology of Microfungi*; Li, D.-W., Ed.; Springer: New York, NY, USA; Cham: Switzerland, 2016; pp. 615–630.
179. Govers, L.L.; in'T Veld, W.A.M.; Meffert, J.P.; Bouma, T.J.; van Rijswick, P.C.J.; Heusinkveld, J.H.T.; Orth, R.J.; van Katwijk, M.M.; van der Heide, T. Marine *Phytophthora* species can hamper conservation and restoration of vegetated coastal ecosystems. *Proc. R. Soc. B Biol. Sci.* **2016**, *283*, 20160812. [CrossRef] [PubMed]
180. Fisher, A.J.; DiTomaso, J.M.; Gordon, T.R.; Aegerter, B.J.; Ayres, D.R. Salt marsh *Claviceps purpurea* in native and invaded *Spartina* marshes in Northern California. *Plant Dis.* **2007**, *91*, 380–386. [CrossRef] [PubMed]
181. Raybould, A.F.; Gray, A.J.; Clarke, R.T. The long-term epidemic of *Claviceps purpurea* on *Spartina anglica* in Poole Harbour: Pattern of infection, effects on seed production and the role of *Fusarium Heterosporum*. *New Phytol.* **1998**, *138*, 497–505. [CrossRef]

182. Hyde, K.D.; Norphanphoun, C.; Chen, J.; Dissanayake, A.J.; Doilom, M.; Hongsanan, S.; Jayawardena, R.S.; Jeewon, R.; Perera, R.H.; Thongbai, B.; et al. Thailand's amazing diversity: Up to 96% of fungi in northern Thailand may be novel. *Fungal Divers.* **2018**, *93*, 215–239. [CrossRef]
183. Devadatha, B.; Jones, E.B.G.; Pang, K.L.; Abdel-Wahab, M.A.; Hyde, K.D.; Sakayaroj, J.; Bahkali, A.H.; Calabon, M.S.; Sarma, V.V.; Sutreong, S.; et al. Occurrence and geographical distribution of mangrove fungi. *Fungal Divers.* **2021**, *106*, 137–227. [CrossRef]
184. Dong, W.; Wang, B.; Hyde, K.D.; McKenzie, E.H.C.; Raja, H.A.; Tanaka, K.; Abdel-Wahab, M.A.; Abdel-Aziz, F.A.; Doilom, M.; Phookamsak, R.; et al. Freshwater Dothideomycetes. *Fungal Divers.* **2020**, *105*, 319–575. [CrossRef]
185. Luo, Z.L.; Hyde, K.D.; Liu, J.K.; Maharachchikumbura, S.S.N.; Jeewon, R.; Bao, D.F.; Bhat, D.J.; Lin, C.G.; Li, W.L.; Yang, J.; et al. Freshwater Sordariomycetes. *Fungal Divers.* **2019**, *99*, 451–660. [CrossRef]
186. Hyde, K.D.; Chethana, K.W.T.; Jayawardena, R.S.; Luangharn, T.; Calabon, M.S.; Jones, E.B.G.; Hongsanan, S.; Lumyong, S. The rise of mycology in Asia. *Sci. Asia* **2020**, *46*, 1–11. [CrossRef]
187. Calabon, M.S.; Jones, E.B.G.; Boonmee, S.; Doilom, M.; Lumyong, S.; Hyde, K.D. Five novel freshwater ascomycetes indicate high undiscovered diversity in lotic habitats in Thailand. *J. Fungi* **2021**, *7*, 1–27. [CrossRef]
188. Saintilan, N. Biogeography of Australian saltmarsh plants. *Austral Ecol.* **2009**, *34*, 929–937. [CrossRef]
189. Spencer, D.M.; Hickling, V.; Spencer, J.F.T. Yeasts from ponds, streams and salt marsh on the Gower Peninsula, Wales. In *Advances in Biotechnology. Proceedings of the Fifth International Yeast Symposium Held in London, Canada, July 20–25*; Stewart, G., Russell, I., Eds.; Elsevier: Amsterdam, The Netherlands, 1981; pp. 515–519.
190. Kurtzman, C.P. *Saturnospora ahearnii*, a new salt marsh yeast from Louisiana. *Antonie Van Leeuwenhoek* **1991**, *60*, 31–34. [CrossRef] [PubMed]

Journal of *Fungi*

Article

Genome and Metabolome MS-Based Mining of a Marine Strain of *Aspergillus affinis*

Micael F. M. Gonçalves [1], Sandra Hilário [1], Marta Tacão [1], Yves Van de Peer [2,3,4,5], Artur Alves [1,*] and Ana C. Esteves [1]

[1] CESAM, Department of Biology, University of Aveiro, 3810-193 Aveiro, Portugal; mfmg@ua.pt (M.F.M.G.); sandra.hilario@ua.pt (S.H.); mtacao@ua.pt (M.T.); acesteves@ua.pt (A.C.E.)
[2] Department of Plant Biotechnology and Bioinformatics, Ghent University, 9052 Ghent, Belgium; yves.vandepeer@psb.vib-ugent.be
[3] Center for Plant Systems Biology, VIB, 9052 Ghent, Belgium
[4] Department of Biochemistry, Genetics and Microbiology, University of Pretoria, Pretoria 0028, South Africa
[5] College of Horticulture, Academy for Advanced Interdisciplinary Studies, Nanjing Agricultural University, Nanjing 210095, China
* Correspondence: artur.alves@ua.pt

Abstract: *Aspergillus* section *Circumdati* encompasses several species that express both beneficial (e.g., biochemical transformation of steroids and alkaloids, enzymes and metabolites) and harmful compounds (e.g., production of ochratoxin A (OTA)). Given their relevance, it is important to analyze the genetic and metabolic diversity of the species of this section. We sequenced the genome of *Aspergillus affinis* CMG 70, isolated from sea water, and compared it with the genomes of species from section *Circumdati*, including *A. affinis*'s strain type. The *A. affinis* genome was characterized considering secondary metabolites biosynthetic gene clusters (BGCs), carbohydrate-active enzymes (CAZymes), and transporters. To uncover the biosynthetic potential of *A. affinis* CMG 70, an untargeted metabolomics (LC-MS/MS) approach was used. Cultivating the fungus in the presence and absence of sea salt showed that *A. affinis* CMG 70 metabolite profiles are salt dependent. Analyses of the methanolic crude extract revealed the presence of both unknown and well-known *Aspergillus* compounds, such as ochratoxin A, anti-viral (e.g., 3,5-Di-tert-butyl-4-hydroxybenzoic acid and epigallocatechin), anti-bacterial (e.g., 3-Hydroxybenzyl alcohol, L-pyroglutamic acid, lecanoric acid), antifungal (e.g., L-pyroglutamic acid, 9,12,13-Trihydroxyoctadec-10-enoic acid, hydroxyferulic acid), and chemotherapeutic (e.g., daunomycinone, mitoxantrone) related metabolites. Comparative analysis of 17 genomes from 16 *Aspergillus* species revealed abundant CAZymes (568 per species), secondary metabolite BGCs (73 per species), and transporters (1359 per species). Some BGCs are highly conserved in this section (e.g., pyranonigrin E and UNII-YC2Q1O94PT (ACR toxin I)), while others are incomplete or completely lost among species (e.g., bikaverin and chaetoglobosins were found exclusively in series *Sclerotiorum*, while asperlactone seemed completely lost). The results of this study, including genome analysis and metabolome characterization, emphasize the molecular diversity of *A. affinis* CMG 70, as well as of other species in the section *Circumdati*.

Keywords: antimicrobial; anti-cancer; comparative genomics; marine fungi; metabolites; whole genome sequencing

Citation: Gonçalves, M.F.M.; Hilário, S.; Tacão, M.; Van de Peer, Y.; Alves, A.; Esteves, A.C. Genome and Metabolome MS-Based Mining of a Marine Strain of *Aspergillus affinis*. *J. Fungi* **2021**, *7*, 1091. https://doi.org/10.3390/jof7121091

Academic Editor: Wei Li

Received: 2 December 2021
Accepted: 17 December 2021
Published: 18 December 2021

1. Introduction

Aspergillus section *Circumdati* encompasses 27 species, many of which are economically, biotechnologically, and medically important, and having vast impacts on human and animal health [1]. *Circumdati* species are notorious for producing highly toxic fungal compounds (e.g., ochratoxin A (OTA)) [2]. In contrast, some *Circumdati* species, such as *A. ochraceus* and *A. slerotiorum*, are used for the biotransformation of steroids and alkaloids, while *A. melleus* is an important source of proteolytic enzymes [3]. The yellow aspergilli *A. affinis* was the first species of the section *Circumdati* isolated from a freshwater environment (submerged decomposing leaf litter), and is able to produce OTA [4], posing a potential danger for marine organisms and for their consumers. However, little is known about *A. affinis*'s ecological role or biotechnological potential. Marine compounds from fungi represent the largest category of marine natural products (MNPs) [5] and are the most widely reported to exhibit a diverse source and remarkable relevant bioactivities, including antibacterial, antifungal, antiviral, anticancer, anti-inflammatory, antioxidant, and cytotoxic activities [6–11]. Recently, during a survey of marine fungi from the Portuguese coast, Gonçalves et al. [12] isolated a strain of *A. affinis* from sea water (strain CMG 70). *Aspergillus affinis* CMG 70 produces amylases, cellulases, chitinases, proteinases and xylanases, and has antimicrobial, cytotoxic and antioxidant activities [13]. The bioactivity profile of *A. affinis* CMG 70 suggests that it has the potential to represent a useful source of novel secondary metabolites. In fact, more than 30% of metabolites isolated from fungi so far are from *Aspergillus* or *Penicillium* [14].

Advances in high-throughput genome sequencing, metabolomic technologies, and bioinformatics have further enabled research on fungal biology, revealing an untapped source of (novel) biosynthetic gene clusters and compounds with a wide range of biotechnological applications [6,15–17]. At present, there is only one BioProject (PRJNA421325) on the whole-genome of *A. affinis* strain ATCC MYA-4773^T (=CBS 129190). Sixteen genomes of other species belonging to the section *Circumdati* are available at the JGI Genome Portal database.

To disclose the biotechnological potential of *A. affinis*, we sequenced and analyzed the genome of *A. affinis* CMG 70. Additionally, to assess the effect of sea salt on the metabolic output of *A. affinis* CMG 70, a metabolomic approach was used. Moreover, given the importance of the section *Circumdati*, a comparative analysis was undertaken using 16 fungal genomes of this section.

2. Materials and Methods

2.1. Culture Conditions and DNA Extraction

Aspergillus affinis CMG 70 was previously isolated from sea water collected at Vagueira (Portugal), during a survey of marine fungi from the Portuguese coast in 2018 [12]. Two mycelium-colonized agar plugs were inoculated in Erlenmeyer flasks containing 50 mL of Potato Dextrose Broth (Merck, Darmstadt, Germany) at 25 °C, without agitation for 7 days, in the dark. Afterwards, mycelium was filtered through sterile filter paper, and was immediately grounded in liquid nitrogen. DNA was extracted according to Pitcher et al. [18]. The quality of the DNA was assessed by agarose gel electrophoresis (0.8%). DNA purity and quantity were determined using a NanoDrop 2000 spectrophotometer (Thermo Fisher Scientific Inc., Waltham, MA, USA).

2.2. Genome Sequencing, Assembly, and Gene Prediction

The *A. affinis* strain CMG 70 genome was sequenced from 100 ng of genomic DNA by Genome Sequencer Illumina HiSeq (2 × 150 bp paired-end reads) with NovaSeq 6000 S2 PE150 XP platform (Eurofins, Brussels, Belgium). Adapter sequences and low-quality reads were removed from output reads using the Trimmomatic software v.0.39 [19]. Quality assessment analysis of the reads was performed with the fastQC program (Babraham, Bioinformatics, 2016). Then, the nuclear genome was assembled using SPAdes v.3.14 [20]. The QUAST web interface (http://cab.cc.spbu.ru/quast/, accessed on 10 January 2021) was used to assess the quality of the assembled genome. Gene prediction of the draft genome assembly was performed using Augustus v.3.3.3 [21] with default parameters and using *A. oryzae* gene models as training set.

2.3. Genome Annotation and Functional Analysis

Several complementary methodologies were used to annotate the sequences. Dispersed Repeat sequences (DRs) were identified in OmicsBox (v.1.4.12) with the Repeat

Masking option (RepeatMasker v.4.0.9) [22]. Tandem Repeat sequences (TRs) were identified by Tandem Repeats Finder (TRF) (http://tandem.bu.edu/cgi-bin/trdb/trdb.exe, accessed on 5 February 2021) [23]. Analyses of noncoding RNAs, such as tRNAs were carried out using the tRNAscan-SE tool (http://lowelab.ucsc.edu/tRNAscan-SE/, accessed on 5 February 2021) with default parameters [24].

Predicted genes were functionally annotated with OmicsBox using Blast2GO [25] against NCBI's nonredundant (Nr) database, Gene Ontology (GO), and Kyoto Encyclopedia of Genes and Genomes (KEGG) with an e-value threshold of 1×10^{-3}. Proteins were classified using InterProScan and the Evolutionary Genealogy of Genes: Non-supervised Orthologous Groups (EggNOG), which also contains the orthologous groups from the original COG/KOG database (euKaryotic cluster of Orthologous Groups of proteins) with an e-value of 1×10^{3}.

Carbohydrate-degrading enzymes (CAZymes) were predicted with the web-based application dbCAN (HMMs 5.0) (http://www.cazy.org/, accessed on 10 February 2021) with default settings (http://bcb.unl.edu/dbCAN2/blast.php, accessed on 10 February 2021) [26]. Fungal secreted proteins, including signal peptides, were predicted using the SignalP [27]. Transporters were identified with the BLAST analysis against the Transporter Classification (TC) Database [28], downloaded in March 2021, with an e-value threshold of 1×10^{-5}, using the Geneious Prime v.2021.0.3 (http://www.geneious.com, accessed on 15 February 2021). The genome was also screened for the presence of Biosynthetic Gene Clusters (BGCs) using the web-based application antiSMASH v.5.0, using strictness 'relaxed' option for detection of well-defined and partials clusters containing the functional parts [29].

2.4. Comparative Analyses

The genome of *A. affinis* CMG 70 was compared to other sequenced and annotated genomes from 16 species in the section *Circumdati* (Table 1). The information available in JGI Genome Portal databases such as genome size, GC content, CAZymes, transporters, and BGCs abundance was used to evaluate genetic and metabolic diversity within the section *Circumdati*. One-way analysis of variance (ANOVA) followed by Student *t*-test ($p < 0.05$) was used to determine significant differences in CAZyme family diversity and transporters abundance between species of the three series within the section *Circumdati*. In addition, a phylogenetic analysis based on Maximum Likelihood using the sequences of the rDNA internal transcribed spacer region (ITS) and tubulin (*tub2*) of the *Aspergillus* strains was performed using MEGA7 [30]. Clade stability was assessed using a bootstrap analysis with 1000 replicates. Sequences were aligned with ClustalX version 2.1 [31] using the parameters described in Gonçalves et al. [12]. All alignments were checked and edited with the BioEdit Alignment Editor version 7.2.5 [32].

2.5. Small-Scale Fermentation and Extraction of Metabolites

A small-scale fermentation was carried out as described in [13]. Briefly, two plugs of mycelium-colonized agar were inoculated into 1-L Erlenmeyer flasks containing 250 mL of PDB (Merck, Germany) in two conditions: with and without 3% sea salt (Sigma-Aldrich, Darmstadt, Germany), with 4 replicates for each condition. The fungus was grown at 25 °C under stationary conditions for 14 days. Culture filtrates were obtained by filtering the mycelium through sterile filter paper. Then, the culture media was filtrated with 0.45 µm cellulose membrane (GN-6 Metricel, Pall Corporation, New York, NY, USA) followed by 0.2 µm nitrate cellulose membrane (Sartorius Stedim Biotech, Gottingen, Germany) in a vacuum system. Cultures media from the 4 replicates were pooled and lyophilized, and dried cultures media were weighed and transferred to tubes. Next, 20 mL of cold 80% MeOH (−80 °C) were added to each tube (containing 2 g of dried sample) and vortexed for 5 min. Each mixture was centrifuged at 14,000× *g* for 10 min at 4 °C to remove precipitated proteins. The supernatant was collected, and the extraction process was repeated. After extraction, the methanolic extracts were filtered using a glass microfiber filter 0.47 mm

(Prat Dumas, Couze-St-Front, France), evaporated in vacuo using a rotary evaporator and lyophilized.

Table 1. List of *Aspergillus* species and strains used in this study. Accessions numbers of ITS and *tub2* are provided.

Species	Strain	Host/Substrate	ITS	*tub2*
Aspergillus affinis	CMG 70	Sea water	MZ230522	MZ254672
Aspergillus affinis	ATCC MYA-4773	Leaf Litter	MN431360	GU721092
Aspergillus crcarbonarius	CBS 556.65	Paper	EF661204	EF661099
Aspergillus cretensis	CBS 112802	Soil	FJ491572	EF661332
Aspergillus elegans	CBS 116.39	Japanese bread	EF661414	EF661349
Aspergillus melleus	CBS 546.65	Soil	EF661425	EF661326
Aspergillus muricatus	CBS 112808	Soil	EF661434	EF661356
Aspergillus ochraceus	AO.MF010	Soil	Genome	Genome
Aspergillus ostianus	CBS 103.07	Unknown	EF661421	EF661324
Aspergillus persi	CBS 112795	Toenail of patient	FJ491580	AY819988
Aspergillus pulvericola	CBS 137327	Indoor house dust	KJ775440	KJ775055
Aspergillus roseoglobulosus	CBS 112800	Decaying leaves	FJ491583	AY819984
Aspergillus sclerotiorum	CBS 549.65	Apple	EF661400	EF661337
Aspergillus sesamicola	CBS 137324	Sesame seed	KJ775437	KJ775063
Aspergillus steynii	IBT 23096	Green coffee bean	EF661416	EF661347
Aspergillus subramanianii	CBS 138230	Shelled brazil nuts	EF661403	EF661339
Aspergillus westerdijkiae	CBS 112803	Plant	EF661427	EF661329
Aspergillus westlandensis	CBS 123905	Air sample	KJ775433	KJ775065

For LC-MS, 5 replicates of dried crude extracts (100 mg) for each condition were used. Metabolite extraction was performed by adding MeOH to each sample and vortexed for 40 min. Then, the samples were centrifuged for 5 min at 20,000× *g* and 400 µL of the methanolic fraction was vacuum dried. Afterwards, 100 µL of cyclohexane/water (1/1, *v*/*v*) was added to each sample and vortexed. Each mixture was centrifuged at 20,000× *g* for 5 min and 90 µL of the aqueous phase was filtered on a 96-filter plate and transferred to a 96-well plate. The samples were 10× diluted in water and 10 µL was analyzed by LC-MS.

2.6. LC-MS Data Analysis, Processing, and Visualization

UHPLC was performed on an ACQUITY UPLC I-Class system (Waters Corporation, Milford, MA, USA) consisting of a binary pump, a vacuum degasser, an autosampler, and a column oven. Chromatographic separation was carried out on an ACQUITY UPLC BEH C18 column (150 × 2.1 mm, 1.7 µm, Waters Corporation), at 40 °C. A gradient of solution A (99:1:0.1 water: acetonitrile: formic acid, pH 3) and solution B (99:1:0.1 acetonitrile: water: formic acid, pH 3) was used: 99% A for 0.1 min decreased to 50% A in 30 min, decreased to 30% in 5 min, decreased to 0% in 2 min. The flow rate was set to 0.35 mL min^{-1}, and the injection volume was 10 µL. The UHPLC system was coupled to a Vion IMS QTOF hybrid mass spectrometer (Waters Corporation). The LockSpray ion source was operated in negative electrospray ionization mode under the following specific conditions: capillary voltage, 2.5 kV; reference capillary voltage, 3 kV; cone voltage, 40 V; source offset, 50 V; source temperature, 120 °C; desolvation gas temperature, 600 °C; desolvation gas flow, 800 L h^{-1}; and cone gas flow, 50 L h^{-1}. Mass range was set from 50 to 1000 Da. The collision energy for full HDMSe was set at 6 eV (low energy) and ramped from 20 to 70 eV (high energy), intelligent data capture intensity threshold was set at 5. Nitrogen (greater than 99.5%) was employed as desolvation and cone gas. Leucin-enkephalin (250 pg μL^{-1} in water: acetonitrile 1:1 [*v*/*v*], with 0.1% formic acid) was used for the lock mass calibration, with scanning every 2 min at a scan time of 0.1 s. Profile data were recorded through a UNIFI Scientific Information System (Waters Corporation). Data processing was performed with Progenesis QI software v.2.4 (Waters Corporation). To understand *A. affinis* metabolome dynamics in response to sea salt, an IQR (interquartile range) filtering was applied due to the large number of significant ions, resulting in a selection of a set of 2500 ions for the data modeling. Principal Component Analysis (PCA), heatmaps, and *t*-test on log-transformed and pareto-scaled (normalized) of the filtered ions were generated and analyzed using online MetaboAnalyst v.4.0 software [33]. Computed *p*-values were adjusted using the Benjamin-Hochberg False Discovery Rate (FDR) correction. Ions having an FDR < 0.01 and a $\log_2$ fold change (FC) > 2 or <-2 were considered differently expressed. For identification purposes, the fragmentation data (ESI negative) of the significant ions were selected and matched against in-house library and 44 external spectral libraries (https://mona.fiehnlab.ucdavis.edu/, accessed on 3 December 2021), using MSsearch software v.2.6. For each ion, the best hit was based on a matching precursor ion ($m/z < 10$ ppm difference) and matching fragments (<50 ppm accuracy), generating five common fragments, including the precursor *m/z*. For each hit, the name of the matching compound followed by the collision energy used, the parent ion as a nominal mass, the chemical formula, a matching factor (MF), a reverse matching factor (RMF), and the name of the library found were obtained (File S1). File S1 contains some positive ionizations, but only ions in negative mode were considered for identification. Thus, annotation was conducted at level 2 of the Metabolomics Standards Initiative (MSI).

3. Results and Discussion

3.1. Sequencing, Assembly Data and Genomic Characteristics

General data related to the draft genome of *A. affinis* strain CMG 70 is presented in Table 2. Briefly, the CMG 70 genome size was estimated at 37.6 Mp, assembled in 421 contigs, with 11,763 predicted coding sequences from which 13.7% encode for hypothetical proteins, and a GC content of 50.21%. The *A. affinis* CMG 70 genome size is larger (0.8%), has a slightly higher GC content (0.2%) and has 5.2% fewer genes than ATCC MYA-4773^{T}.

Table 2. General statistics of the *Aspergillus affinis* CMG 70 genome assembly, and gene prediction.

	General Features
Genome assembled	37.6 Mb
Number of contigs (>500 bp)	421
Largest contig length	919,884 bp
N50	216,796 bp
N75	126,688 bp
L50	51 bp
L75	106 bp
GC content	50.21%
Number of predicted genes	11,763
Total length of predicted genes	18,519,399 bp
Average length of predicted genes	1574 bp
Total length of predicted genes/Genome assembled	49.3%
Average of exons per gene	3
Average of introns per gene	2

3.2. Repetitive Sequences and Prediction of tRNAs

Repetitive sequences are classified in Dispersed Repeats (DRs) and Tandem Repeats (TRs). The total length of the 12,411 DRs in the *A. affinis* CMG 70 genome amounts to 570,106 bp, covering 1.52% of the genome. With respect to the TRs, there are 4491 sequences with a total length of 262,036 bp covering 0.70% of the genome. 251 tRNAs were also predicted, with a total length of 22,032 bp covering 0.06% of the genome (Table 3). Among the tRNAs, 11 are possible pseudogenes and the remaining 240 anti-codon tRNAs correspond to the 20 common amino acid codons.

Table 3. Statistical results of repetitive sequences and noncoding RNAs for the *Aspergillus affinis* CMG 70 genome. SINEs: short interspersed nuclear elements; LINEs: long interspersed nuclear elements; LTRs: long terminal repeats.

Type		Number	Total Length (bp)	Percentage in Genome (%)
Interspersed repeat	SINEs	7	460	0.0012
	LINEs	69	4979	0.0133
	LTRs	124	30,681	0.0817
	DNA transposons	93	9220	0.0245
	Rolling-circles	0	0	0
	Unclassified	17	1433	0.0038
	Small RNA	124	13,588	0.0362
	Satellites	60	4618	0.0123
	Simple repeats	9755	391,258	1.0415
	Low complexicity	2162	113,869	0.3031
	Total	12,411	570,106	1.5176
Tandem repeat		4491	262,036	0.6975
tRNAs		251	22,032	0.0586

3.3. Gene Annotation

The genome of *A. affinis* CMG 70 has 11,584 genes annotated according to the NCBI's nonredundant protein (Nr), UniProt/Swiss-Prot, EggNOG, KEGG, GO, and Pfam databases (Table S1). There are 10,726 (91.2%) cellular proteins and approximately 1037 secreted proteins (8.8%) (Tables S1 and S3). The number of predicted proteins of CMG 70 is similar to that of ATCC MYA-4773^{T}, as well as to what has been reported for other *Aspergillus* species, which vary from 9078 in *A. coremiiformis* to 14,216 in *A. transmontanensis* [34]. Functional analysis (GO, Biological Processes) showed that most genes are involved in cellular (43%) and metabolic process (36%), cellular localization (12%), and biological regulation (9%) (Figure 1, Table S1). Genes classified within the "cellular process" category were mainly classified as being involved in posttranslational modification, protein turnover, chaperones (32%); intracellular trafficking, secretion, and vesicular transport (28%); signal transduction (16%); cell wall and cell cycle control (13%); cytoskeleton (6%); and others (5%), which include defense mechanisms and extracellular structures. Within the "metabolic process" category, *A. affinis* genes are involved in the metabolism and transport of carbohydrates (21%), amino acids (18%), lipids (10%) and inorganic ions (8%); in the biosynthesis of secondary metabolites (19%); and in energy production and conversion (13%). In GO, Molecular Functions, genes are involved in catalytic (51%), binding (39%), and transport (10%) activities (Figure 1, Table S1). These values are in agreement with what has been described in the literature for fungi.

3.4. Carbohydrate-Active Enzymes (CAZymes)

There are 566 genes encoding putative CAZymes, from which 295 carry signal peptides, that were annotated using the HMMER database (Table S2). Among these genes, 279 encode for glycoside hydrolases (GH), 22 for carbohydrate binding modules (CBM), 96 for glycosyltransferases (GT), 107 for oxidoreductases (AA), 39 for carbohydrate esterases (CE), and 23 for pectate lyases (PL) comprising 146 distinct CAZymes families. The main GH family includes β-glucosidades (GH3), chitinases (GH18), cellulases (GH5), β-xylosidades (GH43), polygalaturonases (GH28), and amylases (GH13). Regarding GT, UDP-glucuronosyltransferase (GT1), cellulose/chitin synthases (GT2), and xylanase (GT90) were the most abundant. Carbohydrate binding modules 67, which is a L-rhamnose-binding present in pectin and hemicellulose [35] and CBM20 associated with starch binding [36], were the most CBM abundant. All these enzymes have an important role in the degradation of polysaccharides, such as fucoidan, chitin, pectin, hemicellulose, and starch [37]. This may reveal a certain adaptation for the fungus to obtain carbon sources from different marine substrates, such as the algal fucoidan, pectin, cellulose, and chitin present in some algae and crab and shrimp shells. Glucooligosaccharide/chitooligosaccharide oxidases (AA7), cellobiose dehydrogenase (AA3), and copper-dependent lytic polysaccharide monooxygenases (AA9), which belong to auxiliary activity (AA) family were the most predominant. Carbohydrate esterase families are classified in 18 sub-families and catalyze the de-O or de-N-acylation of substituted saccharides. In *A. affinis*, 11 CEs are present with CE4, the most abundant. CE4 participates in the deacetylation of polysaccharides, such as xylan, chitin, and peptidoglycan [38]. Enzymes acting in the deacetylation of peptidoglycan may be involved in the degradation of bacterial cell wall, being attractive for drug design with potential application in biomedical industry. *Aspergillus affinis* genome encodes PL genes such as pectate lyase (PL1) and rhamnogalacturonan endolyase (PL4). This family is known to be involved in the breakdown of pectin that is synthesized in abundance by terrestrial plants but is not known as a marine polysaccharide [39]. However, pectin-like polysaccharides have been reported in red and green algae, microalgae and in seagrasses [40].

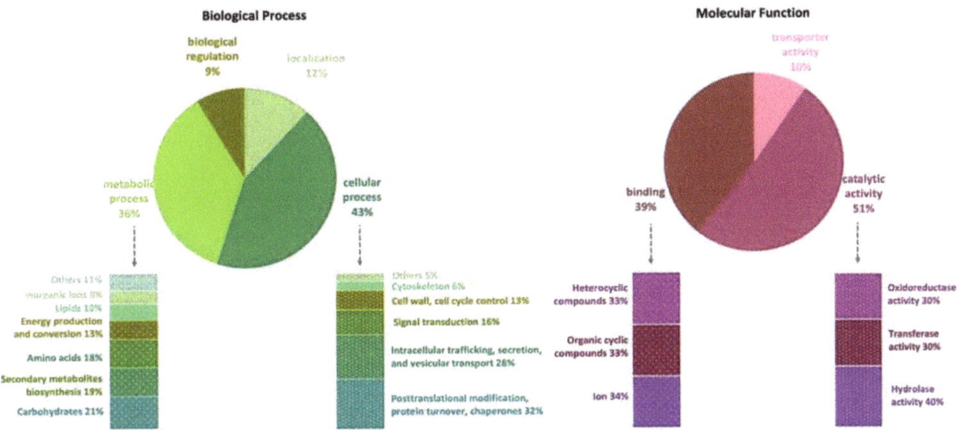

Figure 1. Gene Ontology (GO) functional annotation (pie charts) and EggNOG functional classification (bars charts) of the *Aspergillus affinis* CMG 70 genome.

3.5. Transporter Proteins

We observed transporters from all protein classification (TC) classes: channels and pores (TC 1), electrochemical potential-driven transporters (TC 2), primary active transporters (TC 3), group translocators (TC 4), transmembrane electron carriers (TC 5), accessory factors involved in transport (TC 8), and incompletely characterized transport systems (TC 9). There were 3005 predicted genes annotated as transporters against TC database, accounting for 25.6% of the total predicted genes for *A. affinis* (Table 4 and Table S4). Genes from the TC 2 class are the most abundant transporters of *A. affinis* CMG 70 genome (32.7%), followed by genes from the classes TC 1 (19.6%), TC 9 (16.1%), and TC 3 (15.3%). The *A. affinis* genome encodes transporters involved in the transport of zinc, sugar/H+, florfenicol, pantothenate, and various MFS (Major Facilitator Superfamily) transporters. It is known that MFS transporters in fungi play an important role in multidrug resistance [41] and are required for fungal growth under stress conditions [42]. Furthermore, we identified several genes coding for glycerol, inositol, sodium, and chloride transporters. By increasing these compounds' production and accumulation and others such as erythritol, arabitol, xylitol, mannitol, mycosporines and nitrogen-containing compounds (e.g., glycine, betaine, and free amino acids), the cell is able to maintain a positive turgor pressure [43]. Two mechanisms explain how fungi tolerate high salinity levels: their high affinity transport systems and osmoregulatory capacity. Marine fungi produce and accumulate specific solutes that allow them to function in saltwater [44]. For example, in the halophyte *Mesembryanthemum crystallinum*, the myo-inositol and its transporters play a major role in the tolerance to salt stress [45]. Moreover, Kogej et al. [43] and Plemenitaš et al. [46] also showed the production of glycerol, erythritol, arabitol, and mannitol and the involvement of alkali metal transporters (K^+/Na^+) by the halophilic fungus *Hortaea werneckii* as osmoadaptation. All four polyols have also been detected in *A. flavus* and *A. parasiticus* in response to osmotic stress [47]. We annotated many genes involved in glycerol, mannitol, inositol, trehalose, sorbitol, glycine, and betaine biosynthetic process, suggesting that *A. affinis* has adaptability mechanisms to thrive in saltwater. Furthermore, genes essential for the MAPK high osmolarity (*Sln1-Ypd1-Ssk1-Ssk2-Pbs2-Hog1* and *Sho1-Cdc42-Ste20(or Cla4)-Ste11-Pbs2-Hog1*) and cell wall stress cascades (*Wsc1-Rom2-Rho1-Pkc1-Bck1-Mkk1-Slt2*) were identified, resulting in glycerol accumulation to reduce the osmotic pressure and in cell wall remodeling. Gladfelter et al. [48] suggests that the high-osmolarity-glycerol signaling pathway seems in part to be linked to the water balance, cell stability and turgor in fungi. Also, transporters associated to ionic homeostasis (TC 1) encoding for calcium channels, nucleoporins, and heat shock 70 proteins' transporters were also detected, allowing rapid changes in the cell physiology of *A. affinis*.

Table 4. Genes predicted to code for transporters in the genome of *Aspergillus affinis* CMG 70.

Transporter Class	Number of Genes (n)
Channels and pores (TC 1)	586
Electrochemical potential-driven transporters (TC 2)	983
Primary active transporters (TC 3)	460
Group translocators (TC 4)	109
Transmembrane electron carriers (TC 5)	42
Accessory factors involved in transport (TC 8)	342
Incompletely characterized transport systems (TC 9)	483
Total	3005

3.6. Biosynthetic Gene Clusters

Seventy-two biosynthetic gene clusters (BGCs) involved in the secondary metabolism of *A. affinis* CMG 70 were predicted (Table S5). Biosynthetic gene clusters encode a form of machinery that produce bioactive compounds. In addition to biosynthetic genes, BGC typically include genes for expression control, self-resistance, and export of the compounds they encode [49].

The BGCs identified encode 7 terpenes, 5 indoles, 23 t1PKs (type 1 polyketide synthases), 8 NRPS (non-ribosomal peptide synthase), 7 NRPS-t1PKs, 14 NRPS-like, 3 NRPS-like-t1PKs, 1 NRPS-like-indole, 3 NRPS-indole, and 1 betalactone. From the BGCs identified, 9 BGCs have 100% similarity with known BGCs, such as asperlactone (anti-fungal), serinocyclin A (anti-insect), UNII-YC2Q1O94PT (ACR toxin I), pyranonigrin E (antimicrobial), biotin (vitamin), clavaric acid (antitumor), pseurotin (antibacterial), 6-methylsalicyclic acid (mammal-toxic) and AbT1 (anti-fungal). Other BGCs, such as cluster 33, shared 75% gene similarity with nidulanin A BGC, cluster 61 is likely to be an ochratoxin A BGC (60% of genes show similarity), and cluster 41 acts as aspergillic acid coding BGC (57% of genes show similarity). Cluster 14, 22 and 34 have 50% similarity with notoamide A, hexadehydroastechrome and asperphenamate, respectively. Other genes probably involved in BGC of squalestatin S1, ankaflavin, patulin, shearinine D, NG-391 and ochrindole A, were also detected.

3.7. Phylogenetic Analyses

Recently, Houbraken et al. [50] created three series within the section *Circumdati* to distinguish the species, namely *Circumdati*, *Sclerotiorum*, and *Steyniorum*. As can be seen in Figure 2, *A. affinis* CMG 70 groups in the same clade as the type species of *A. affinis* strain ATCC MYA-4773 (=CBS 129190), which belongs to ser. *Circumdati*. This series forms a sister clade with ser. *Steyniorum*, which is phylogenetic related with ser. *Sclerotiorum*.

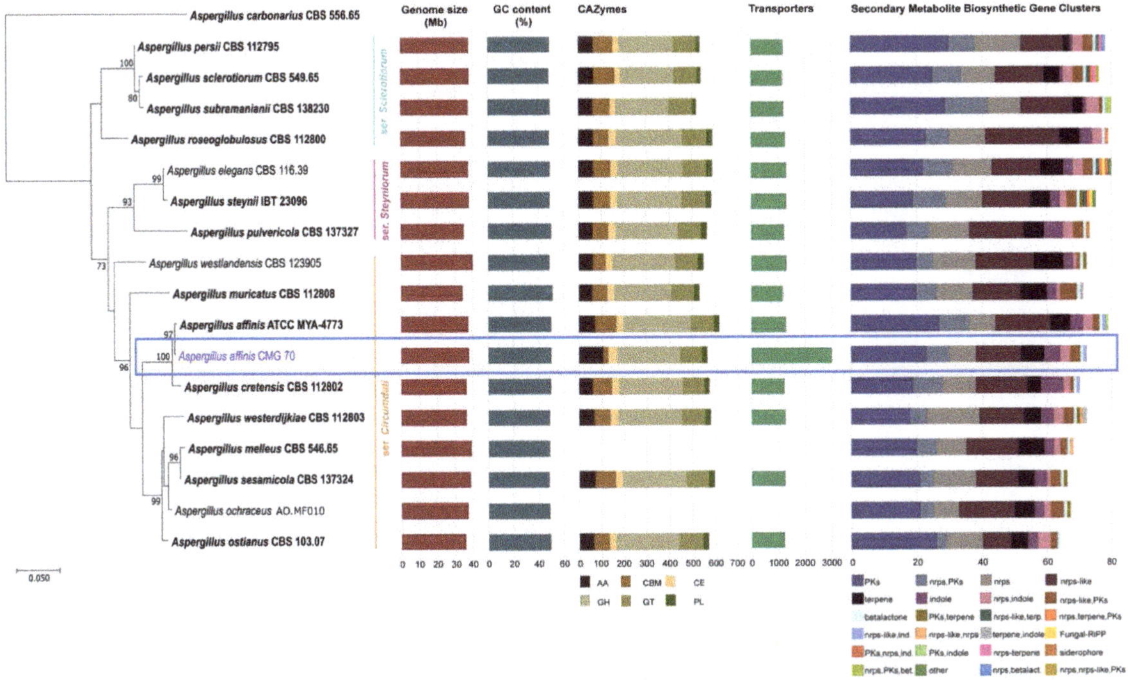

Figure 2. Phylogenetic relationship of 16 *Aspergillus* species from the section *Circumdati* based on ITS and *tub2* sequence data and inferred using the Maximum Likelihood method under Tamura 3-parameter model. The tree is drawn to scale, with branch lengths measured in the number of substitutions per site and rooted to *Aspergillus carbonarius* (CBS 556.65). Bootstrap values (≥70%) are shown at the nodes. Ex-type strains are in bold and the strain under study is in blue. The number of carbohydrate-active enzymes, transporters and biosynthetic gene clusters are indicated using bar graphs.

3.8. Comparative Analyses

3.8.1. General Features

The analysis showed that the genomic features of *A. affinis* strain CMG 70 and of ATCC MYA-4773^T are similar, regarding genome size, GC content and number of predicted genes. The average size of all genomes from the section *Circumdati* is 37.08 Mb. *Aspergillus muricatus* CBS 112808 has the smallest genome, i.e., 1.5% smaller than *A. westlandensis* CBS 123905. Moreover, *Circumdati* species genomes have a moderate GC content around 49.37% (Figure 2, Table S6).

3.8.2. CAZymes

Differences in the number of CAZymes between both strains of *A. affinis* were observed with more 8.6% CAZymes in ATCC MYA-4773^T (Figure 2, Table S7). In addition, some CAZymes appeared to be strain-specific, such as those from families AA2, CBM46, CBM67, CBM87, CE2, CE18, GH146, GH 109, and GT109 for strain CMG 70, while AA5, CBM1, CBM13, CBM18, CBM 32, CBM35, CBM48, CBM50, GH23, GH74, GH115, GH133, GT31, and GT41 for strain ATCC MYA-4773^T (Table S7). These differences may be related to the availability and type of carbohydrates in the environment where the strains where isolated from.

An average of 568 CAZymes per species was predicted (Figure 2, with the exception of *A. melleus* CBS 546.65 and *A. ochraceus* AO.MF010 since no information is available at the Mycocosm portal). This is similar to what has been reported for 23 species of the section *Flavi* (598/species) [34]. Within the section *Circumdati*, there is a clear difference between the type and number of CAZymes among the three series of this section (Figure 3A). Within this section, the ser. *Steyniorum* showed the lower abundance of CAZymes, while series *Sclerotiorum* showed the highest. Carbohydrate esterases, GT, PL and GH were more prevalent in ser. *Sclerotiorum*, in opposition to ser. *Circumdati* in which AA and CBM are more prevalent.

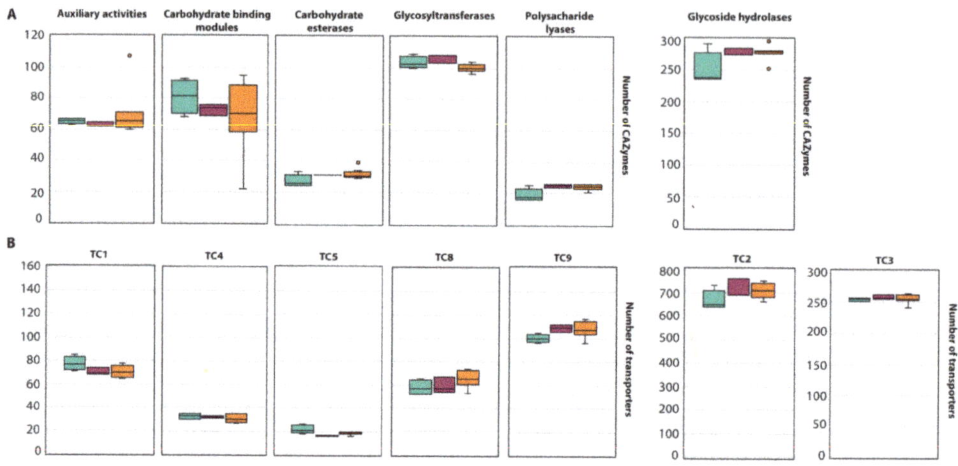

Figure 3. Boxplot representing the diversity of CAZyme family (**A**) and the abundance of transporters (**B**) in the section *Circumdati*: series *Sclerotiorum* (blue), series *Steyniorum* (pink) and series *Circumdati* (orange). In the boxplot, the midline represents the median and the upper and lower limit of the box represents the third and first quartile. One way analysis of variance (ANOVA) followed by Student *t*-test was used. No significant differences ($p > 0.05$) were observed.

3.8.3. Transporter Proteins

The genome of *A. affinis* CMG 70 contains 57.2% more transporters compared with ATCC MYA-4773^T (Figure 2, Table S6). The highest number of transporters predicted in *A. affinis* CMG 70 might be associated with to the salinity control, since this strain was isolated from a marine environment.

A total of 20,387 transporters were predicted for the 15 *Circumdati* strains analyzed (approximately 1360 transporters per strain) (Table S6, Figure 2). The number of predicted transporters is similar in most species. Exceptions are *A. melleus* CBS 546.65 and *A. ochraceus* AO.MF010 with no annotated transporters and *A. affinis* CMG 70, with 2.2 times the number of transporters than the rest of the genomes analyzed. TC2 transporters' family is the most represented in all *Circumdati* species although slightly more abundant in ser. *Steyniorum* (Figure 3B). The second most abundant class in the section *Circumdati* was the TC 3. According to our analyses, there is no difference in the distribution of this transporter family among the three series of this section (Figure 3B). No variation was also observed for TC 1, TC 4 and TC 5. In contrast, TC 8 and TC 9 are the most prevalent in the ser. *Circumdati*, followed by ser. *Steyniorum* and ser. *Sclerotiorum* (Figure 3B).

3.8.4. BGCs

Aspergillus affinis CMG 70 contains 8.9% fewer BGCs than *A. affinis* ATCC MYA-4773^T (Figure 2, Table S6). Overall, genomes of *Aspergillus* species from the section *Circumdati* are rich in gene clusters involved in the synthesis of secondary metabolites (average 73/species). Type 1 polyketide synthases were the most abundant type of gene clusters, followed by NRPS and NRPS-like, PKs-NRPS hybrid clusters, terpenes, and indoles (Figure 2). *Aspergillus elegans* and *A. subramanianii* have the highest number of BGCs (80), while *A. ostianus* has the lowest (64). Figure 4 shows the list and similarity of known secondary metabolite BGCs of *Circumdati* species genomes. The pyranonigrin E and UNII-

YC2Q1O94PT (ACR toxin I) BGCs were detected in all genomes with 100% similarity showing a high degree of conservation in this section. Pyranonigrin E is a PKs-NRPS hybrid metabolite from *A. niger* isolated from a marine source. Pyranonigrins are of considerable interest as potent antioxidants [51]. ACR toxin I is responsible for brown spot of rough lemon disease by the rough lemon pathotype of *Alternaria alternata* [52]. This suggests that *Circumdati* species may be able to also cause lemon leaf spot disease, but more studies are needed to understand the effect of this toxin in other plants.

Additionally, aspergillic acid, asperphenamate, and hexadehydroastechrome/terezine-D/astechrome BGCs were detected in all genomes but with similarity above 28%, 50%, and 37%, respectively, suggesting that some genes may be partially incomplete or lost. Nidulanin A, sequalestatin S1, notoamide A and ochrindole A were also detected in all genomes with exception of *A. ostianus*, *A. elegans*, *A. pulvericola*/*A. roseoglobulosus*, and *A. subramanianii*/*A. sclerotiorum* respectively (Figure 4). Interestingly, bikaverin and chaetoglobosins BGCs were detected exclusively in *Aspergillus* series *Sclerotiorum* (with the exception of *A. roseoglobulosus*). On the other hand, asperlactone BGC were detected in all species of series *Circumdati* and *Steyniorum* and curiously only in *A. roseoglobulosus*, which belong to series *Sclerotiorum* (Figure 4). Asperlactone belongs to methylsalicylic acid (MSA) type polyketide group and is produced by *A. westerdijkiae*. It has been reported that asperlactone has strong antibacterial and antifungal activities [53,54].

Aspergillic acid is a hydroxamic acid-containing pyrazinone isolated from *A. flavus* that exhibits antibiotic properties and toxicity for mammals [55]. Lebar et al. [56] reported that *Circumdati* species do not produce aspergillic acid, but neoaspergillic acid and its hydroxylated analog neohydroxyaspergillic acid, indicating that the cluster responsible for these is a homolog of aspergillic acid BGC. This six-gene cluster is constituted by *AsaA* (ankyrin domain protein), *AsaB* (GA4 desaturase family protein), *AsaC* (NRPS-like), *AsaD* (cytochrome P450 oxidoreductase), *AsaR* (C6 transcription factor) and *AsaE* (MFS transporter). This gene architecture was found in both strains of *A. affinis*, with the *AsaR* gene incomplete. However, the C6 transcription factor is not essential for the synthesis of aspergillic acid and its derivatives [56].

Nidulanin A is a cyclic tetrapeptide isolated from *A. nidulans*. The nidulanin A gene cluster is conserved in all *Aspergillus* and *Penicillium* spp. and its biological functions are not yet known [57]. Recently, Raffa and Keller [58] mentioned that this compound is being tested for antimicrobial or virulence-related properties. The presence of the four genes encoding nidulanin A (MFS and ABC multidrug transporter, NRPS and conserved hypothetical protein) was observed in both strains of *A. affinis*. Although we did not detect nidulanin in its metabolome, we cannot overrule the hypothesis of nidulanin being produced—or another very similar compound—by *Circumdati* species.

Notoamides are alkaloids with the pyranoindole ring common to stephacidins (antitumor alkaloids) found in *A. ochraceus* and in several members of the paraherquamide family. These prenylated indole alkaloids were obtained and characterized from a culture of a marine *Aspergillus* sp. isolated from the mussel *Mytilus edulis* [59]. Currently, there is no well-known property or function for notoamides, although some studies showed cytotoxicity against tumor cell lines, insecticidal, antibiotic and antiparasitic activities [60–62]. The genetic architecture of a notoamide BGC comprises 18 genes (*notA–R*). It was not possible to detect *notK–R* genes in both strains of *A. affinis* (Figure 5A). The cluster is identical only in 10 *not* genes (*notA–J*) and the pattern of the exon/intron arrangement in the corresponding genes is also highly similar between strains, including the 2 genes that were not described in the *not* gene cluster—the cold shock protein and the ubiquitin carbon terminal hydrolase genes. Li et al. [63] stated that the sequence similarity from *notK* to *notR* is quite reduced, and that the gene architecture differs drastically, suggesting that the previously assigned *not* gene cluster probably ends at *notJ* and the other *not* genes are unlikely involved in notoamide biosynthesis. Since *notK–R* genes were not also detected and considering the hypothesis of the *notK–R* not being involved in notoamide synthesis, it is possible that *Circumdati* species produce notoamide or a notoamide related compound.

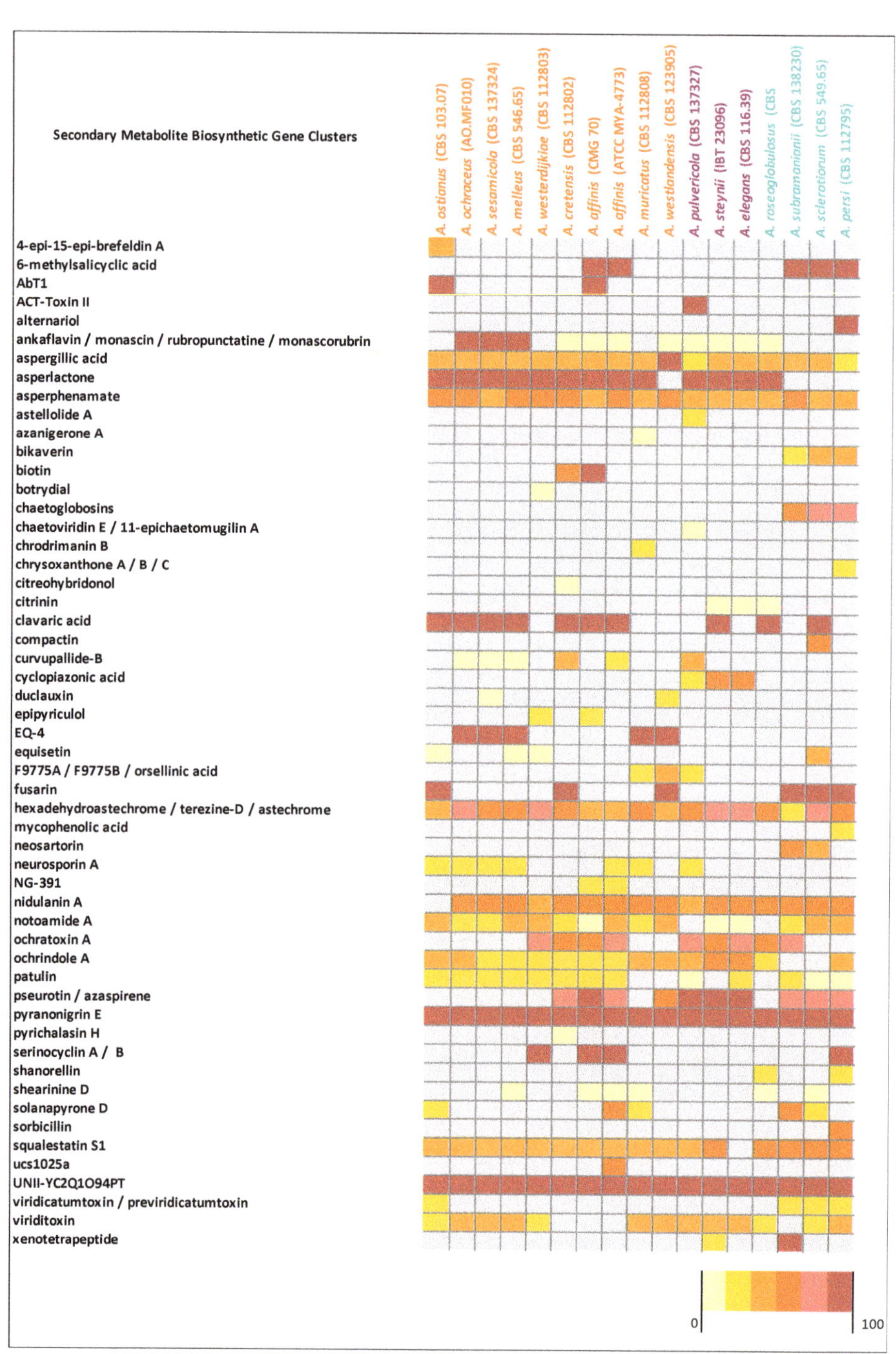

Figure 4. Matrix indicating the similarity of secondary metabolite gene clusters of *Circumdati* genomes (series *Sclerotiorum* in blue, series *Steyniorum* in pink, and series *Circumdati* in orange) in relation to known clusters from the antiSMASH. The color key is given as a percentage.

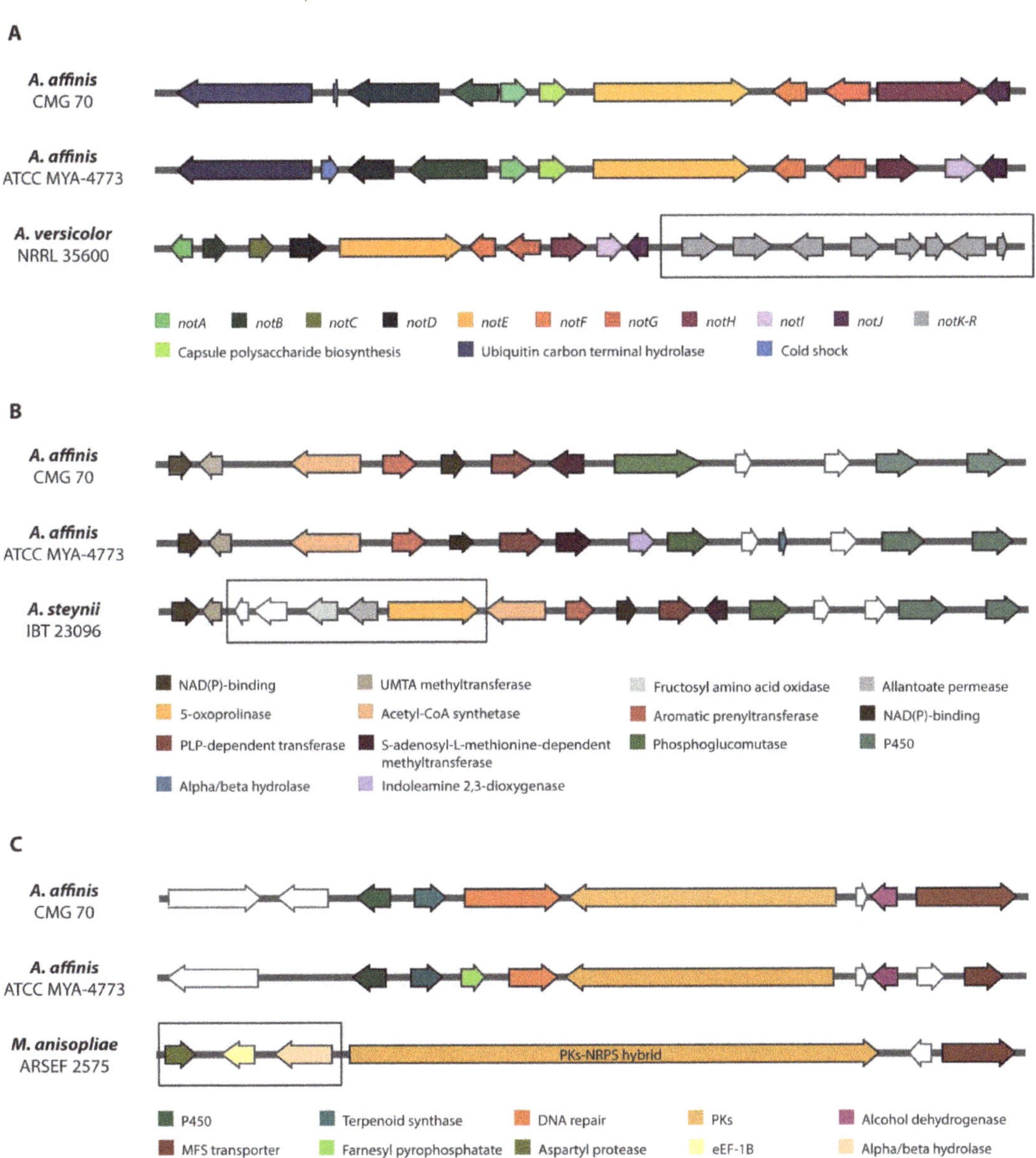

Figure 5. Comparison of three biosynthetic gene regions in *Aspergillus affinis* strains CMG 70 and ATCC MYA-4773^T with: (**A**) Notoamide BGC of *Aspergillus versicolor* NRRL 35600; (**B**) Ochrindole A BGC of *Aspergillus steynii* IBT 23096; and (**C**) NG-391 BGC of *Metarhizium anisopliae* ARSEF 2575. The genes encoding hypothetical proteins are represented as white arrows. The square box shows the missing genes in relation to the *A. affinis* cluster.

Ochratoxin A (OTA) is a problematic toxic metabolite that is widely distributed in food products, such as cereals, rice, soya, coffee, cocoa, beans, peas, peanuts, fresh grapes, and dry fruits, posing risks to human and animal health [2]. It was first reported in *A. ochraceus*, but many other *Aspergillus* and *Penicillium* species and other molds have been reported as OTA-producing species. Recently, Gil-Serna et al. [2] showed that the genomic regions that encode for OTA widely differ in *Circumdati* species. Some species, including *A. affinis*, contain a potentially functional OTA biosynthetic cluster suggesting that these species have the potential to synthetize the toxin, and others contain partial regions which might be related to their inability to produce OTA. In *A. affinis* CMG 70 we found the cluster region containing five genes known to be involved in OTA biosynthesis: halogenase, bZIP transcription factor, cytochrome P450 monooxygenase, NRPS, and a PKs. In fact, when we analyzed the dried crude extracts of *A. affinis* CMG 70 (see Section 3.9), ochratoxin A was detected as one of the most expressed compounds.

Ochrindoles (A–D) are prenylated bis-indolyl benzoid/quinone and ochrindole A is the most common of the four ochrindole compounds known [64]. Ochrindoles are known for their anti-insect properties, making ochrindoles (or derivatives)-producing species, of interest for the pesticide industry. Kjærbølling et al. [34] already identified candidate-genes for the ochrindole cluster in *A. steynii*, a member of the section *Circumdati*. Both strains of *A. affinis* shared 12 genes within ochrindole A BGC which is comprised of 17 genes (Figure 5B). The lack of the five genes (hypothetical protein, fructosyl amino acid oxidase, allantoate permease and 5-oxoprolinase coding genes) might (or not!) compromise the synthesis of ochrindole A. A deeper investigation on this subject is needed.

Patulin is a carcinogenic mycotoxin produced by several species found in fruit and vegetable-based products, posing a serious health risk to consumers [65]. Patulin production has been doubtfully reported in several species, including some *Penicillium* and *Aspergillus* spp., such as *A. ochraceus* [66]. Confirmed and efficient production of patulin has been found only in *A. clavatus*, *A. giganteus* and *A. longivesica* (section *Clavati*). The biosynthesis of patulin and its gene cluster are well known. We identified 3 of the 15 *pat* genes: *patC* (MFS transporter), *patD* (dehydrogenase) and *patE* (oxidoreductase), in both strains of *A. affinis*, suggesting that this species and probably all the others *Circumdati* species do not produce patulin. In fact, we did not detect patulin in the extracts of *A. affinis*. Nielsen et al. [67] showed that although *Penicillium roqueforti* has most of the *pat* genes needed for production of patulin, some genes are lacking and therefore it is unable to produce it.

Squalestatin S1 (also known as zaragozic acid) is a potent inhibitor of squalene synthase, an important enzyme for sterol biosynthesis [68]. Squalestatin S1 exhibits antifungal activity [69] and was found in some ascomycetes [70,71]. More recently the squalestatin S1 producing BGC from *Aspergillus* sp. Z5 was reported in *Paecilomyces penicillatus* [72] and halophilic marine fungus *Eurotium rubrum* [73]. The cluster of both *A. affinis* strains shared three out of four genes of squalestatin S1 BGC: the core enzyme farnesyl-diphosphate farnesyltransferase (squalene synthase), a DnaJ domain protein and other one conserved hypothetical protein.

When observing the predicted BGCs of the two strains of *A. affinis* (CMG 70 and ATCC MYA-4773^{T}) (Figure 4) we were able to detect some differences in the diversity of the BGCs present: the AbT1, biotin and epipyriculol BGCs were detected only in CMG 70, while curvupallide-B, neurosporin A, solanapyrone D and ucs1025a in ATCC MYA-4773^{T}. Furthermore, we found that the NG-391 BGC was exclusive of *A. affinis* suggesting that this cluster region is species specific. NG-391 was firstly identified in an insect pathogen *Metarhizium robertsii* [74] with similar structure to the mutagenic and carcinogenic mycotoxin fusarin C [75]. However, Donzelli et al. [74] reported that NG-391 does not contribute significantly to *M. robertsii* virulence. Recently, Kato et al. [76] isolated a lucilactaene compound from *Fusarium* sp. RK97-94 which is structurally related to NG-391. The same authors reported that lucilactaene and NG-391 do not have the 7-methyl group present in fusarins and show antimalarial activity and moderate growth inhibitory activity against cancer cells. With a similar core biosynthetic gene and the MFS transporter, the NG-391 cluster in *A. affinis* also has a cytochrome P450, a terpenoid synthase, a DNA repair protein, an alcohol dehydrogenase, and a farnesyl pyrophosphate synthase (Figure 5C).

3.9. Metabolome Analysis

We profiled the metabolomes of *A. affinis* CMG 70 grown with and without sea salt. Quintuplicate profiles were combined for each condition for comparative analysis. The full list of ions is given in Table S8. Despite the presence of unknown compounds, the major classes identified were polyketides, phenolic compounds, terpenes, amino acids, drugs, mycotoxins, carbohydrates, carboxylic acids, fatty acids, alkaloids, and indoles.

The scores of PCA on all filtered ions clearly revealed dissimilarities in the metabolome of the salted and non-salted extracts of *A. affinis* (Figure 6). These results show that *A. affinis* produces different compounds in response to osmotic stress and may adapt to salinity oscillations.

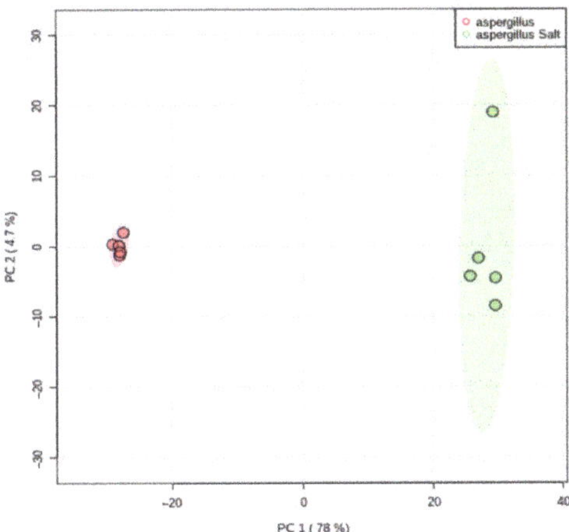

Figure 6. Principal Component Analysis (PCA) scores plot of salted and non-salted extracts of *Aspergillus affinis* CMG 70. Green represents salted extracts and in red non-salted extracts.

Subsequently, statistical testing on the filtered ions was conducted using a *t*-test. Computed *p*-values were adjusted using the false discovery rate (FDR) correction. 749 ions (FDR < 0.01 and a $\log_2$ fold change (FC) > 2 or <−2) were significantly more (523) or less (226) abundant when the fungus was grown with sea salt (Table S9, Figure 7). Due to the lack of information, many of the predicted compounds of *A. affinis* CMG 70 remain unidentified. From the identified compounds, the most abundant in both conditions were, for example, ochratoxin A, daunomycinone, maltose, maltotriose, laminaritetraose, 3,5-Di-tert-butyl-4-hydroxybenzoic acid, methyldopa, 2-hydroxypentyl glucosinolate, 2-Chloro-N6-cyclopentyladenosine, N-Fructosyl pyroglutamate, and lecanoric acid. Gonçalves et al. [13] reported different biological activities of *A. affinis* CMG 70 under salted and non-salted conditions, suggesting that salt induces an alteration to the metabolic profile. Compounds such as 19R-Hydroxy-PGF2alpha, catechin, ferulic acid sulfate, 9S,11R-Dihydroxy-15-oxo-13E-prostenoic acid, N-Caffeoyl-O-methyltyramine, 7-(2-Cyclopentylidenehydrazinyl)-7-oxoheptanoic acid, and D-myo-Inositol-1,3,4,5,6-pentaphosphate, and hydroxyferulic acid were most abundant under salted conditions, while 4-O-β-Galactopyranosyl-D-mannopyranose, lactobionic acid, (−)-Gallocatechin 3-gallate, quercetin-3-O-xyloside, 1,6-Anhydro-β-D-glucose, kelampayoside A, (+)-Fluprostenol, maltotetraose, isoreserpin, daunomycinone, (−)-Catechin gallate, aleuritic acid, maltose, palatinose, maltotriose, and laminaritetraose under non-salted conditions. As demonstrated by Overy et al. [77] media supplemented with sea salt applies a selective pressure on the metabolic profile of the fungus, regardless of the marine or terrestrial origin of the isolates. It is noteworthy that salt induces an increase of lipids, peptides, polyketides, and phenolic compounds in *A. affinis*. This suggests possible physiological mechanisms due to the accumulation of osmolytes and mechanical strengthening of the cells to adapt and tolerate different salinity levels.

At this point, it should be stressed that, as stated by Drabinska et al. [78], the presence of salt changes the nature of the molecular interactions between compounds. It might affect the quality of the extraction and induce the decrease of the intensity of the detected ions. Though some differences might be due to the presence of salts and not to differential expression by the fungus grown in the presence of salt, the water–cyclohexane extraction step should ensure that the amount of salt in the sample is reduced to a non-significant level.

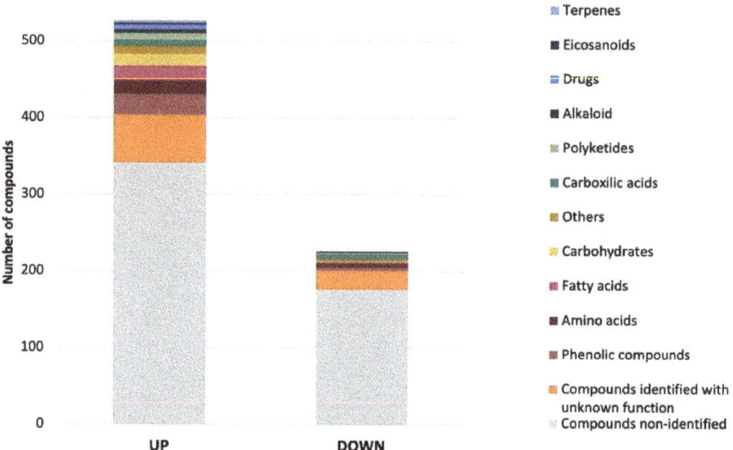

Figure 7. Structural classification of up and down regulated ($p < 0.01$) metabolites produced by *Aspergillus affinis* CMG 70, grown in the presence of sea salt, in comparison to *A. affinis* CMG 70, grown in the absence of sea salt.

Analysis of *A. affinis* extracts by LC-MS proved effective in detecting bioactive compounds that have been reported for their multiple activities, such as anti-bacterial, anti-fungal, anti-viral, anti-cancer, anti-inflammatory, and pesticides (Table 5). Despite significant efforts, new drugs are required to combat the increase in drug-resistance and the emergence of new viral infections. In this regard, we identified 3,5-Di-tert-butyl-4-hydroxybenzoic acid, that has been reported recently as a precursor of anti-viral compounds [79]. In addition, epigallocatechin, which has been associated with anti-viral properties [80], was also found in our crude extracts. To the best of our knowledge, this is the first report of these bioactive compounds in a fungus.

The dissemination of antibiotic resistance in clinical and non-clinical environments is a serious, difficult to control problem, and a risk to public health [81]. Therefore, the discovery and production of new anti-bacterial compounds is crucial and constitutes a breakthrough for medicine. In the present study, we identified anti-bacterial compounds, such as 3-Hydroxybenzyl alcohol, L-Pyroglutamic acid, and lecanoric acid. 3-Hydroxybenzyl alcohol was detected for the first time in *A. nidulans* isolated from a forest soil sample (India) [82]. We also identified genes involved in the biosynthesis of carbapenem, streptomycin, novobiocin, penicillin, and cephalosporin (Table S1). Apart from anti-viral and anti-bacterial compounds, we also detected antifungal compounds, such as 9,12,13-Trihydroxyoctadec-10-enoic, hydroxyferulic acid, L-pyroglutamic acid, lecanoric acid, scopoletin and 4,6-dihydroxy-4-(hydroxymethyl)-3,4a,8,8-tetramethyl-5,6,7,8a-tetrahydronaphthalen-1-one. This last compound was also identified in the marine-derived fungus *A. insuetus*, which was isolated from the Mediterranean sponge *Psammocinia* sp. [83] and in *Pleosporales* sp. from marine sediments of Bohai Sea [84]. Furthermore, lecanoric acid has been detected in a marine strain of *A. versicolor* [85].

Although marine fungi are less explored compared to terrestrial fungi, some marine species have yielded a wide range of diverse compounds with anti-cancer properties [7].

In this context, some known chemotherapeutic metabolites, such as daunomycinone and mitoxantrone, were also found in the crude extract of *A. affinis*. To our knowledge, this is first report of these compounds in fungi.

Table 5. Metabolites with biotechnological potential of *Aspergillus affinis* CMG70 belonging to various chemical classes and related functions. Metabolites were annotated at MSI-level 2. *m/z*—ratio mass/charge detected; Rt—retention time (min). PubChem was used for class identification.

Putative Metabolite	Molecular Formula	m/z	Rt	Adduct	Class	Function
3-Hydroxybenzyl alcohol	$C_7H_9O_2$	123.0445	9.22	$[M–H]^-$	Benzyl alcohol	Anti-bacterial [82]
9,12,13-Trihydroxyoctadec-10-enoic acid	$C_{18}H_{34}O_5$	329.2324	23.40	$[M–H]^-$	Fatty Acid	Anti-fungal [86]
Aleuritic acid	$C_{16}H_{32}O_5$	303.2169	19.84	$[M–H]^-$	Fatty Acid	Main component of shellac, a natural resin with applications in food, pharmaceutics and coatings [87]
3,5-Di-tert-butyl-4-hydroxybenzoic acid	$C_{15}H_{22}O_3$	249.1489	25.17	$[M–H]^-$	Phenolic Compound	Antioxidant and anti-inflammatory activities. Also, it is used as a precursor to anti-viral compounds and to cyclooxygenase inhibitors [79,88]
Carbidopa	$C_{10}H_{14}N_2O_4$	193.0498	10.04	$[M–H–H_4N_2]^-$	Catecholamine	Used in Parkinson's disease treatment [89]
Catechin	$C_{15}H_{14}O_6$	289.0013	5.34	$[M–H]^-$	Phenol	Used as carbon source for growth [90]
4,6-dihydroxy-4-(hydroxymethyl)-3,4a,8,8-tetramethyl-5,6,7,8a-tetrahydronaphthalen-1-one	$C_{15}H_{24}O_4$	267.1589	17.50	$[M–H]^-$	Naphthalene	Anti-fungal [83,84]
Daunomycinone	$C_{21}H_{18}O_8$	379.0825	1.05	$[M–H–H_2O]^-$	Naphthacene	Antibiotic with anti-cancer activity [91]
Epigallocatechin	$C_{15}H_{14}O_7$	611.1352	2.55	$[2M–H]^-$	Phenol	Anti-viral, antimicrobial, antitoxin and antitumor [80]
Folinic acid	$C_{20}H_{23}N_7O_7$	472.1561	4.14	$[M–H]^-$	Polyketide	Used in combination with other chemotherapy drugs [92]
Hydroxyferulic acid	$C_{20}H_{10}O_5$	209.0443	8.77	$[M–H]^-$	Coumaric Acid	Anti-fungal, involved in lignin biosynthesis [93]
Guanosine	$C_{10}H_{13}N_5O_5$	282.0838	2.27	$[M–H]^-$	Nucleoside	Antioxidant, neuroprotective, cardiotonic and immuno-modulatory properties [94]
Inosine	$C_{10}H_{12}N_4O_5$	267.0719	6.31	$[M–H]^-$	Nucleoside	Antioxidant, neuroprotective, cardiotonic and immuno-modulatory properties [94]
Isofraxidin	$C_{11}H_{10}O_5$	221.0444	10.81	$[M–H]^-$	Coumarin	Antioxidant, anti-malarial and neuprotective [95]
L-Pyroglutamic acid	$C_5H_7NO_3$	257.0768	1.80	$[2M–H]^-$	Imino Acid	Anti-fungal and anti-bacterial [96]
Lecanoric acid	$C_{16}H_{14}O_7$	167.0343	9.21	$[M–H–C_8H_6O_3]^-$	Polyphenol	Anti-bacterial, anti-fungal, anthelmintic and antioxidant properties [97,98]
Mitoxantrone	$C_{22}H_{28}N_4O_6$	443.1945	5.66	$[M–H]^-$	Anthraquinone	Anti-cancer [99]
Ochratoxin A	$C_{20}H_{18}ClNO_6$	402.0746	27.01	$[M–H]^-$	Carboxylic Acid	Mycotoxin, nephrotoxic, immunotoxic, carcinogenic and teratogenic [100]
Saccharopine	$C_{11}H_{20}N_2O_6$	275.1239	2.03	$[M–H]^-$	Amino Acid	Plays a role in the metabolism of lysine and swainsonine, which is a potential chemotherapy drug [101]
Scopoletin	$C_{10}H_8O_4$	191.0336	14.95	$[M–H]^-$	Coumarin	Anti-fungal [102]

4. Conclusions

This study discloses the genome sequence of *A. affinis* CMG 70 and analyses the biosynthetic potential among *Aspergillus* species from the section *Circumdati*. Overall, the present study has illustrated high similarity in genome size, GC content and transporters. Furthermore, we have also shown that members of the section *Circumdati* are a rich source of CAZymes, with different abundances between the three series of this section. We have shown that the pyranonigrin E and UNII-YC2Q1O94PT (ACR toxin I) BGCs are highly conserved in all genomes of the section *Circumdati*. Moreover, we also observed that some BGCs that are incomplete or truncated. In addition, the asperlactone cluster was detected only in series *Circumdati* and *Steyniorum* while it seemed to be completely lost in series *Sclerotiorum*. Contrarily, the bikaverin and chaetoglobosins clusters were found exclusively in *Sclerotiorum*.

The *A. affinis* CMG 70 genome has some clusters, transporters and CAZymes' genes that appear to be strain-specific. These features might be related to fungal adaptation to the marine environment, maintaining osmotic potential through: (1) the production and accumulation of specific solutes (osmolytes) that allow them to function in saltwater; (2) increase of transporters that allow ion exchange; (3) activation of signaling pathways allowing the water balance, cell stability, and positive turgor; (4) high affinity CAZymes to marine polysaccharides enabling the efficient degradation of the available carbon sources in the marine food web. Combining genome analysis with metabolites profiling showed a variety of gene components and secondary metabolites. Additionally, efforts should also be taken to determine the properties of both known and especially unknown molecules to unravel its promising potential. We cannot rule out that many of these molecules may play an important role in the fungus' osmoregulatory capacity to thrive in the marine environment. Moreover, different fermentation culture conditions should be used to amplify the production of specific compounds, evidencing the remarkable plasticity of fungal secondary metabolism.

Supplementary Materials: The following are available online at https://www.mdpi.com/article/10.3390/jof7121091/s1, Table S1: Gene annotation, Table S2: Carbohydrate active enzymes prediction, Table S3: Secreted proteins, Table S4: Transporter's prediction, Table S5: Biosynthetic Gene Clusters, Table S6: Summary of genomic features of *Circumdati* genomes, Table S7: Comparison of CAZymes families between *A. affinis* CMG 70 and ATCC MYA-4773, Table S8: Full list of compounds, Table S9: List of the significantly differential compounds, File S1: matched spectral library compounds.

Author Contributions: Conceptualization: M.F.M.G., A.C.E. and A.A.; data curation: M.F.M.G., S.H. and M.T.; funding acquisition: A.A.; formal analysis: M.F.M.G. and S.H.; investigation: M.F.M.G. and S.H.; methodology: M.F.M.G., S.H., M.T., A.C.E. and A.A.; resources: A.A.; supervision: A.A., A.C.E. and Y.V.d.P.; writing—original draft: M.F.M.G.; writing—review and editing: A.A., A.C.E., Y.V.d.P., M.T. and S.H. All authors have read and agreed to the published version of the manuscript.

Funding: The authors acknowledge financial support from the Portuguese Foundation for Science and Technology (FCT) to CESAM (UIDB/50017/2020+UIDP/50017/2020), Marta Tacão (CEECIND/00977/2020) and the PhD grants of M. Gonçalves (SFRH/BD/129020/2017) and S. Hilário (SFRH/BD/137394/2018).

Institutional Review Board Statement: Not applicable.

Informed Consent Statement: Not applicable.

Data Availability Statement: This Whole-Genome Shotgun project has been deposited in the GenBank database under the accession number JAGXNN000000000. The genome raw sequencing data and the assembly reported in this paper is associated with NCBI BioProject: PRJNA723818 and BioSample: SAMN18830867 within GenBank. The SRA accession number is SRR14465410. Data generated or analyzed during this study are included in this published article and its supplementary information files.

Acknowledgments: We want to acknowledge the VIB Metabolomics Core (VIB-UGent, Belgium) for performing the metabolic profiling, LC-MS data processing and basic statistical analysis and Scott Baker at Pacific Northwest National Laboratory, at Department of Environmental Molecular Sciences Laboratory, Richland, USA, for permitting to access the *Aspergillus* genomes that are deposited in the JGI Genome Portal database.

Conflicts of Interest: The authors declare no conflict of interest.

References

1. Visagie, C.M.; Varga, J.; Houbraken, J.; Meijer, M.; Kocsubé, S.; Yilmaz, N.; Fotedar, R.; Seifert, K.A.; Frisvad, J.C.; Samson, R.A. Ochratoxin production and taxonomy of the yellow aspergilli (*Aspergillus* section *Circumdati*). *Stud. Mycol.* **2014**, *78*, 1–61. [CrossRef] [PubMed]
2. Gil-Serna, J.; Vázquez, C.; Patiño, B. The Genomic regions that contain ochratoxin A biosynthetic genes widely differ in *Aspergillus* Section *Circumdati* Species. *Toxins* **2020**, *12*, 754. [CrossRef] [PubMed]
3. Luisetti, M.; Piccioni, P.D.; Dyne, K.; Donnini, M.; Bulgheroni, A.; Pasturenzi, L.; Donnetta, A.M.; Peonaal, V. Some properties of the alkaline proteinase from *Aspergillus melleus*. *Int. J. Tissue React.* **1991**, *13*, 187–192. [PubMed]
4. Davolos, D.; Persiani, A.M.; Pietrangeli, B.; Ricelli, A.; Maggi, O. *Aspergillus affinis* sp. nov., a novel ochratoxin A-producing *Aspergillus* species (section *Circumdati*) isolated from decomposing leaves. *Int. J. Syst. Evol. Microbiol.* **2012**, *62*, 1007–1015. [CrossRef]
5. Carroll, A.R.; Copp, B.R.; Davis, R.A.; Keyzers, R.A.; Prinsep, M.R. Marine natural products. *Nat. Prod. Rep.* **2019**, *36*, 122–173. [CrossRef]
6. Han, J.; Liu, M.; Jenkins, I.D.; Liu, X.; Zhang, L.; Quinn, R.J.; Feng, Y. Genome-inspired chemical exploration of marine fungus *Aspergillus fumigatus* MF071. *Mar. Drugs* **2020**, *18*, 352. [CrossRef]
7. Deshmukh, S.K.; Prakash, V.; Ranjan, N. Marine fungi: A source of potential anticancer compounds. *Front. Microbiol.* **2018**, *8*, 2536. [CrossRef]
8. Jans, P.E.; Mfuh, A.M.; Arman, H.D.; Shaffer, C.V.; Larionov, O.V.; Mooberry, S.L. Cytotoxicity and mechanism of action of the marine-derived fungal metabolite trichodermamide B and synthetic analogues. *J. Nat. Prod.* **2017**, *80*, 676–683. [CrossRef]
9. Chen, C.J.; Zhou, Y.Q.; Liu, X.X.; Zhang, W.J.; Hu, S.S.; Lin, L.P.; Huo, G.M.; Jiao, R.H.; Tan, R.X.; Ge, H.M. Antimicrobial and anti-inflammatory compounds from a marine fungus *Pleosporales* sp. *Tetrahedron Lett.* **2015**, *56*, 6183–6189. [CrossRef]
10. Moghadamtousi, S.Z.; Nikzad, S.; Kadir, H.A.; Abubakar, S.; Zandi, K. Potential antiviral agents from marine fungi: An overview. *Mar. Drugs* **2015**, *13*, 4520–4538. [CrossRef]
11. Sun, H.H.; Mao, W.J.; Chen, Y.; Guo, S.D.; Li, H.Y.; Qi, X.H.; Chen, Y.-L.; Xu, J. Isolation, chemical characteristics and antioxidant properties of the polysaccharides from marine fungus *Penicillium* sp. F23-2. *Carbohydr. Polym.* **2009**, *78*, 117–124. [CrossRef]
12. Gonçalves, M.F.M.; Esteves, A.C.; Alves, A. Revealing the hidden diversity of marine fungi in Portugal with the description of two novel species, *Neoascochyta fuci* sp. nov. and *Paraconiothyrium salinum* sp. nov. *Int. J. Syst. Evol. Microbiol.* **2020**, *70*, 5337–5354. [CrossRef] [PubMed]
13. Gonçalves, M.F.M.; Paço, A.; Escada, L.F.; Albuquerque, M.S.F.; Pinto, C.A.; Saraiva, J.; Duarte, A.S.; Rocha-Santos, T.A.P.; Esteves, A.C.; Alves, A. Unveiling biological activities of marine fungi: The effect of sea salt. *Appl. Sci.* **2021**, *11*, 6008. [CrossRef]
14. Berdy, J. Bioactive microbial metabolites. *J. Antibiot.* **2005**, *58*, 1–26. [CrossRef] [PubMed]
15. Han, J.; Zhang, J.; Song, Z.; Zhu, G.; Liu, M.; Dai, H.; Hsiang, T.; Liu, X.; Zhang, L.; Quinn, R.J.; et al. Genome-based mining of new antimicrobial meroterpenoids from the phytopathogenic fungus *Bipolaris sorokiniana* strain 11134. *Appl. Microbiol. Biotechnol.* **2020**, *104*, 3835. [CrossRef] [PubMed]
16. Han, J.; Zhang, J.; Song, Z.; Liu, M.; Hu, J.; Hou, C.; Zhu, G.; Jiang, L.; Xia, X.; Quinn, R.J.; et al. Genome-and MS-based mining of antibacterial chlorinated chromones and xanthones from the phytopathogenic fungus *Bipolaris sorokiniana* strain 11134. *Appl. Microbiol. Biotechnol.* **2019**, *103*, 5167–5181. [CrossRef] [PubMed]
17. Doroghazi, J.R.; Albright, J.C.; Goering, A.W.; Ju, K.-S.; Haines, R.R.; Tchalukov, K.A.; Labeda, D.P.; Kelleher, N.L.; Metcalf, W.W. A roadmap for natural product discovery based on large-scale genomics and metabolomics. *Nat. Chem. Biol.* **2014**, *10*, 963. [CrossRef]
18. Pitcher, D.G.; Saunders, N.A.; Owen, R.J. Rapid extraction of bacterial genomic DNA with guanidium thiocyanate. *Lett. Appl. Microbiol.* **1989**, *8*, 151–156. [CrossRef]
19. Bolger, A.M.; Lohse, M.; Usadel, B. Trimmomatic: A flexible trimmer for Illumina sequence data. *Bioinformatics* **2014**, *30*, 2114–2120. [CrossRef]
20. Bankevich, A.; Nurk, S.; Antipov, D.; Gurevich, A.A.; Dvorkin, M.; Kulikov, A.S.; Lesin, V.M.; Nikolenko, S.I.; Pham, S.; Prjibelski, A.D.; et al. SPAdes: A new genome assembly algorithm and its applications to single-cell sequencing. *J. Comput. Biol.* **2012**, *19*, 455–477. [CrossRef]
21. Stanke, M.; Steinkamp, R.; Waack, S.; Morgenstern, B. AUGUSTUS: A web server for gene finding in eukaryotes. *Nucleic Acids Res.* **2004**, *32*, W309–W312. [CrossRef] [PubMed]
22. Smit, A.F.A.; Hubley, R.; Green, P. RepeatMasker Open-4.0. 2018. Institute for Systems Biology. 2015. Available online: http://www.repeatmasker.org (accessed on 5 February 2021).
23. Gelfand, Y.; Rodriguez, A.; Benson, G. TRDB—The tandem repeats database. *Nucleic Acids Res.* **2007**, *35*, D80–D87. [CrossRef]

24. Lowe, T.M.; Eddy, S.R. tRNAscan-SE: A program for improved detection of transfer RNA genes in genomic sequence. *Nucleic Acids Res.* **1997**, *25*, 955–964. [CrossRef] [PubMed]
25. Götz, S.; García-Gómez, J.M.; Terol, J.; Williams, T.D.; Nagaraj, S.H.; Nueda, M.J.; Robles, M.; Talón, M.; Dopazo, J.; Conesa, A. High-throughput functional annotation and data mining with the Blast2GO suite. *Nucleic Acids Res.* **2008**, *36*, 3420–3435. [CrossRef] [PubMed]
26. Yin, Y.; Mao, X.; Yang, J.; Chen, X.; Mao, F.; Xu, Y. dbCAN: A web resource for automated carbohydrate-active enzyme annotation. *Nucleic Acids Res.* **2012**, *40*, W445–W451. [CrossRef] [PubMed]
27. Petersen, T.N.; Brunak, S.; Von Heijne, G.; Nielsen, H. SignalP 4.0: Discriminating signal peptides from transmembrane regions. *Nat. Methods* **2011**, *8*, 785–786. [CrossRef]
28. Saier Jr, M.H.; Reddy, V.S.; Tsu, B.V.; Ahmed, M.S.; Li, C.; Moreno-Hagelsieb, G. The transporter classification database (TCDB): Recent advances. *Nucleic Acids Res.* **2016**, *44*, D372–D379. [CrossRef]
29. Blin, K.; Shaw, S.; Steinke, K.; Villebro, R.; Ziemert, N.; Lee, S.Y.; Medema, M.H.; Weber, T. antiSMASH 5.0: Updates to the secondary metabolite genome mining pipeline. *Nucleic Acids Res.* **2019**, *47*, W81–W87. [CrossRef]
30. Kumar, S.; Stecher, G.; Tamura, K. MEGA7: Molecular evolutionary genetics analysis version 7.0 for bigger datasets. *Mol. Biol. Evol.* **2016**, *33*, 1870–1874. [CrossRef]
31. Thompson, J.D.; Gibson, T.J.; Plewniak, F.; Jeanmougin, F.; Higgins, D.G. The CLUSTAL_X windows inter- face: Flexible strategies for multiple sequence alignment aided by quality analysis tools. *Nucleic Acids Res.* **1997**, *25*, 4876–4882. [CrossRef]
32. Hall, T.A. BioEdit a user-friendly biological sequence alignment editor and analysis program for windows 95/98/ NT. *Nucleic Acids Symp. Ser.* **1999**, *41*, 94–98.
33. Chong, J.; Soufan, O.; Li, C.; Caraus, I.; Li, S.; Bourque, G.; Wishart, D.S.; Xia, J. MetaboAnalyst 4.0: Towards more transparent and integrative metabolomics analysis. *Nucleic Acids Res.* **2018**, *46*, W486–W494. [CrossRef] [PubMed]
34. Kjærbølling, I.; Vesth, T.C.; Frisvad, J.C.; Nybo, J.L.; Theobald, S.; Kuo, A.; Bowyer, P.; Matsuda, Y.; Mondo, S.; Lyhne, E.K.; et al. Linking secondary metabolites to gene clusters through genome sequencing of six diverse *Aspergillus* species. *Proc. Natl. Acad. Sci. USA* **2018**, *115*, E753–E761. [CrossRef]
35. Koivistoinen, O.M.; Arvas, M.; Headman, J.R.; Andberg, M.; Penttilä, M.; Jeffries, T.W.; Richard, P. Characterisation of the gene cluster for L-rhamnose catabolism in the yeast *Scheffersomyces* (Pichia) *stipitis*. *Gene* **2012**, *492*, 177–185. [CrossRef]
36. Shen, J.; Zheng, L.; Chen, X.; Han, X.; Cao, Y.; Yao, J. Metagenomic analyses of microbial and carbohydrate-active enzymes in the rumen of dairy goats fed different rumen degradable starch. *Front. Microbiol.* **2020**, *11*, 1003. [CrossRef]
37. Barbosa, A.I.; Coutinho, A.J.; Costa Lima, S.A.; Reis, S. Marine polysaccharides in pharmaceutical applications: Fucoidan and chitosan as key players in the drug delivery match field. *Mar. Drugs* **2019**, *17*, 12. [CrossRef]
38. Oberbarnscheidt, L.; Taylor, E.J.; Davies, G.J.; Gloster, T.M. Structure of a carbohydrate esterase from *Bacillus anthracis*. *Proteins Struct. Funct. Bioinform.* **2007**, *66*, 250–252. [CrossRef]
39. Hehemann, J.H.; Truong, L.V.; Unfried, F.; Welsch, N.; Kabisch, J.; Heiden, S.E.; Junker, S.; Becher, D.; Thürmer, A.; Daniel, R.; et al. Aquatic adaptation of a laterally acquired pectin degradation pathway in marine gammaproteobacteria. *Environ. Microbiol.* **2017**, *19*, 2320–2333. [CrossRef] [PubMed]
40. Popper, Z.A.; Michel, G.; Herve, C.; Domozych, D.S.; Willats, W.G.; Tuohy, M.G.; Kloareg, B.; Stengel, D.B. Evolution and diversity of plant cell walls: From algae to flowering plants. *Annu. Rev. Plant. Biol.* **2011**, *62*, 567–590. [CrossRef] [PubMed]
41. Chen, L.H.; Tsai, H.C.; Yu, P.L.; Chung, K.R. A major facilitator superfamily transporter-mediated resistance to oxidative stress and fungicides requires Yap1, Skn7, and MAP kinases in the citrus fungal pathogen *Alternaria alternata*. *PLoS ONE* **2017**, *12*, e0169103. [CrossRef]
42. Xu, X.; Chen, J.; Xu, H.; Li, D. Role of a major facilitator superfamily transporter in adaptation capacity of *Penicillium funiculosum* under extreme acidic stress. *Fungal Genet. Biol.* **2014**, *69*, 75–83. [CrossRef] [PubMed]
43. Kogej, T.; Stein, M.; Volkmann, M.; Gorbushina, A.A.; Galinski, E.A.; Gunde-Cimerman, N. Osmotic adaptation of the halophilic fungus *Hortaea werneckii*: Role of osmolytes and melanization. *Microbiology* **2007**, *153*, 4261–4273. [CrossRef]
44. Rojas-Jimenez, K.; Rieck, A.; Wurzbacher, C.; Jürgens, K.; Labrenz, M.; Grossart, H.P. A salinity threshold separating fungal communities in the Baltic Sea. *Front. Microbiol.* **2019**, *10*, 680. [CrossRef]
45. Niño-González, M.; Novo-Uzal, E.; Richardson, D.N.; Barros, P.M.; Duque, P. More transporters, more substrates: The *Arabidopsis* major facilitator superfamily revisited. *Mol. Plant* **2019**, *12*, 1182–1202. [CrossRef]
46. Plemenitaš, A.; Lenassi, M.; Konte, T.; Kejžar, A.; Zajc, J.; Gostinčar, C.; Gunde-Cimerman, N. Adaptation to high salt concentrations in halotolerant/halophilic fungi: A molecular perspective. *Front. Microbiol.* **2014**, *5*, 199. [CrossRef]
47. Nesci, A.; Etcheverry, M.; Magan, N. Osmotic and matric potential effects on growth, sugar alcohol and sugar accumulation by *Aspergillus* section *Flavi* strains from Argentina. *J. Appl. Microbiol.* **2004**, *96*, 965–972. [CrossRef] [PubMed]
48. Gladfelter, A.S.; James, T.Y.; Amend, A.S. Marine fungi. *Curr. Biol.* **2019**, *29*, R191–R195. [CrossRef]
49. Martinet, L.; Naômé, A.; Deflandre, B.; Maciejewska, M.; Tellatin, D.; Tenconi, E.; Smargiasso, N.; Pauw, E.; van Wezel, G.P.; Rigali, S. A single biosynthetic gene cluster is responsible for the production of bagremycin antibiotics and ferroverdin iron chelators. *Mbio* **2019**, *10*, e01230-19. [CrossRef]
50. Houbraken, J.; Kocsubé, S.; Visagie, C.M.; Yilmaz, N.; Wang, X.C.; Meijer, M.; Kraak, B.; Hubka, V.; Bensch, K.; Samson, R.A.; et al. Classification of *Aspergillus, Penicillium, Talaromyces* and related genera (Eurotiales): An overview of families, genera, subgenera, sections, series and species. *Stud. Mycol.* **2020**, *95*, 5–169. [CrossRef]

51. Awakawa, T.; Yang, X.L.; Wakimoto, T.; Abe, I. Pyranonigrin E: A PKS-NRPS hybrid metabolite from *Aspergillus niger* identified by genome mining. *ChemBioChem* **2013**, *14*, 2095–2099. [CrossRef] [PubMed]
52. Izumi, Y.; Ohtani, K.; Miyamoto, Y.; Masunaka, A.; Fukumoto, T.; Gomi, K.; Tada, Y.; Ichimura, K.; Peever, T.L.; Akimitsu, K. A polyketide synthase gene, ACRTS2, is responsible for biosynthesis of host-selective ACR-toxin in the rough lemon pathotype of *Alternaria alternata*. *Mol. Plant Microbe Interact.* **2012**, *25*, 1419–1429. [CrossRef] [PubMed]
53. Bacha, N.; Dao, H.P.; Atoui, A.; Mathieu, F.; O'Callaghan, J.; Puel, O.; Liboz, T.; Dobson, A.D.W.; Lebrihi, A. Cloning and characterization of novel methylsalicylic acid synthase gene involved in the biosynthesis of isoasperlactone and asperlactone in *Aspergillus westerdijkiae*. *Fungal Genet. Biol.* **2009**, *46*, 742–749. [CrossRef] [PubMed]
54. Torres, M.; Balcells, M.; Sala, N.; Sanchis, V.; Canela, R. Bactericidal and fungicidal activity of *Aspergillus ochraceus* metabolites and some derivatives. *Pestic. Sci.* **1998**, *53*, 9–14. [CrossRef]
55. White, E.C.; Hill, J.H. Studies on antibacterial products formed by molds: I. aspergillic acid, a product of a strain of *Aspergillus flavus*. *J. Bacteriol.* **1943**, *45*, 433–443. [CrossRef]
56. Lebar, M.D.; Mack, B.M.; Carter-Wientjes, C.H.; Gilbert, M.K. The aspergillic acid biosynthetic gene cluster predicts neoaspergillic acid production in *Aspergillus* section *Circumdati*. *World Mycotoxin J.* **2019**, *12*, 213–222. [CrossRef]
57. Andersen, M.R.; Nielsen, J.B.; Klitgaard, A.; Petersen, L.M.; Zachariasen, M.; Hansen, T.J.; Blicher, L.H.; Gotfredsen, C.H.; Larsen, T.O.; Nielsen, K.F.; et al. Accurate prediction of secondary metabolite gene clusters in filamentous fungi. *Proc. Natl. Acad. Sci. USA* **2013**, *110*, E99–E107. [CrossRef]
58. Raffa, N.; Keller, N.P. A call to arms: Mustering secondary metabolites for success and survival of an opportunistic pathogen. *PLoS Pathog.* **2019**, *15*, e1007606. [CrossRef]
59. Kato, H.; Yoshida, T.; Tokue, T.; Nojiri, Y.; Hirota, H.; Ohta, T.; Williams, R.M.; Tsukamoto, S. Notoamides A–D: Prenylated Indole Alkaloids Isolated from a Marine-Derived Fungus, *Aspergillus* sp. *Angew. Chem. Int. Ed.* **2007**, *46*, 2254–2256. [CrossRef]
60. Ding, Y.; Wet, J.R.D.; Cavalcoli, J.; Li, S.; Greshock, T.J.; Miller, K.A.; Finefield, J.M.; Sunderhaus, J.D.; McAfoos, T.J.; Tsukamoto, S.; et al. Genome-based characterization of two prenylation steps in the assembly of the stephacidin and notoamide anticancer agents in a marine-derived *Aspergillus* sp. *J. Am. Chem. Soc.* **2010**, *132*, 12733–12740. [CrossRef] [PubMed]
61. Greshock, T.J.; Grubbs, A.W.; Tsukamoto, S.; Williams, R.M. A concise, biomimetic total synthesis of stephacidin A and notoamide B. *Angew. Chem. Int. Ed.* **2007**, *46*, 2262–2265. [CrossRef] [PubMed]
62. Liesch, J.M.; Wichmann, C.F. Novel antinematodal and antiparasitic agents from *Penicillium charlesii*. II. Structure determination of paraherquamides B, C, D, E, F, and G. *J. Antibiot.* **1990**, *43*, 1380–1386. [CrossRef] [PubMed]
63. Li, S.; Srinivasan, K.; Tran, H.; Yu, F.; Finefield, J.M.; Sunderhaus, J.D.; McAfoos, T.J.; Sachiko Tsukamoto, S.; Williams, R.M.; Sherman, D.H. Comparative analysis of the biosynthetic systems for fungal bicyclo [2.2.2] diazaoctane indole alkaloids: The (+)/(−)-notoamide, paraherquamide and malbrancheamide pathways. *MedChemComm* **2012**, *3*, 987–996. [CrossRef] [PubMed]
64. de Guzman, F.S.; Bruss, D.R.; Rippentrop, J.M.; Gloer, K.B.; Gloer, J.B.; Wicklow, D.T.; Dowd, P.F. Ochrindoles AD: New bis-indolyl benzenoids from the sclerotia of *Aspergillus ochraceus* NRRL 3519. *J. Nat. Prod.* **1994**, *57*, 634–639. [CrossRef] [PubMed]
65. Moake, M.M.; Padilla-Zakour, O.I.; Worobo, R.W. Comprehensive review of patulin control methods in foods. *Compr. Rev. Food Sci. Food Saf.* **2005**, *4*, 8–21. [CrossRef]
66. Frisvad, J.C. A critical review of producers of small lactone mycotoxins: Patulin, penicillic acid and moniliformin. *World Mycotoxin J.* **2018**, *11*, 73–100. [CrossRef]
67. Nielsen, J.C.; Grijseels, S.; Prigent, S.; Ji, B.; Dainat, J.; Nielsen, K.F.; Frisvad, J.C.; Workman, M.; Nielsen, J. Global analysis of biosynthetic gene clusters reveals vast potential of secondary metabolite production in *Penicillium* species. *Nat. Microbiol.* **2017**, *2*, 17044. [CrossRef]
68. Baxter, A.; Fitzgerald, B.J.; Hutson, J.L.; McCarthy, A.D.; Motteram, J.M.; Ross, B.C.; Sapra, M.; Snowden, M.A.; Watson, N.S.; Williams, R.J. Squalestatin 1, a potent inhibitor of squalene synthase, which lowers serum cholesterol in vivo. *J. Biol. Chem.* **1992**, *267*, 11705–11708. [CrossRef]
69. Hasumi, K.; Tachikawa, K.; Sakai, K.; Murakawa, S.; Yoshikawa, N.; Kumazawa, S.; Endo, A. Competitive inhibition of squalene synthetase by squalestatin 1. *J. Antibiot.* **1993**, *46*, 689–691. [CrossRef]
70. Bills, G.F.; Peláez, F.; Polishook, J.D.; Diez-Matas, M.T.; Harris, G.H.; Clapp, W.H.; Dufresne, C.; Byrne, K.M.; Nallin-Omstead, M.; Jenkins, R.G.; et al. Distribution of zaragozic acids (squalestatins) among filamentous ascomycetes. *Mycol. Res.* **1994**, *98*, 733–739. [CrossRef]
71. Dawson, M.J.; Farthing, J.E.; Marshall, P.S.; Middleton, R.F.; O'neill, M.J.; Shuttleworth, A.; Stylli, C.; Tait, R.M.; Taylor, P.M.; Wildman, H.G.; et al. The squalestatins, novel inhibitors of squalene synthase produced by a species of *Phoma* I. Taxonomy, fermentation, isolation, physico-chemical properties and biological activity. *J. Antibiot.* **1992**, *45*, 639–647. [CrossRef] [PubMed]
72. Wang, X.; Peng, J.; Sun, L.; Bonito, G.; Guo, Y.; Li, Y.; Fu, Y. Genome sequencing of *Paecilomyces penicillatus* provides insights into its phylogenetic placement and mycoparasitism mechanisms on morel mushrooms. *Pathogens* **2020**, *9*, 834. [CrossRef]
73. Bandapali, O.R.; Hansen, F.T.; Parveen, A.; Phule, P.; Goud, E.S.K.; Acharya, S.K.; Sørensen, J.L.; Kumar, A. Identification and characterization of biosynthetic gene clusters from halophilic marine fungus *Eurotium rubrum*. *Preprints* **2020**, 2020020388. [CrossRef]
74. Donzelli, B.G.; Krasnoff, S.B.; Churchill, A.C.; Vandenberg, J.D.; Gibson, D.M. Identification of a hybrid PKS-NRPS required for the biosynthesis of NG-391 in *Metarhizium robertsii*. *Curr. Genet.* **2010**, *56*, 151–162. [CrossRef] [PubMed]

75. Gelderblom, W.C.; Thiel, P.G.; Marasas, W.F.; Van der Merwe, K.J. Natural occurrence of fusarin C, a mutagen produced by *Fusarium moniliforme*, in corn. *J. Agric. Food Chem.* **1984**, *32*, 1064–1067. [CrossRef]
76. Kato, S.; Motoyama, T.; Futamura, Y.; Uramoto, M.; Nogawa, T.; Hayashi, T.; Hirota, H.; Tanaka, A.; Takahashi-Ando, N.; Kamakura, T.; et al. Biosynthetic gene cluster identification and biological activity of lucilactaene from *Fusarium* sp. RK97-94. *Biosci. Biotechnol. Biochem.* **2020**, *84*, 1303–1307. [CrossRef] [PubMed]
77. Overy, D.; Correa, H.; Roullier, C.; Chi, W.C.; Pang, K.L.; Rateb, M.; Shang, Z.; Capon, R.; Bills, G.; Kerr, R. Does osmotic stress affect natural product expression in fungi? *Mar. Drugs* **2017**, *15*, 254. [CrossRef]
78. Drabińska, N.; Młynarz, P.; de Lacy Costello, B.; Jones, P.; Mielko, K.; Mielnik, J.; Persad, R.; Ratcliffe, N.M. An optimization of liquid–liquid extraction of urinary volatile and semi-volatile compounds and its application for gas chromatography-mass spectrometry and proton nuclear magnetic resonance spectroscopy. *Molecules* **2020**, *25*, 3651. [CrossRef]
79. Vachy, R. Novel Glucopyranose Derivatives, Preparation Thereof, and Biological Uses Thereof. U.S. Patent Application No. 12/737,748, 18 February 2010.
80. Matsumoto, Y.; Kaihatsu, K.; Nishino, K.; Ogawa, M.; Kato, N.; Yamaguchi, A. Antibacterial and antifungal activities of new acylated derivatives of epigallocatechin gallate. *Front. Microbiol.* **2012**, *3*, 53. [CrossRef]
81. O'Neill, J. Tackling drug-resistant infections globally: Final report and recommendations. *Rev. Microb. Resist.* **2016**. Available online: https://amr-review.org/sites/default/files/160518_Final%20paper_with%20cover.pdf (accessed on 18 May 2021).
82. Kumar, C.G.; Mongolla, P.; Pombala, S.; Bandi, S.; Babu, K.S.; Ramakrishna, K.V.S. Biological evaluation of 3-hydroxybenzyl alcohol, an extrolite produced by *Aspergillus nidulans* strain KZR-132. *J. Appl. Microbiol.* **2017**, *122*, 1518–1528. [CrossRef] [PubMed]
83. Cohen, E.; Koch, L.; Thu, K.M.; Rahamim, Y.; Aluma, Y.; Ilan, M.; Yarden, O.; Carmeli, S. Novel terpenoids of the fungus *Aspergillus insuetus* isolated from the Mediterranean sponge *Psammocinia* sp. collected along the coast of Israel. *Bioorg. Med. Chem.* **2011**, *19*, 6587–6593. [CrossRef]
84. Cao, F.; Zhao, D.; Chen, X.Y.; Liang, X.D.; Li, W.; Zhu, H.J. Antifungal drimane sesquiterpenoids from a marine-derived *Pleosporales* sp. fungus. *Chem. Nat. Compd.* **2017**, *53*, 1189–1191. [CrossRef]
85. Orfali, R.; Aboseada, M.A.; Abdel-Wahab, N.M.; Hassan, H.M.; Perveen, S.; Ameen, F.; Alturki, E.; Abdelmohsen, U.R. Recent updates on the bioactive compounds of the marine-derived genus *Aspergillus*. *RSC Adv.* **2021**, *11*, 17116–17150. [CrossRef]
86. Masui, H.; Kondo, T.; Kojima, M. An antifungal compound, 9, 12, 13-trihydroxy-(E)-10-octadecenoic acid, from *Colocasia antiquorum* inoculated with *Ceratocystis fimbriata*. *Phytochemistry* **1989**, *28*, 2613–2615. [CrossRef]
87. Heredia-Guerrero, J.A.; Goldoni, L.; Benítez, J.J.; Davis, A.; Ceseracciu, L.; Cingolani, R.; Bayer, I.S.; Heinze, T.; Koschella, A.; Heredia, A.; et al. Cellulose-polyhydroxylated fatty acid ester-based bioplastics with tuning properties: Acylation via a mixed anhydride system. *Carbohydr. Polym.* **2017**, *173*, 312–320. [CrossRef] [PubMed]
88. Kindra, D.R.; Evans, W.J. Bismuth-based cyclic synthesis of 3, 5-di-tert-butyl-4-hydroxybenzoic acid via the oxyarylcarboxy dianion, $(O_2CC_6H_2{}^tBu_2O)^{2-}$. *Dalton Trans.* **2014**, *43*, 3052–3054. [CrossRef]
89. Zhu, H.; Lemos, H.; Bhatt, B.; Islam, B.N.; Singh, A.; Gurav, A.; Huang, L.; Browning, D.D.; Mellor, A.; Fuzeles, S.; et al. Carbidopa, a drug in use for management of Parkinson disease inhibits T cell activation and autoimmunity. *PLoS ONE* **2017**, *12*, e0183484. [CrossRef]
90. Maia, M.; Ferreira, A.E.; Nascimento, R.; Monteiro, F.; Traquete, F.; Marques, A.P.; Cunha, J.; Eiras-Dias, J.E.; Cordeiro, C.; Figueiredo, A.; et al. Integrating metabolomics and targeted gene expression to uncover potential biomarkers of fungal/oomycetes-associated disease susceptibility in grapevine. *Sci. Rep.* **2020**, *10*, 15688. [CrossRef]
91. Howlett, D.R.; George, A.R.; Owen, D.E.; Ward, R.V.; Markwell, R.E. Common structural features determine the effectiveness of carvedilol, daunomycin and rolitetracycline as inhibitors of Alzheimer β-amyloid fibril formation. *Biochem. J.* **1999**, *343*, 419–423. [CrossRef]
92. Hsu, H.T.; Wu, L.M.; Lin, P.C.; Juan, C.H.; Huang, Y.Y.; Chou, P.L.; Chen, J.L. Emotional distress and quality of life during folinic acid, fluorouracil, and oxaliplatin in colorectal cancer patients with and without chemotherapy-induced peripheral neuropathy: A cross-sectional study. *Medicine* **2020**, *99*, e19029. [CrossRef]
93. Kim, Y.S.; Lim, C.H. Isolation and structure elucidation of antifungal compounds from the antarctic lichens, *Stereocaulon alpinum* and *Sphaerophorus globosus*. *Korean J. Agric. Sci.* **2020**, *47*, 183–191.
94. Ledesma-Amaro, R.; Buey, R.M.; Revuelta, J.L. Increased production of inosine and guanosine by means of metabolic engineering of the purine pathway in *Ashbya gossypii*. *Microb. Cell Factories* **2015**, *14*, 58. [CrossRef] [PubMed]
95. Majnooni, M.B.; Fakhri, S.; Shokoohinia, Y.; Mojarrab, M.; Kazemi-Afrakoti, S.; Farzaei, M.H. Isofraxidin: Synthesis, biosynthesis, isolation, pharmacokinetic and pharmacological properties. *Molecules* **2020**, *25*, 2040. [CrossRef]
96. Bilska, K.; Stuper-Szablewska, K.; Kulik, T.; Buśko, M.; Załuski, D.; Perkowski, J. Resistance-related L-pyroglutamic acid affects the biosynthesis of trichothecenes and phenylpropanoids by *F. graminearum sensu stricto*. *Toxins* **2018**, *10*, 492. [CrossRef]
97. Luo, H.; Yamamoto, Y.; Kim, J.A.; Jung, J.S.; Koh, Y.J.; Hur, J.S. Lecanoric acid, a secondary lichen substance with antioxidant properties from *Umbilicaria antarctica* in maritime Antarctica (King George Island). *Polar Biol.* **2009**, *32*, 1033–1040. [CrossRef]
98. Nugraha, A.S.; Untari, L.F.; Laub, A.; Porzel, A.; Franke, K.; Wessjohann, L.A. Anthelmintic and antimicrobial activities of three new depsides and ten known depsides and phenols from Indonesian lichen: *Parmelia cetrata* Ach. *Nat. Prod. Res.* **2020**, *35*, 5001–5010. [CrossRef]

99. Khan, S.N.; Lal, S.K.; Kumar, P.; Khan, A.U. Effect of mitoxantrone on proliferation dynamics and cell-cycle progression. *Biosci. Rep.* **2010**, *30*, 375–381. [CrossRef] [PubMed]
100. Meucci, V.; Armani, A.; Tinacci, L.; Guardone, L.; Battaglia, F.; Intorre, L. Natural occurrence of ochratoxin A (OTA) in edible and not edible tissue of farmed gilthead seabream (*Sparus aurata*) and European seabass (*Dicentrarchus labrax*) sold on the Italian market. *Food Control* **2021**, *120*, 107537. [CrossRef]
101. Lu, P.; Li, X.; Wang, S.; Gao, F.; Yuan, B.; Xi, L.; Du, L.; Jiya, H.; Zhao, L. Saccharopine reductase influences production of Swainsonine in *Alternaria oxytropis*. *Sydowia* **2021**, *73*, 69–74.
102. Sun, H.; Wang, L.; Zhang, B.; Ma, J.; Hettenhausen, C.; Cao, G.; Sun, G.; Wu, J.; Wu, J. Scopoletin is a phytoalexin against *Alternaria alternata* in wild tobacco dependent on jasmonate signalling. *J. Exp. Bot.* **2014**, *65*, 4305–4315. [CrossRef]

Journal of *Fungi*

MDPI

Article

Mold and Yeast-Like Fungi in the Seaside Air of the Gulf of Gdańsk (Southern Baltic) after an Emergency Disposal of Raw Sewage

Małgorzata Michalska [1,*,†], Monika Kurpas [1], Katarzyna Zorena [1,†], Piotr Wąż [2] and Roman Marks [3]

1 Department of Immunobiology and Environment Microbiology, Faculty of Health Sciences with Institute of Maritime and Tropical Medicine, Medical University of Gdańsk, 80-210 Gdańsk, Poland; monika.kurpas@gumed.edu.pl (M.K.); kzorena@gumed.edu.pl (K.Z.)

2 Department of Nuclear Medicine, Faculty of Health Sciences with Institute of Maritime and Tropical Medicine, Medical University of Gdańsk, 80-210 Gdańsk, Poland; piotr.waz@gumed.edu.pl

3 Institute of Marine and Environmental Sciences, University of Szczecin, 70-453 Szczecin, Poland; roman.marks@usz.pl

* Correspondence: malgorzata.michalska@gumed.edu.pl

† These authors have contributed equally to this work.

Abstract: The aim of this study was to determine the correlation between the meteorological factors and the number of molds and yeast-like fungi in the air in the five coastal towns in the years 2014–2017, and in 2018, after emergency disposal of raw sewage to the Gdańsk Gulf. In the years 2014–2018, a total number of 88 air samples were collected in duplicate in the five coastal towns of Hel, Puck, Gdynia, Sopot, and Gdańsk-Brzeźno. After the application of the (PCA) analysis, this demonstrated that the first principal component (PC1) had a positive correlation with the water temperature, wind speed, air temperature, and relative humidity. The second principal component (PC2) had a positive correlation with the relative humidity, wind speed, wind direction, and air temperature. In 2018, potentially pathogenic mold and yeast-like fungi (*Candida albicans*, *Stachybotrys chartarum* complex, *Aspergillus* section *Fumigati*) were detected in the seaside air. While the detected species were not observed in the years 2014–2017. We suggest that it is advisable to inform residents about the potential health risk in the event of raw sewage disposal into the water. Moreover, in wastewater treatment plants, tighter measures, including wastewater disinfection, should be introduced.

Keywords: emergency disposal of raw of sewage; seaside air; bioaerosol; mold; yeast-like fungi

Citation: Michalska, M.; Kurpas, M.; Zorena, K.; Wąż, P.; Marks, R. Mold and Yeast-Like Fungi in the Seaside Air of the Gulf of Gdańsk (Southern Baltic) after an Emergency Disposal of Raw Sewage. *J. Fungi* **2021**, *7*, 219. https://doi.org/10.3390/jof7030219

Academic Editor: Federico Baltar

Received: 31 December 2020
Accepted: 16 March 2021
Published: 17 March 2021

1. Introduction

Biological aerosols are a subset of atmospheric particles consisting of both living and non-living organisms, including bacteria, viruses, pollens, molds, yeast-like fungi, and their metabolic products (e.g., mycotoxins) [1,2]. Most molds are not harmful to humans, however, previous studies, including ours, have shown that yeast-like fungi and spores of mold fungi are the etiological factors of many diseases, including allergy, pneumonia, bronchitis, neoplastic diseases, and type 1 diabetes [3–6]. The storage and sorting of organic waste, composting, agricultural production, food processing, and wastewater treatment systems emit large volumes of bioaerosols, which lead to significant exposure to biological factors [7–9]. Studies have demonstrated that not only emergency and uncontrolled wastewater disposals but also outflows from treatment plants and heavy rains cause an increase in the number of microorganisms, such as mold and yeast-like fungi in coastal seawater and sand [8–12]. Moreover, the mold and yeast-like fungi survive in salt seawater and in beach sand for many months [13–15]. More than 30 years ago, Anderson (1979) discovered that pathogenic fungi, for instance, *Trichosporon cutaneum*, *Candida albicans*, *Microsporum gypseum*, and *Trichophyton mentagrophytes*, could survive in the sand for over a month [16]. A similar study was conducted by other authors who demonstrated that several species of dermatophytes (*Epidermophyton floccosum, Microsporum canis, M. gypseum, T. mentagrophytes, T. rubrum*), and *Scopulariopsis brevicaulis* can survive in the sand from 25 to 360 days [17,18]. Therefore, Vogel et al. suggested that pathogenic yeast-like fungi found in seawater, sewage and beach sand could be a good additional mycological indicator in assessing the safety of marine bathing waters [18].

In terms of mycology, clean air and seawater are of key importance to the health of seaside town inhabitants. This is especially true of holiday resorts that dispose of treated sewage into seawaters. A typical example is seaside bathing areas located along the Gulf of Gdańsk—a bay in the south-eastern part of the Baltic Sea, located between Poland and Russia. Wastewater from the two largest sewage treatment plants is disposed of into the Gulf of Gdańsk [19]. The first (Gdańsk-Wschód) disposes of sewage into the Gulf of Gdańsk to a distance of 2.5 km from the shoreline through a deep water collector, the second (Gdynia-Dębogórze) uses a collector over a distance of 2.3 km to dispose of wastewater to the Bay of Puck (western part of the Gulf of Gdańsk). What is more, from the 15th to 18th May 2018, there was an emergency raw sewage disposal into the Motława River, which flows into the Gulf of Gdańsk (Figure 1). We have adopted a research hypothesis that emergency disposal of raw sewage into the Gulf of Gdańsk could have led to microbiological contamination of the water in the Gulf of Gdańsk and the local air. The preliminary results of our study showed an increase in the number of coliform bacteria and *Escherichia coli* in the seawater and air as a result of emergency disposal of raw sewage to the Gulf of Gdańsk in 2018 [7]. Thus far, the results of mycological tests performed after the disposal have not been presented. Therefore, the aim of this study was to determine the correlation between the meteorological factors and the number of molds and yeast-like fungi in the air in the five coastal towns in the years 2014–2017, and in 2018, after emergency disposal of raw of sewage to the Gdańsk Gulf.

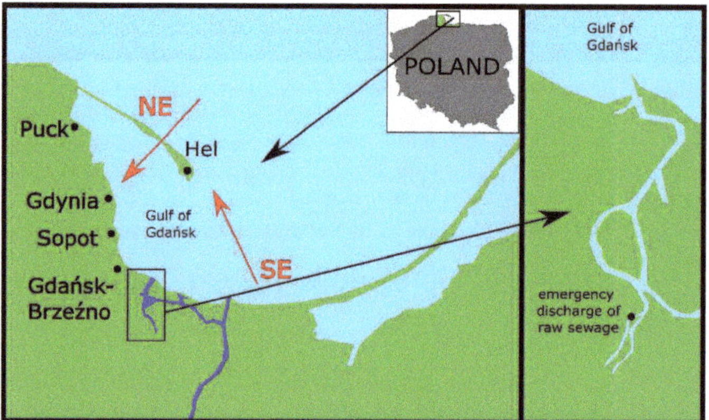

Figure 1. The emergency discharge of raw sewage into the Motława River to the Gulf of Gdańsk in 2018. Map showing the five coastal towns: Hel, Puck, Gdynia, Sopot, and Gdańsk-Brzeźno, where microbiological coastal air were samples collected. The wind direction marked on the map were Southeast (SE) and Northeast (NE) (red arrow). Place of emergency discharge of raw sewage into the Motława River to the Gulf of Gdańsk (black arrow).

2. Materials and Methods

2.1. Collection of Air Samples of the Gulf of Gdańsk

In the years 2014–2018, a total of 88 air samples were collected in duplicate, in 5 coastal towns, i.e., Hel, Puck, Gdynia, Sopot, and Gdańsk-Brzeźno on the Gulf of Gdansk (Figure 1). In the years 2014–2017, 62 air samples were collected between 14th May and 14th July, every 28 days between 9:00 a.m. and 2:00 p.m. In 2018 after the emergency disposal of raw sewage, the 26 air samples were collected between 14th May and 23rd July, every 14 days between 9:00 a.m. and 2:00 p.m. The air samples were not collected in the rain and heavy rainfall.

Air samples were collected at a height of 50 cm and about 100 cm from the waterline. In all cases, the samples were collected for 10 min over the Gulf of Gdansk in the 5 coastal towns of Hel, Puck, Gdynia, Sopot, and Gdańsk-Brzeźno. The air samples were collected by impaction with a SAS Super ISO 100 (Milan, Italy) sampler. The nozzle of the sampler was positioned perpendicular to the wind direction. The sampler automatically collected 100-L samples of air. The extracted air was then transported through small holes to a head with a Petri dish containing a Sabouraud dextrose agar medium. The maximum efficiency of the collection was for particulate matter of d50 = 2–4 μm. The flow rate was 90 Lpm. All removable parts of the air sampler were sterilized by autoclaving before sampling, and the sterilized sampler head was cleaned between samples with 70% ethanol.

2.2. Mould Fungi Incubation and Identification

The fungi were counted after a 120-hour incubation at 28 °C on Sabouraud Dextrose Agar medium by Merck (Darmstadt, Germany). Yeast-like fungi were identified by CHROMagar Candida, Graso Biotech (Starogard Gdańskicity, Poland).

Mold fungi were identified based on their macro- and microscopic features, with the use of a Nicon Eclipse E2000 microscope at 400, 600, 1000× magnification and a key for the identification of fungi [20,21]. Mould colonies were identified on the basis of the color, texture, topography of the culture surface, smell of the colony, color of the reverse of the colony, and the presence of the diffuse pigment. Microscopic features of the fungal colonies were identified based on their microscopic features, i.e., the presence of macroconidia and microconidia, their shape, and appearance [20,21].

2.3. Sample Analysis of Mould Fungi

The number of colonies of fungi were expressed as a colony-forming unit (CFU) per 1 m^3 of the air (CFU/m^3). When applying the impact method, we used the Feller table attached to the manual of the air sampler [5,22].

The colonies collected should be revised by the equation:

$$\mathrm{Pr} = \mathrm{N}[1/\mathrm{N} + 1/\mathrm{N} - 1 + 1/\mathrm{N} - 2 + 1/\mathrm{N} - \mathrm{r} + 1] \tag{1}$$

where Pr is the revised colony in stage, N is the number of sieve pores, and r is the number of viable particles counted on the agar plate.

The number of colonies of fungi (CFU/m^3) was calculated using the following equation:

$$\mathrm{C(CFU/m^3)} = \mathrm{T} \times 1000\mathrm{t(min)} \times \mathrm{F(L/min)} \tag{2}$$

where C—airborne fungi concentration; CFU—colony-forming unit; T—total colonies after application of the Pr statistical correction; t—sampling time and F—airflow rate.

2.4. Characterisation of Meteorological Conditions

During the collection of air samples between spring and summer during 2014–2017 and in 2018, we recorded air and water temperature, humidity, wind speed, and wind direction using a GMH 3330 thermo-hygrometer by Greisinger (Remscheidcity, Germany). The air temperature during 2014–2017 ranged from 26 °C to 3 °C (spring season) and from 20 °C to 16 °C (summer season), respectively. Relative humidity in the spring season was between 30% and 88%, and from 59% to 82% in the summer season. Wind speed in the spring season varied between 0 and 31 km/h and between 7 and 25 km/h in the summer. The air temperature in 2018 fluctuated between 27 °C and 10 °C in the spring and between 27 °C to 15 °C in the summer. Relative humidity in the spring season was between 39% and 93%, and from 44% to 70% in the summer season. Wind speed in the spring season varied between 0 and 15 km/h and between 2.6 and 32 km/h in the summer. Air samples were not collected when it rained.

2.5. Statistical Analysis

To search for hidden relationships and regularities between meteorological factors and the number of mold and yeast-like fungi in the air, we used one of the numerous methods of factor analysis. For data analysis, Principal Component Analysis (PCA) ready-made procedures from the "ggfortify", "FactoMineR," and "factoextra" packages [23–26] were used. The algorithm of the PCA method allowed the data to be transformed (using the features of the eigenvalues and eigenvectors of the covariance matrix or the correlation matrix for this purpose) in order to obtain the greatest possible differences in standard deviations in the new variables. The first major component of PC1 is related to most of the variability of the original data set, and the second major component of PC2 is related to the second largest component, and so on [27]. Assuming that the first few components contained a significant amount of variation in the original data set, together they may account for almost all of the variability in the data and thus simplify the interpretation of the results. The obtained results of the statistical analysis made it possible to determine the correlation between meteorological factors, the number of molds, the number of yeast-like fungi in the coastal air in the researched locations on the Gulf of Gdańsk, and PC1 and PC2. The PCA statistical analysis was performed for 2 research periods, 2014–2017 and 2018.

3. Results

3.1. The Principal Component Analysis of Mold and Yeast-Like Fungi Detected in Air Samples in the Five Seaside Towns in the Study Period of 2014–2017 and in 2018

The first three main principal components explained almost 78% of the total variance. The first principal component (PC1) explained 33.54% of the total variance. The second principal component (PC2) explained 27.23% of the total variance. The third principal component (PC3) explained 17.15% of the variance (Table 1). The PCA loading plot of the first two principal components compared the numbers of mold and yeast-like fungi and the meteorological factors in the coastal towns of Hel, Puck, Gdynia, Sopot, and Gdańsk-Brzeźno in the years 2014–2017, and in 2018 (Figure 2).

Table 1. Eigenvalues and values of the explained variance for the model from the mold and yeast-like fungi license.

	Eigenvalue	Explained Part of Multivariate Variability of Accessions [%]	Cumulative Part of Multivariate Variability [%]
PC1	2.0127	33.5458	33.5458
PC2	1.6341	27.2358	60.7816
PC3	1.0294	17.1567	77.9383
PC4	0.8123	13.5377	91.4760
PC5	0.2931	4.8847	96.3607
PC6	0.2184	3.6393	100.0000

Table 2 shows the values of the correlation between the variables used in the model and the main components shown in Figure 2. For each of the determined correlation values, the p-value was given. Table 2 includes only those variables that, at the assumed significance level, gave a statistically significant result $p < 0.00001$.

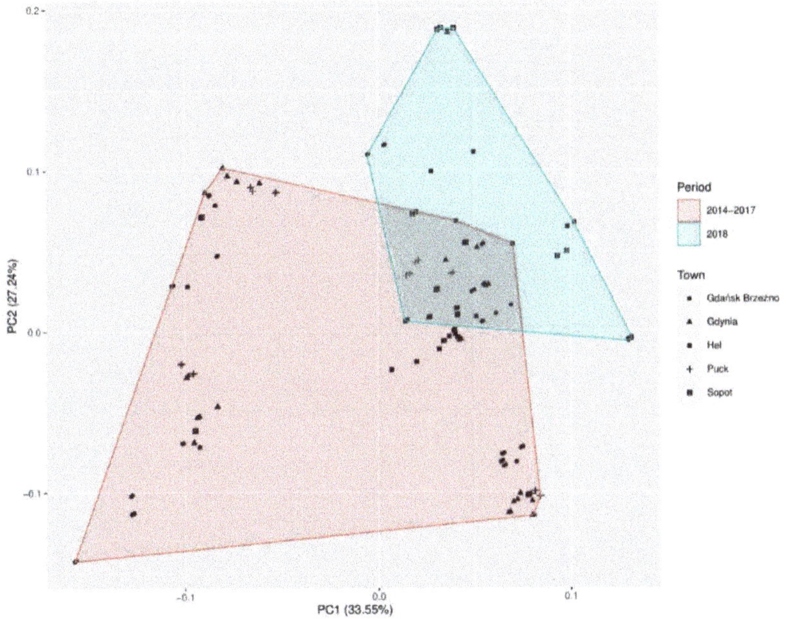

Figure 2. Principal component analysis (PCA) of mold and yeast-like fungi detected in air samples in the seaside towns and meteorological factors. The plot score of the first two principal components contains almost 60.78% of the explained variance. Two clusters can be distinguished: The red one is composed in the study period of 2014–2017 and the blue cluster in 2018.

Table 2. The values of the correlation between PC1, PC2, and the variables used to build the model.

Variables	PC1		PC2	
	Correlation	*p*-Value	Correlation	*p*-Value
water temperature (°C)	0.8220	<0.00001	−0.3612	<0.00001
wind speed (km/h)	0.7567	<0.00001	0.5178	<0.00001
air temperature (°C)	0.6263	<0.00001	−0.6215	<0.00001
relative humidity %	0.4883	<0.00001	0.8031	<0.00001
the number of mold and yeast-like fungi (CFU/m^3)	0.3608	<0.00001	−0.2039	0.00665
wind direction	-	-	0.4035	<0.00001

The first principal component (PC1) had a positive correlation ($p < 0.00001$) with the water temperature, wind speed, air temperature, and relative humidity, explaining 33.54% of the total variance observed. The second principal component (PC2) had a positive correlation ($p < 0.00001$) with the relative humidity, wind speed, and wind direction, and a negative correlation with the air temperature, explaining 27.23% of the variance (Table 2).

3.2. The Correlation between the Number of Mold Fungi *Aspergillus sp., Penicillium sp., Cladosporium sp.* in the Research Period of 2014–2017 and in 2018

In the next stage of the study, statistical analysis based on PCA was used to evaluate the number of mold fungi *Aspergillus* sp. (Tables 3 and 4), *Penicillium* sp., (Tables 5 and 6) and *Cladosporium* sp. (Tables 7 and 8) in the years 2014–2017 and in 2018.

Table 3. Eigenvalues and values of the explained variance for the model from the *Aspergillus* sp. license.

	Eigenvalue	Explained Part of Multivariate Variability of Accessions [%]	Cumulative Part of Multivariate Variability [%]
PC1	1.9859	33.0988	33.0988
PC2	1.6496	27.4930	60.5919
PC3	1.1141	18.5688	79.1606
PC4	0.7382	12.3037	91.4644
PC5	0.2952	4.9196	96.3840
PC6	0.2170	3.6160	100.0000

Table 4. The values of the correlation between PC1, PC2, and the variables used to build the model.

Variables	PC1		PC2	
	Correlation	*p*-Value	Correlation	*p*-Value
water temperature (°C)	0.8264	<0.00001	−0.3557	<0.00001
wind speed (km/h)	0.7576	<0.00001	0.5263	<0.00001
air temperature (°C)	0.6393	<0.00001	−0.6195	<0.00001
relative humidity %	0.4825	<0.00001	0.8094	<0.00001
the number of *Aspergillus* sp. (CFU/m^3)	0.2921	0.00008	−0.2827	0.00014

Table 5. Eigenvalues and values of the explained variance for the model from the *Penicillium* sp. license.

	Eigenvalue	Explained Part of Multivariate Variability of Accessions [%]	Cumulative Part of Multivariate Variability [%]
PC1	1.9918	33.1967	33.1967
PC2	1.6671	27.7855	60.9822
PC3	0.9693	16.1555	77.1377
PC4	0.8699	14.4989	91.6367
PC5	0.2942	4.9041	96.5408
PC6	0.2076	3.4592	100.0000

Table 6. The values of the correlation between PC1, PC2, and the variables used to build the model.

Variables	PC1		PC2	
	Correlation	*p*-Value	Correlation	*p*-Value
wind speed (km/h)	0.8688	<0.00001	0.2811	0.00016
water temperature (°C)	0.6987	<0.00001	−0.5693	<0.00001
relative humidity %	0.6545	<0.00001	0.6085	<0.00001
air temperature (°C)	0.4711	<0.00001	−0.7575	<0.00001
the number of *Penicillium* sp. (CFU/m^3)	0.3130	<0.00001	0.3216	0.00001
wind direction	-		0.4653	<0.00001

Table 7. Eigenvalues and values of the explained variance for the model from the *Cladosporium* sp. license.

	Eigenvalue	Explained Part of Multivariate Variability of Accessions [%]	Cumulative Part of Multivariate Variability [%]
PC1	1.9655	32.7577	32.7577
PC2	1.7050	28.4173	61.1751
PC3	1.0191	16.9845	78.1596
PC4	0.8001	13.3358	91.4954
PC5	0.2921	4.8681	96.3634
PC6	0.2182	3.6366	100.0000

Table 8. The values of the correlation between PC1, PC2, and the variables used to build the model.

Variables	PC1		PC2	
	Correlation	*p*-Value	Correlation	*p*-Value
water temperature (°C)	0.8287	<0.00001	−0.3116	<0.00001
wind speed (km/h)	0.7492	<0.00001	0.5426	<0.00001
air temperature (°C)	0.6621	<0.00001	−0.6036	<0.00001
relative humidity %	0.4723	<0.00001	0.7982	<0.00001
the number of *Cladosporium* sp. (CFU/m^3)	0.2199	0.00336	−0.4275	<0.00001
wind direction			0.3597	<0.00001

3.2.1. The Correlation between the Number of Mold Fungi Aspergillus sp., the Meteorological Elements in the Research Period of 2014–2017 and in 2018

The analysis of the PCA eigenvalues of the correlation matrix (Table 3) revealed the three main principal components, which could explain 79.16% of the total variance. The first principal components (PC1) explained 33.10% of the total variance. The second principal component (PC2) explained 27.49% of the total variance. The third principal component (PC3) explained 18.57% of the variance.

The PCA loading plot of the first two principal components comparing the numbers of *Aspergillus* sp. and the meteorological factors in the five coastal towns in the years 2014–2017, and in 2018 (Figure 3).

Table 4 shows the values of the correlation between the variables used in the model and the principal components shown in Figure 3, along with the *p*-value. Table 4 presents only those variables that, at the assumed significance level, gave a statistically significant result $p < 0.00001$.

The first principal component was a significant correlation ($p < 0.00001$) with the water temperature, wind speed, and air temperature, explaining 33.10% of the total variance observed (Table 4). The second principal component (PC2) was correlated with the relative humidity, wind speed, and air temperature explaining 27.49% of the variance.

3.2.2. The Correlation between the Number of Mold Fungi Penicillium sp. the Meteorological Elements in the Research Period of 2014–2017 and in 2018

Analysis of the eigenvalues of the correlation matrix (Table 5) revealed the three main principal components, which could explain 77.14% of the total variance. The first principal components (PC1) explained 33.20% of the total variance. The second principal component (PC2) explained 27.79% of the total variance. The third principal component (PC3) explained 16.16% of the variance.

The PCA loading plot of the first two principal components comparing the numbers of *Penicillium* sp. and the meteorological factors in the five coastal towns in the years 2014–2017, and in 2018 (Figure 4).

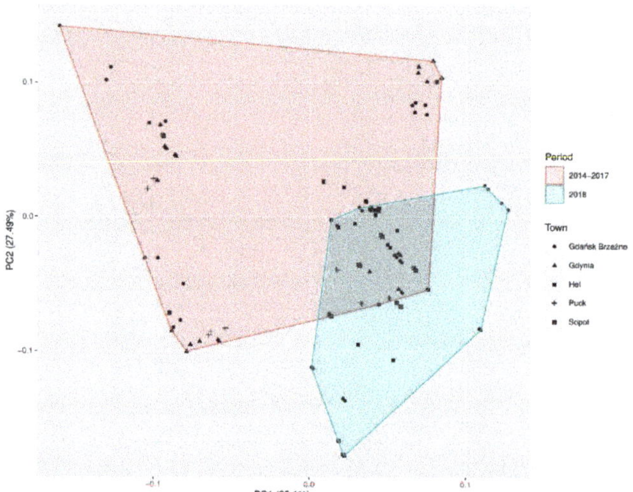

Figure 3. Principal component analysis (PCA) of *Aspergillus* sp. detected in air samples in the seaside towns (Hel, Puck, Gdynia, Sopot, and Gdańsk Brzeźno). The plot score of the first two principal components contained almost 60.59% of the explained variance. Two clusters can be distinguished: The red one is composed in the study period of 2014–2017 and the blue cluster in 2018.

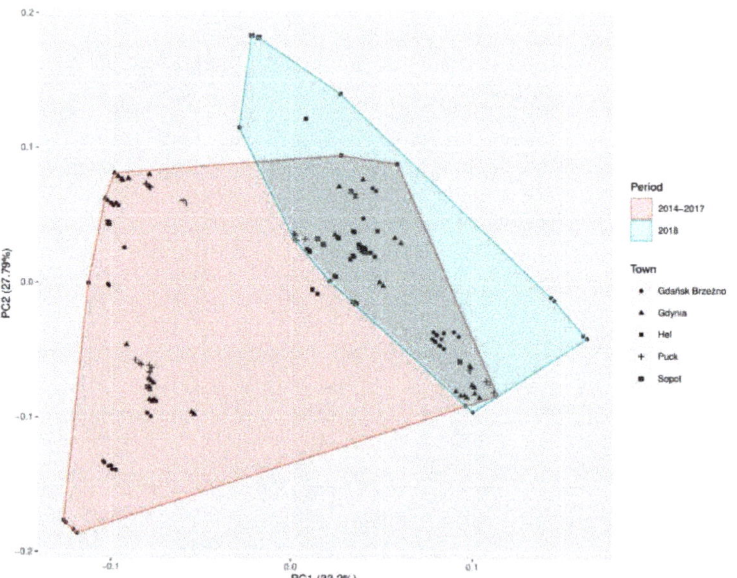

Figure 4. Principal component analysis (PCA) of *Penicillium* sp. detected in air samples in the seaside towns (Hel, Puck, Gdynia, Sopot, and Gdańsk Brzeźno). Here the plot score of the first two principal components is reported. It contains almost 60.98% of the explained variance. Two clusters can be distinguished: The red one is composed in the study period of 2014–2017 and the blue cluster in 2018.

Table 6 presents the values of the correlation between the variables used in the model and the main components presented in Figure 4. For each of the determined correlation values, the p-value was given. Table 6 contains only those variables which, at the assumed significance level, gave a statistically significant result.

The first principal component (PC1) was the combination of the wind speed and water temperature and then relative humidity and air temperature, explaining 33.10% of the total variance observed (Table 6). The principal component (PC2) was correlated with the air temperature, relative humidity, water temperature ($p < 0.00001$), and wind direction ($p = 0.00016$), explaining 27.49% of the variance.

3.2.3. The Correlation between the Number of Mold Fungi Cladosporium sp. and the Meteorological Factors in the Research Period of 2014–2017 and in 2018

The PCA analysis of the eigenvalues of the correlation matrix (Table 7) revealed the three main principal components, which could explain 78.2% of the total variance. The first principal components (PC1) explained 32.76% of the total variance. The second principal component (PC2) explained 28.42% of the total variance. The third principal component (PC3) explained 17.0% of the variance.

Table 8 presents the values of the correlation between the variables used in the model and the main components presented in Figure 5. For each of the determined correlation values, the p-value was given. Table 8 lists only those variables that, at the assumed significance level, gave a statistically significant result.

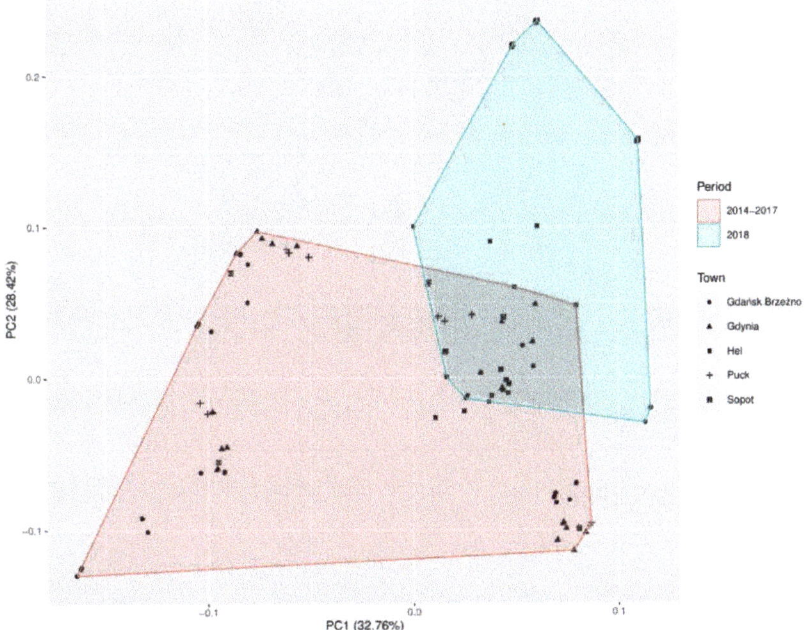

Figure 5. Principal component analysis (PCA) of *Cladosporium* sp. detected in air samples in the seaside towns (Hel, Puck, Gdynia, Sopot, and Gdańsk Brzeźno). The plot score of the first two principal component contains almost 61.17% of the explained variance. Two clusters can be distinguished: The red one is composed in the study period of 2014–2017 and the blue cluster in 2018.

The most important principal component (PC1) of mold fungi *Cladosporium* sp. was the significant correlation ($p < 0.00001$) with the water temperature, wind speed, and air temperature, explaining 32.76% of the total variance observed (Table 8). The second principal component (PC2) was the significant correlation with the relative humidity, air temperature, and wind speed, explaining 28.42% of the variance.

3.3. The Qualitative Assessment of Mold and Yeast-Like Fungi in the Atmospheric Air of the Seaside Towns in the Years 2014–2017

In the air samples collected in the years 2014–2017 in the towns of Hel, Puck, Gdynia, Sopot, and Gdańsk-Brzeźno, *Ascomycota* (98.29%), *Basidiomycota* (1.52%), and *Zygomycota*, (*Fungi Incertae Sedis*) were detected (0.19%) (Figure 3).

Three classes of *Ascomycota* fungi were found—*Eurotiomycetes* (77.50%), *Dothideomycetes* (18.44%), and *Saccharomycetes* (2.34%). In the *Eurotiomycetes* class, the following genus was isolated: *Penicillium* (63.24%), *Aspergillus* (13.56%), and *Trichophyton* (0.70%). Within the *Penicillium* genus, *Penicillium* section *Viridicata* (40.11%) and *Penicillium* section *Chrysogena* (23.13%) were detected. Within the *Aspergillus* genus, there was *Aspergillus* section *Nigri* (13.56%). Within the *Trichophyton* genus, the *Trichophyton mentagrophytes* complex (0.70%) was isolated. In the class of *Dothideomycetes*, the *Cladosporium* genus was found, with the *Cladosporium herbarum* complex (16.79%) and the *Aureobasidium* genus, with the *Aureobasidium pullulans* complex (1.65%). In the *Saccharomycetes* class, the *Saccharomycetes* genus was isolated (2.34%). In the *Basidiomycota* phylum, the *Cystobasidiomycetes* class, the *Rhodotorula* genus, with the *Rhodotorula* sp. (1.52%) was detected. In the *Zygomycota* phylum, the *Mucoromycotina* class, with the *Mucor mucedo* group (0.19%) was detected. The percentage of mold and yeast-like fungi in the seaside air is shown in Figure 6.

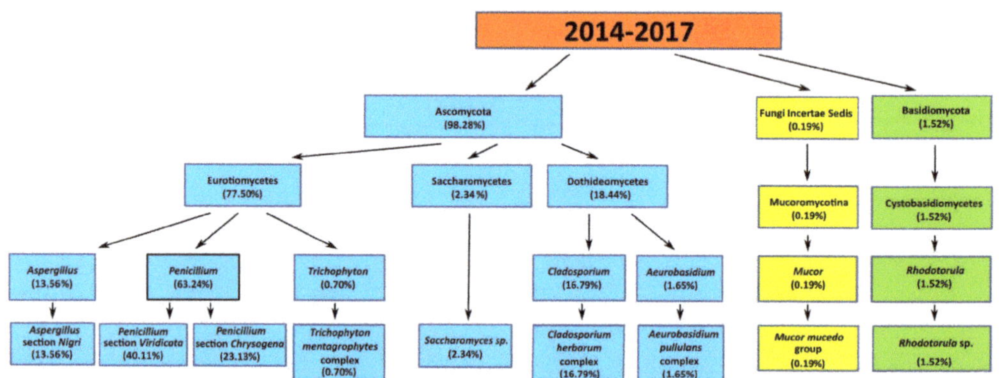

Figure 6. The percentage share of mold and yeast-like fungi in the air of the seaside towns (Hel, Puck, Gdynia, Sopot, and Gdańsk-Brzeźno) in the years 2014–2017.

3.4. The Qualitative Assessment of Mold and Yeast-Like Fungi in the Air of Seaside Towns in 2018

Ascomycota (96.17%), *Basidiomycota* (1.21%), and *Zygomycota* (*Fungi Incertae Sedis*) (2.62%) were detected in the air samples collected in 2018 in the seaside towns of Hel, Puck, Gdynia, Sopot, and Gdańsk-Brzeźno (Figure 4). In the *Ascomycota* phylum, four classes of fungi were found: *Eurotiomycetes* (60.07%), *Dothideomycetes* (4.60%), *Saccharomycetes* (27.26%), and *Sordariomycetes* (4.24%).

The genus isolated within the *Eurotiomycetes* class were: *Penicillium* (20.72%), *Aspergillus* (38.5%), and *Trichophyton* (0.85%). Within the *Penicillium* genus, *Penicillium* section *Viridicata* (11.87%) and *Penicillium* section *Chrysogena* (8.84%) were detected. Within the *Aspergillus* genus, the species: *Aspergillus* section *Nigri* (37.05%) and *Aspergillus* section *Fumigati* (1.45%) were isolated. Within the *Trichophyton* genus, the *Trichophyton mentagrophytes* complex was detected (0.85%). In the *Dothideomycetes* class, there was *Cladosporium* genus, the *Cladosporium herbarum* species complex (4.31%), and the *Alternaria* genus with the *Alternaria alternata* complex (0.29%). In the *Sordariomycetes* class, the *Stachobytris* genus with the *Stachybotrys chartarum* complex (4.24%) was found. In the *Saccharomycetes* class, the *Candida* genus was isolated with *Candida albicans* (27.26%). In the *Basidiomycota* phylum, the *Cystobasidiomycetes* class, the *Rhodotorula* genus, and the *Rhodotorula* sp. (1.21%) were detected. In the *Zygomycota* phylum, the *Mucoromycotina* class with the *Mucor mucedo* group (2.62%) was found. In 2018, potentially pathogenic and allergenic mold and yeast-like fungi were detected in the seaside air, such as *A.* section *Fumigati* (1.45%), *S. chartarum* complex (4.24%), and *C. albicans* (27.26%). The species were not observed in the years 2014–2017. The percentage share of these mold and yeast-like fungi in the samples of seaside air is shown in Figure 7.

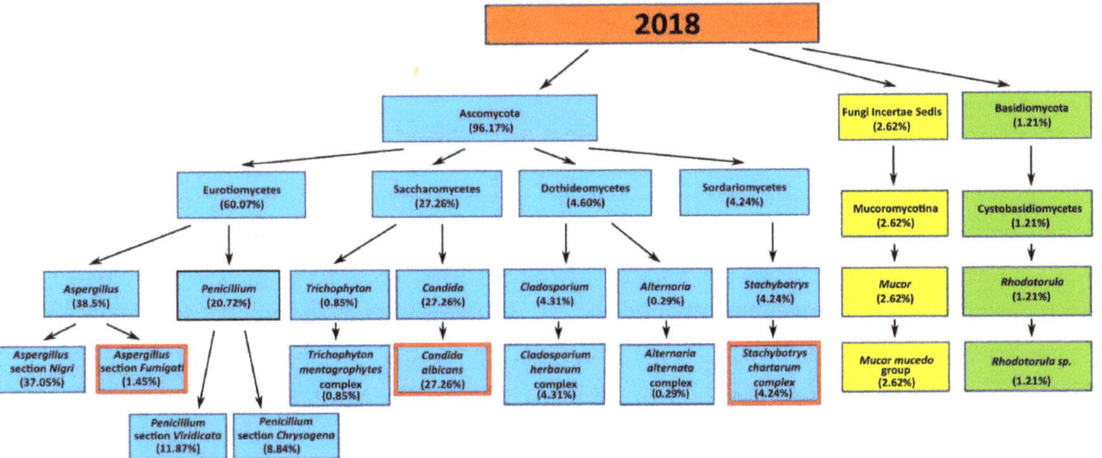

Figure 7. The percentage share of mold and yeast-like fungi in the air of seaside towns (Hel, Puck, Gdynia, Sopot, and Gdańsk-Brzeźno) in 2018. The red frame marks the species were not observed in the years 2014–2017.

4. Discussion

The analysis based on the PCA correlation of mold and yeast-like fungi detected in air samples in the seaside towns and meteorological factors showed a statistically significant relationship ($p < 0.00001$). The principal component (PC1) was correlated with the air and water temperature, and wind speed explains 33.54% of the total variance. The second principal component (PC2) was correlated with the relative air humidity, air temperature, direction, and wind speed, and explains 27.23% of the total variance (Table 2). Similar research results were obtained by Grinn-Gofron and Bosiacka [28]. Their 4-year study showed that air temperature, dew point, relative humidity, and average wind speed had the greatest influence on the composition of spores in the air [28]. Previous interdisciplinary studies too, including our research, showed that the direction and speed of the wind are one of the most important meteorological factors affecting the formation of aerosols at the water-air interface [29–32]. In our study, the important meteorological factors affecting airborne spore of *Aspergillus* sp. and *Penicillium* sp. concentrations were relative humidity and wind speed. The same results were noted by Grinn-Gofroń in Szczecin. The daily values of relative humidity and average wind speed were positively correlated for $p = 0.001$ and $p = 0.01$, respectively [33]. In addition, the principal component (PC1) and (PC2) of *Cladosporium* sp. was significantly correlated ($p < 0.00001$) with the wind speed, air temperature, and relative humidity. In the same way, air temperature, wind speed, and relative humidity were associated with *Cladosporium* spore dispersal in Morocco [34]. In addition, our PCA statistical study of the seaside air of 2018 revealed that, compared to the number of fungi detected in the years 2014–2017, the highest number of mold and yeast-like fungi, after emergency disposal of sewage into the Gulf of Gdańsk, was detected in Hel, Sopot, and Gdańsk-Brzeźno (Figure 2). We believe that the greater number of mold fungi in the air samples in these seaside towns could have been influenced not only by water and air temperature, wind speed, and humidity but also by the direction of the wind (SE) blowing from the Bay of Gdańsk. We suggest that it is advisable to inform residents about the potential health risk in the event of raw sewage disposal into the water.

On the other hand, in Gdynia, in 2018, neither mold nor yeast-like fungi were found compared to the period of 2014–2017. The lack of mold and yeast-like fungi in the air of Gdynia in 2018 could be caused by the direction of the wind blowing from the sea (NE). The Gulf of Gdańsk is partially separated from the Baltic Sea by the Hel Peninsula and the city of Hel. In Hel, the dominant wind direction is west, and in Gdynia, a north-east direction. In the city of Gdynia, due to the presence of the Hel Peninsula, the wind blowing from the direction of the sea is less frequently observed.

To our knowledge, no results of mycological studies on the quality of seaside air after emergency disposal of raw sewage are available. Therefore, the outcomes of 2018 were compared to the quality of air at the sewage treatment plant in order to indicate the likely origin of the species. Filamentous fungi of the genus *Aspergillus*, *Cladosporium*, and *Mucor* and yeast-like fungi, for example, *Candida*, were detected in domestic human and animal sewage [35–38]. Michałkiewicz et al. found that the majority of yeast-like fungi isolated from the air of the four wastewater treatment plants was *Candida* [39]. Other studies demonstrated that the following mold and yeast-like fungi were predominant: *Cladosporium* sp., *A. fumigatus*, *A. alternata*, *C. albicans*, and *Rhodotorula* sp. [40–44]. Potentially pathogenic fungi, such as *Olpidium*, *Paecilomyces*, *Aspergillus*, *Rhodotorula*, *Penicillium*, *Candida*, *Synchytrium*, *Phyllosticta*, and *Mucor* have been detected in three wastewater treatment plants located in the Gauteng province of the Republic of South Africa [45]. In Portuguese studies, mold fungi *Aspergillus*, *Fusarium*, and yeast-like fungi *Candida* were found in beach sand contaminated with leaking toilet sewage [12].

In turn, other researchers conducted a sanitary evaluation of sand and water from 16 beaches of São Paulo State, Brazil [46]. Ninety-six samples each of wet and dry sand and seawater were collected and analyzed for fecal indicator bacteria. Correlation analysis indicated a significant relationship between fecal indicator densities in wet sand and seawater. There was a significant correlation between the densities of fecal coliforms and fecal streptococci for both types of sand, and this correlation was higher in wet sand. These data suggest the necessity of some criteria for microbiological control [46]. Transmission of infectious diseases in terrestrial beach environments can occur via direct exposure to microbes found in sand or through the flux of microbes from water to sand within the swash or intertidal zone. In addition to direct exposure, sand can also serve as a vehicle for transferring pathogenic microbes to and from the adjacent water [10].

Recent research suggests that being in and using the beach may be a risk factor for infectious diseases, thus monitoring of both seawater, beach sand, and coastal air is warranted [7,10,46,47].

In conclusion, the study results of 2018 indicate that untreated wastewater associated with emergency disposal to the Gulf of Gdańsk was a likely source of mold and yeast-like fungi in the seaside air. The analysis of PCA data demonstrated a statistically significant relationship between the meteorological factors and the number of mold and yeast-like fungi reported in the period of 2014–2017 and in 2018. Although failures of sewage treatment plant collectors, heavy rainfall, and floods occur quite often, they should not lead to an increase in the number of potentially pathogenic bacteria and mold and yeast-like fungi in the coastal seawater and air. Therefore, it is important to build new wastewater treatment plants, and expand and modernize the existing ones, thus that pathogenic microorganisms are effectively eliminated in the process of wastewater treatment. Moreover, tighter measures, including wastewater disinfection, should be introduced in wastewater treatment plants.

Author Contributions: Conceptualization, M.M. and K.Z.; methodology, M.M. and K.Z.; validation, R.M., P.W. and K.Z.; formal analysis, M.M. and P.W.; investigation, M.M.; statistical analysis, P.W.; data curation, K.Z.; R.M. and P.W.; writing—original draft preparation, M.M. and K.Z.; writing—

review and editing, K.Z., M.M., M.K., R.M. and P.W.; supervision, K.Z., R.M. All authors have read and agreed to the published version of the manuscript.

Funding: This research was funded by Medical University of Gdańsk, grant number ST-02-0108/07/780.

Institutional Review Board Statement: Not applicable.

Informed Consent Statement: Not applicable.

Data Availability Statement: Not applicable.

Conflicts of Interest: The authors declare no conflict of interest.

References

1. Walser, S.M.; Gerstner, D.G.; Brenner, B.; Bünger, J.; Eikmann, T.; Janssen, B.; Kolb, S.; Kolk, A.; Nowak, D.; Raulf, M.; et al. Evaluation of exposure-response relationships for health effects of microbial bioaerosols-A systematic review. *Int. J. Hygen Environ. Health* **2015**, *218*, 577–589. [CrossRef] [PubMed]
2. Schlosser, O.; Robert, S.; Debeaupuis, C. Aspergillus fumigatus and mesophilic moulds in air in the surrounding environment downwind of non-hazardous waste landfill sites. *Int. J. Hygen Environ. Health* **2016**, *219*, 239–251. [CrossRef] [PubMed]
3. Kowalewska, B.; Kawko, M.; Zorena, K.; Myśliwiec, M. Yeast-like fungi in the gastrointestinal tract in children and adolescents with diabetes type 1. *Pediatr. Endocrinol. Diabetes Metab.* **2014**, *22*, 170–177. [CrossRef] [PubMed]
4. Kowalewska, B.; Zorena, K.; Szmigiero-Kawko, M.; Wąż, P.; Myśliwiec, M. Higher diversity in fungal species discriminates children with type 1 diabetes mellitus from healthy control. *Patient Prefer. Adherence* **2016**, *10*, 591–599.
5. Michalska, M.; Wąż, P.; Zorena, K.; Bartoszewicz, M.; Korzeniowska, K.; Krawczyk, S.; Myśliwiec, M. Potential effects of microbial air quality on the number of new cases of diabetes type 1 in children in two regions of Poland: A pilot study. *Infect. Drug Resist.* **2019**, *12*, 2323–2334. [CrossRef]
6. Zhang, I.; Pletcher, S.D.; Goldberg, A.N.; Barker, B.M.; Cope, E.K. Fungal Microbiota in Chronic Airway Inflammatory Disease and Emerging Relationships with the Host Immune Response. *Front. Microbiol.* **2017**, *8*, 2477. [CrossRef]
7. Michalska, M.; Zorena, K.; Bartoszewicz, M. Analysis of faecal bacteria isolated from air and sweater samples following an emergency sewage discharge into the gulf of Gdansk in 2018–preliminary study. *Int. Marit. Health* **2019**, *70*, 239–243. [CrossRef]
8. Noble, R.T.; Weisberg, S.B.; Leecaster, M.K.; McGee, C.D.; Dorsey, J.H.; Vainik, P.; Orozco-Borbon, V. Storm effects on regional beach water quality along the southern California shoreline. *J. Water Health* **2003**, *1*, 23–31. [CrossRef]
9. Lyons, B.P.; Devlin, M.J.; Hamid, S.A.; Al-Otiabi, A.F.; Al-Enezi, M.; Massoud, M.S.; Al-Zaidan, A.; Smith, A.; Morris, S.; Bersuder, P.; et al. Microbial water quality and sedimentary faecal sterols as markers of sewage contamination in Kuwait. *Mar. Pollut. Bull.* **2015**, *100*, 689–698. [CrossRef] [PubMed]
10. Solo-Gabriele, H.; Harwood, V.; Kay, D.; Fujioka, R.; Sadowsky, M.; Whitman, R.; Brandão, J. Beach sand and the potential for infectious disease transmission: Observations and recommendations. *J. Mar. Biol. Assoc. UK* **2016**, *96*, 101–120. [CrossRef]
11. O'Mullan, G.D.; Dueker, E.M.; Juhl, A.R. Challenges to Managing Microbial Fecal Pollution in Coastal Environments: Extra-Enteric Ecology and Microbial Exchange Among Water, Sediment, and Air. *Curr. Pollut. Rep.* **2017**, *3*, 1. [CrossRef]
12. Brandão, J.; Albergaria, I.; Albuquerque, J.; José, S.; Grossinho, J.; Ferreira, F.C.; Raposo, A.; Rodrigues, R.; Silva, C.; Jordao, L.; et al. Untreated sewage contamination of beach sand from a leaking underground sewage system. *Sci. Total Environ.* **2020**, *740*, 140237. [CrossRef]
13. Velonakis, E.; Dimitriadi, D.; Papadogiannakis, E.; Vatopoulos, A. Present status of effect of microorganisms from sand beach on public health. *J. Coast. Life Med.* **2014**, *2*, 746–756.
14. Whitman, R.; Harwood, V.J.; Edge, T.A.; Nevers, M.B.; Byappanahalli, M.N.; Vijayavel, K.; Brandão, J.; Sadowsky, M.J.; Alm, E.W.; Crowe, A.; et al. Microbes in Beach Sands: Integrating Environment, Ecology and Public Health. *Rev. Environ. Sci. Biotechnol.* **2014**, *13*, 329–368. [CrossRef]
15. Pougnet, R.; Pougnet, L.; Allio, I.; Lucas, D.; Dewitte, J.-D.; Loddé, B. Maritime environment health risks related to pathogenic microorganisms in seawater. *Int. Marit. Health* **2018**, *69*, 35–45. [CrossRef]
16. Carrillo-Muñoz, A.; Rodriguez, J.T.; Madrenys-Brunet, N.; Dronda-Ayza, A. Comparative study on the survival of 5 species of dermatophytes and S. brevicaulis in beach sand, under laboratory conditions. *Rev. Iberoam. Micol.* **1990**, *7*, 36–38.
17. Vogel, C.; Rogerson, A.; Schatz, S.; Laubach, H.; Tallman, A.; Fell, J. Prevalence of yeast in beach sand at three bathing beaches in South Florida. *Water Res.* **2007**, *41*, 1915–1920. [CrossRef] [PubMed]
18. Anderson, J.H. In vitro survival of human pathogenic fungi in Hawaiian beach sand. *Sabouraudia* **1979**, *17*, 13–22. [CrossRef] [PubMed]
19. Quant, B.; Bray, R.; Olańczuk-Neyman, K.; Jankowska, K.; Kulbat, E.; Łuczkiewicz, A.; Sokołowska, A.; Fudala, S. *Badania Nad Dezynfekcją Ścieków Oczyszczonych Odprowadzanych do Wód Powiserzchniowych*; Monografie Komitetu Inżynierii Środowiska Polskiej Akademii Nauk: Lublin, Poland, 2009; Volume 61, pp. 19–29. (In Polish)
20. Käärik, A.; Keller, J.; Kiffer, E.; Perreau, J.; Reisinger, O.; Nilsson, S. *Atlas of Airborne Fungal Spores in Europe*; Springer: Berlin/Heidelberg, Germany, 1983.

21. Campbell, C.K.; Johnson, E.M.; Warnock, D.W. *Identification of Pathogenic Fungi*; Wiley-Blackwell Chichester, West Sussex: Hoboken, NJ, USA, 2013.
22. Feller, W. *An Introduction to Probability Theory and Its Applications*; John Wiley and Sous Inc.: New York, NY, USA, 1950; Volume II.
23. Tang, Y.; Horikoshi, M.; Li, W. ggfortify: Unified Interface to Visualize Statistical Result of Popular R Packages. *R J.* **2016**, *8*, 474. [CrossRef]
24. Horikoshi, M.; Tang, Y. ggfortify: Data Visualization Tools for Statistical Analysis Results. 2016. Available online: https://CRAN.R-project.org/package=ggfortify (accessed on 10 February 2020).
25. Alboukadel Kassambara and Fabian Mundt, factoextra: Extract and Visualize the Results of Multivariate Data Analyses. R Package Version 1.0.7. 2020. Available online: https://CRAN.R-project.org/package=factoextra (accessed on 10 February 2020).
26. Le, S.; Josse, J.; Husson, F. FactoMineR: An R Package for Multivariate Analysis. *J. Stat. Softw.* **2008**, *25*, 1–18. [CrossRef]
27. Chao, H.J.; Schwartz, J.; Milton, D.K.; Burge, H.A. Populations and determinants of airborne fungi in large office buildings. *Environ. Health Perspect.* **2002**, *110*, 777–782. [CrossRef]
28. Grinn-Gofroń, A.; Bosiacka, B. Effects of meteorological factors on the composition of selected fungal spores in the air. *Aerobiologia (Bologna)* **2015**, *31*, 63–72. [CrossRef]
29. Michalska, M.; Bartoszewicz, M.; Cieszyńska, M.; Nowacki, J. Bioaerozol morski w rejonie Zatoki Gdańskiej. *Med. Srod.* **2011**, *14*, 24–28.
30. Rahlff, J.; Stolle, C.; Giebel, H.A.; Brinkhoff, T.; Ribas-Ribas, M.; Hodapp, D.; Wurl, O. High wind speeds prevent formation of a distinct bacterioneuston community in the sea-surface microlayer. *FEMS Microbiol. Ecol.* **2017**, *93*, 93. [CrossRef] [PubMed]
31. Marks, R.; Górecka, E.; McCartney, K.; Borkowski, W. Rising bubbles as mechanism for scavenging and aerosolization of diatoms. *J. Aerosol. Sci.* **2019**, *128*, 79–88. [CrossRef]
32. Ruiz-Gil, T.; Acuña, J.J.; Fujiyoshi, S.; Tanaka, D.; Noda, J.; Maruyama, F.; Jorquera, M.A. Airborne bacterial communities of outdoor environments and their associated influencing factors. *Environ. Int.* **2020**, *145*, 106156. [CrossRef] [PubMed]
33. Grinn-Gofroń, A. Airborne Aspergillus and Penicillium in the atmosphere of Szczecin, (Poland) (2004–2009). *Aerobiologia (Bologna)* **2011**, *27*, 67–76. [CrossRef] [PubMed]
34. Sidel, F.F.B.; Bouziane, H. Fungal Spores of Cladosporium in the Air of Tetouan Meteorological Parameters and Forecast Models. *Int. J. Environ. Sci. Nat. Res.* **2017**, *3*, 555609.
35. Park, J.H.; Ryu, S.H.; Lee, J.Y.; Kim, H.J.; Kwak, S.H.; Jung, J.; Lee, J.; Sung, H.; Kim, S.-H. Airborne fungal spores and invasive aspergillosis in hematologic units in a tertiary hospital during construction: A prospective cohort study. *Antimicrob. Resist. Infect. Control* **2019**, *8*, 88. [CrossRef]
36. Korzeniewska, E. Emission of bacteria and fungi in the air from wastewater treatment plants-A review. *Front. Biosci. (Schol. Ed.)* **2011**, *1*, 393–407. [CrossRef]
37. Niazi, S.; Hassanvand, M.S.; Mahvi, A.H.; Nabizadeh, R.; Alimohammadi, M.; Nabavi, S.; Faridi, S.; Dehghani, A.; Hoseini, M.; Moradi-Joo, M.; et al. Assessment of bioaerosol contamination (bacteria and fungi) in the largest urban wastewater treatment plant in the Middle East. *Environ. Sci. Pollut. Res. Int.* **2015**, *22*, 16014–16021. [CrossRef]
38. Zhang, H.; Feng, J.; Chen, S.; Li, B.; Sekar, R.; Zhao, Z.; Jia, J.; Wang, Y.; Kang, P. Disentangling the Drivers of Diversity and Distribution of Fungal Community Composition in Wastewater Treatment Plants Across Spatial Scales. *Front. Microbiol.* **2019**, *9*, 1291. [CrossRef]
39. Michałkiewicz, M.; Pruss, A.; Dymaczewski, Z.; Jeż-Walkowiak, J.; Kwaśna, S. Microbiological Air Monitoring around Municipal Wastewater Treatment Plants. *Pol. J. Environ. Stud.* **2011**, *20*, 1243–1250.
40. Cyprowski, M.; Sowiak, M.; Soroka, P.M.; Buczyńska, A.; Kozajda, A.; Szadkowska-Stańczyk, I. Ocena zawodowej ekspozycji na aerozole grzybowe w oczyszczalniach ścieków. *Med. Pr.* **2008**, *59*, 365–371.
41. Filipkowska, Z.; Gotkowska-Płachta, A.; Korzeniewska, E. Grzyby pleśniowe oraz drożdże i grzyby drożdżoidalne w powietrzu atmosferycznym na terenie oczyszczalni ścieków z systemem stawów napowietrzanych i stabilizacyjnych. *Woda-Środowisko-Obsz. Wiej.* **2008**, *8*, 69–82. (In Polish)
42. Małecka-Adamowicz, M.; Kubera, Ł.; Donderski, W.; Kolet, K. Microbial air contamination on the premises of the sewage treatment plant in Bydgoszcz (Poland) and antibiotic resistance of Staphylococcus spp. *Arch. Environ. Prot.* **2017**, *43*, 58–65. [CrossRef]
43. Bednarz, A.; Pawłowska, S. A fungal spore calendar for the atmosphere of Szczecin, Poland. *Acta Agrobot.* **2016**, *69*, 1669. [CrossRef]
44. Kowalski, M.; Wolany, J.; Pastuszka, J.S.; Płaza, G.; Wlazło, A.; Ulfig, K.; Malina, A. Characteristics of airborne bacteria and fungi in some Polish wastewater treatment plants. *Int. J. Environ. Sci. Technol.* **2017**, *14*, 2181–2192. [CrossRef]
45. Assress, H.A.; Selvarajan, R.; Nyoni, H.; Ntushelo, K.; Mamba, B.B.; Msagati, T.A.M. Diversity, Co-occurrence and Implications of Fungal Communities in Wastewater Treatment Plants. *Sci. Rep.* **2019**, *9*, 14056. [CrossRef]
46. Sato, M.I.Z.; Di Bari, M.; Lamparelli, C.C.; Truzzi, A.C.; Coelho, M.C.L.S.; Hachich, E.M. Sanitary quality of sands from marine recreational beaches of São Paulo, Brazil. *Braz. J. Microbiol.* **2005**, *36*, 321–326. [CrossRef]
47. Sabino, R.; Rodrigues, R.; Costa, I.; Carneiro, C.; Cunha, M.; Duarte, A.; Faria, N.; Ferreira, F.; Gargaté, M.J.; Júlio, C.; et al. Routine screening of harmful microorganisms in beach sands: Implications to public health. *Sci. Total Environ.* **2014**, *15*, 1062–1069. [CrossRef] [PubMed]

Journal of *Fungi*

Article

The Culturable Mycobiome of Mesophotic *Agelas oroides*: Constituents and Changes Following Sponge Transplantation to Shallow Water

Eyal Ben-Dor Cohen [1,2], Micha Ilan [1] and Oded Yarden [2,*]

[1] School of Zoology, George S Wise Faculty of Life Sciences, Tel Aviv University, Tel Aviv 6997801, Israel; eyalbendor@gmail.com (E.B.-D.C.); milan@tauex.tau.ac.il (M.I.)
[2] Department of Plant Pathology and Microbiology, The RH Smith Faculty of Agriculture, Food and Environment, The Hebrew University of Jerusalem, Rehovot 7610001, Israel
* Correspondence: oded.yarden@mail.huji.ac.il

Abstract: Marine sponges harbor a diverse array of microorganisms and the composition of the microbial community has been suggested to be linked to holo-biont health. Most of the attention concerning sponge mycobiomes has been given to sponges present in shallow depths. Here, we describe the presence of 146 culturable mycobiome taxa isolated from mesophotic niche (100 m depth)-inhabiting samples of *Agelas oroides*, in the Mediterranean Sea. We identify some potential in vitro interactions between several *A. oroides*-associated fungi and show that sponge meso-hyl extract, but not its predominantly collagen-rich part, is sufficient to support hyphal growth. We demonstrate that changes in the diversity of culturable mycobiome constituents occur following sponge transplantation from its original mesophotic habitat to shallow (10 m) waters, where historically (60 years ago) this species was found. We conclude that among the 30 fungal genera identified as associated with *A. oroides*, *Aspergillus*, *Penicillium* and *Trichoderma* constitute the core mycobiome of *A. oroides*, and that they persist even when the sponge is transplanted to a suboptimal environment, indicative of the presence of constant, as well as dynamic, components of the sponge mycobiome. Other genera seemed more depth-related and appeared or disappeared upon host's transfer from 100 to 10 m.

Keywords: mycobiome; marine sponge; marine fungi; *Agelas*; mesophotic

Citation: Ben-Dor Cohen, E.; Ilan, M.; Yarden, O. The Culturable Mycobiome of Mesophotic *Agelas oroides*: Constituents and Changes Following Sponge Transplantation to Shallow Water. *J. Fungi* **2021**, *7*, 567. https://doi.org/10.3390/jof7070567

Academic Editor: Federico Baltar

Received: 7 April 2021
Accepted: 28 June 2021
Published: 16 July 2021

1. Introduction

Sponges (phylum Porifera) are a diverse and abundant filter-feeding phylum of sessile marine invertebrates, with close to 10,000 known species [1]. They inhabit hard and soft bottom habitats, while a few species (ca. 150) occupy fresh water [2]. Sponges are an important component of the benthic fauna, with a significant impact on biogeochemical cycling of key nutrients [3]. In some ecosystems sponges provide three-dimensional habitats in the benthic-exposed environment [4]. Thus, they influence the composition, diversity, and abundance of the epibenthic faunal community [5] and the pelagic fauna [6]. Marine sponges are known to harbor a large diversity of microorganisms [7–9]. These associated microorganisms are thought to be involved in a variety of ecological functions including production of secondary metabolites [10,11]. While experimental studies suggest that the composition of the microbial community is linked to holo-biont health, the potential role of the microbiome as involved in sponge acclimation and adaptation to environmental change is unknown [3]. While sponges host microorganisms from all domains of life [12], sponge-associated eukaryotes have been the most recent to gain attention. These, and in particular, fungi, have been shown to be both common and diverse in almost every class of sponges [11,13–18]. While host genetics, environmental factors and geography have been shown to influence the sponge prokaryotic microbiome [19,20], the influence of these factors on the mycobiome have yet to be characterized.

Agelas oroides (Demospongiae) is a massive sponge, with size ranges between 5 cm to 25 cm. It is well known for its secondary metabolites production [21–23]. It can be found throughout the Mediterranean Sea, along a depth gradient of 4–115 m, indicating that this sponge can inhabit mesophotic niches. Over the last decade, mesophotic coral ecosystems (MCEs), found at ocean depths of 30–150 m, have gained increasing levels of attention [24–26]. Typically, this niche is characterized by lower temperatures and light intensities than the shallow water, and is an environment richer in nutrients. This habitat has been considered to be part of the "deep reef refuge hypothesis", which suggests that, following some adjustment, macro-organisms inhabiting this niche can find refuge from biotic and a-biotic stresses common in shallower habitats [27]. In spite of its reported broad depth range, *A. oroides* was last recorded in Israeli Levant region in shallow waters (<7 m) 50 years ago [28], and since then considered as lost to this region. It was recently re-found as flourishing along the Israeli coastline, at mesophotic (ca. 100 m) depths [23,29]. A recent transplantation experiment supports the suggestion that *A. oroides* disappeared from the Levant shallow water due to the increase in shallow water summer temperatures, and the extended period of these higher water temperatures during the past 50 years [29].

Members of the genus *Agelas* have been described as high microbial abundance (HMA) sponges [30]. One suggested role for the large microbial consortia present in these sponges is in nutritional supplementation [31]. The number of reports on *Agelas*-associated fungi is limited and has mainly focused on several *Aspergillus* spp. that have been isolated from *A. oroides* and their capacity to produce natural products [32,33].

Here, we report on culturable constituents of the mesophotic *A. oroides* mycobiome, identify some potential interactions between several *A. oroides*-associated fungi and describe changes in mycobiome constituents following sponge transplantation from their mesophotic habitat to shallow (10 m) waters, where historically this species was found.

2. Materials and Methods

2.1. Sponge Sample Collection

Samples (n = 10) of the sponge *Agelas oroides* were collected (with a permit from the Israel Nature and Parks Authority) from the Mediterranean Sea by a remotely operated vehicle (ROV) at ~100 m depth from two sites, using the R/V Mediterranean Explorer (EcoOcean) (Figure 1), as described in detail by Idan et al. [29]. Two samples (A, B) were collected from Haifa Rosh HaCarmel (32°52.44′ N, 34°51.47′ E) and eight samples (C–J) were collected approximately 16 km off-shore at Herzliya, Israel (32°10.62′ N, 34°37.98′ E). Sponge samples A and B were collected in the autumn (September 2017), C-G during the winter (January 2018), H in the summer (July 2018) and samples I and J also in the winter (February 2019). The approximate size of each sponge sample was ~350 cm^3. Samples were kept in sealed plastic containers with seawater and treated in the laboratory within three hours post collection. Using a sterile blade, the sponges were cut in the center to expose inner parts. These core sponge sections (~5.0 cm^3) were dipped twice in sterile double distilled water (SDDW) prior to plating on fungal growth media.

For the transplantation experiment, two sponge samples (~430 cm^3) were collected from the mesophotic habitat using the ROV. The samples were maintained, for two weeks, in open water tables at the marine facility of Ruppin Academic Center (Mikhmoret). Then, the sponges were stabilized by being attached to a brick (400 cm^2) and placed (by SCUBA) in a cage at 10 m depth in the Mediterranean Sea ("Playground" site, Mikhmoret beach, 32°25.71′ N, 34°.52.39′ E). The cage was covered with a plastic net (3 cm mesh) to prevent animals from feeding on the sponges and to reduce light intensity. Every two weeks three sponge fragments (~5.0 cm^3) were collected from different parts of the sponges, and determined, on the basis of visual inspection, as either healthy, visibly unhealthy or dead. The fragments were compressed and fungal isolation and identification were carried out as described below.

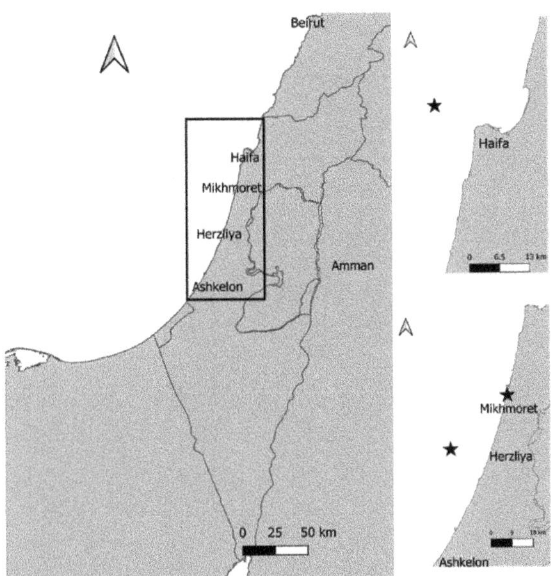

Figure 1. Map of *Agelas oroides* sampling and transplantation points (marked by asterisks).

2.2. Isolation and Identification of *Agelas oroides*-Derived Fungi

After rinsing the sponge core sections, they were either compressed using an autoclaved mortar and pestle, as previously described [17] or cut to smaller fragments. Following the compression procedure, 100 μL of sponge extract were plated on potato dextrose agar (PDA, Difco, Franklin Lakes, NJ, USA) Petri dishes amended with 250 mg/L chloramphenicol. Petri dishes were incubated at 25 °C, in the dark, until a mycelium emerged (3–60 days). Following fungal growth, disks (0.4 cm diameter) from the edge of the colony were transferred to fresh medium, until pure cultures were obtained. In parallel, the smaller, not compressed, sponge fragments were placed on identical, slanted, medium in 13 mm test tubes until fungal growth was evident. Pure cultures from these tubes were purified as described above.

To increase the variety of isolated fungi, sponge samples were plated on additional dishes, amended with either cycloheximide (500 μg mL^{-1}, Sigma, St. Louis, MO, USA), the benzimidazole fungicide Benomyl (10 μg mL^{-1}, Dupont, Wilmington, NC, USA) or the pyrimidine sulfamate fungicide Bupirimate (10 μg mL^{-1}, Makhteshim-Agan Group, Airport City, Israel), as previously described by Paz et al. [17].

For DNA-based molecular identification of the isolated fungi, fungal mycelium was scraped from the culture plates and suspended in 2 mL tubes containing 200 μL deionized water (NANOpure, Barnstead Co., Newton, MA, USA) and an equal amount of 0.5 mm glass beads (acid washed, Sigma-Aldrich, St. Louis, MO, USA). The tube was agitated using a Mini Bead Beater (Biospec Products Inc., Bartlesville, OK, USA) for 100 s, followed by 10 min boiling to inactivate endogenous nucleases. The samples were cooled to room temperature and subsequent DNA isolation was carried out using the GenElute™ Plant Genomic DNA miniprep kit (Merck, Herzliya Pituach, Israel) as described in the user manual.

PCR was carried out with S1000™ Thermal Cycler (Bio-Rad, Hercules, CA, USA) using PCRBIO VeriFi mix (PCR BIOSYSTEMS, Wayne, PA, USA). Primers used in this study are listed in Table 1. These included internal transcribed sequences of rDNA, as well as specific partial gene sequences (on the basis of different genera) used to identify species. Initial denaturation of DNA was carried out at 95 °C for 1 min followed by 35 cycles of three-step PCR amplifications consisting of denaturation at 95 °C for 0.5 min, primer annealing at the appropriate temperature (Table 1) for 15 s, and extension at 72 °C for 30 s. PCR products were separated by agarose gel electrophoresis. Amplicons were purified (Wizard SV Gel and PCR Clean Up System, Promega, San Luis Obispo, CA, USA), sequenced (MacLab, South San Francisco, CA, USA) and assembled with the Fragment Merger Tool [34]. Fungal strain identification was based on BLAST comparison of the sequences obtained with available databases.

Table 1. Primers used in this study.

Gene Name and Targeted Templates	Primer Designation (Annealing Temp.)	Sequence
Internal Transcribed Spacer (ITS) (all fungal templates)	ITS-5F ITS-4R (62 °C)	GGAAGTAAAAGTCGTAACAAGG TCCTCCGCTTATTGATATGC
ACTINa (ACT) (*Cladosporium* spp.)	ACT-512F ACT-783R (60 °C)	ATGTGCAAGGCCGGTTTCGC TACGAGTCCTTCTGGCCCAT
β-tubulin (Tub2) (*Acremonium* spp.; *Chaetomium* sp.; *Nigrospora* sp.; *Parengydontium* spp.; *Penicillium* spp.; *Talaromyces* sp.; *Zygosporium* spp.) (benA) (*Aspergillus* spp.; *Penicillium* spp.)	Bt2a-F Bt2b-R (68.9 °C) Ben2f Bt2b-R (55 °C)	GGTAACCAAATCGGTGCTGCTTTC ACCCTCAGTGTAGTGACCCTTGGC TCCAGACTGGTCAGTGTGTAA ACCCTCAGTGTAGTGACCCTTGGC
Calmodulin (CaM) (Aspergillus spp.)	CMD 5 CMD 6 (55 °C)	CCGAGTACAAGGAGGCCTTC CCGATAGAGGTCATAACGTGG
Glyceraldehyde-3-phosphate dehydrogenase (GPD) (*Alternaria* spp.)	gpd1-F gpd2-R (58 °C)	CAACGGCTTCGGTCGCATTG GCCAAGCAGTTGGTTGTGC
RNA polymerase II subunit 2 (RPB2) (*Acremonium* spp.; *Penicillium* sp.; *Pichia* sp.)	fRPB2-5F fRPB2-7cR (58.5 °C)	GATGATAGAGATCATTTTGG ATGGGTAAACAAGCTATGGG
Large Subunit (D1/D2) of rRNA (*Crocicreas* spp.; *Arthrographis* sp.; *Sarocladium* sp.)	NL-1F NL-4R (55 °C)	GCATATCAATAAGCGGAGGAAAAG GGTCCGTGTTTCAAGACGG
Translation elongation factor 1-alpha (TEF1) (*Trichoderma* spp.) (*Fusarium* spp.)	EF1-728F TEF1LLEr (62 °C) ef1 ef2 (55 °C)	CATCGAGAAGTTCGAGAAGG AACTTGCAGGCAATGTGG ATGGGTAAGGARGACAAGAC GGARGTACCAGTSATCATGTT

2.3. Fungal-Fungal Interactions and Assays of Growth on Sponge Tissue

Interactions between three of the strains isolated from mesophotic sponge samples (*Alternaria alternata* strain CP-02-2; accession number MZ568118; *Parengydontium album* strain CP-03-1; accession number MZ568223 and *Zygosporium masonii* (strain CP-03-2; accession number MZ568332) were analyzed in dual cultures. Agar plugs (4 mm diameter) from the edges of growing colonies were placed either in the center (single cultures) or near the periphery (dual cultures) of PDA dishes and incubated at 25 °C for periods ranging 9–18 days, until colony fronts were close enough to each other to visualize the presence or absence of an interaction. To determine growth rates, images of the various colonies (n = 4) obtained at different time points were analyzed using ImageJ [35]. Two way-ANOVA statistical analysis was used to determine the significance of differences in the growth rates.

To determine the temperature-dependent growth rates of *Penicillium stekii* (strain EP-14-1; accession number MZ568306), mycelial plugs (4 mm) of the fungus were first placed in the center of an SWA-containing petri dish. The dishes were incubated at either 18 °C, 25 °C, 28 °C, 31 °C or 34 °C. The temperatures in the mesophotic habitat were stable between January to October 2018 (19.5 °C–18.1 °C) while the temperatures in the shallow water ranged between 17.5 °C–31 °C. Hence, the temperatures tested represented temperatures prevalent in the Mediterranean Sea (excluding 34 °C, which is higher than previously measured temperatures; based on Idan et al. [29]). Growth rates were measured (n = 4) as described in the dual culture experiments.

In order to examine if *A. oroides* sponge fragments could support fungal growth, one specimen of *A. oroides* was collected from the mesophotic habitat and divided into two. One part was compressed ("pressed sponge") and the other did not undergo any mechanical treatment ("native sponge"). Both subsamples were autoclaved and placed on pre-moistened Whatman filter paper in standard petri dishes. Disks (3 mm) of *P. album* cultured on SWA were placed on the sponge subsamples (two replicates each, of pressed and native sponge samples). After seven days at 18 °C, fungal growth was imaged using a Stemi SV 6 stereoscope (Zeiss, Oberkochen, Germany). A follow-up experiment, using the same samples, included adding 200 µL of Potato dextrose broth (PDB) on one side of the pressed sponge and observing fungal growth at various time points after the PDB amendment.

3. Results

3.1. Fungal Diversity in Agelas oroides

Overall, 146 taxa, represented by 237 sponge-derived culturable fungal colonies, were isolated from the 10 *A. oroides* specimens originally collected from the mesophotic sites (Figure 2). Among the cultivated strains, 96% (230 taxa) were identified as Ascomycota (Table 2). Approximately half of them were designated as belonging to four orders (Euritales, Capnodiales, Hypocreales and Pleosporales). Overall, 30 genera were identified on the basis of ITS1 sequencing and a majority of the species comprising the culturable mycobiome were identified using additional molecular markers (Tables 1 and 2). The most common genera were *Penicillium*, *Cladosporium* and *Aspergillus* spp. (33%, 22% and 11%, respectively) and were present in most of the sponges collected from both mesophotic sites. Amending the isolation medium with cyclohexmide, Benlate or Bupirimate increased the diversity of the obtained taxa. This included 11, 10 and 4 taxa, respectively, which were not isolated on the standard, chloramphenicol-amended, PDA medium (Figure 3). While most identified taxa were found to reoccur in the different mesophotic sponge samples, members of one genus, *Fusarium* spp. were found to occur uniquely in sponge specimens following their transplantation to shallow water (see below).

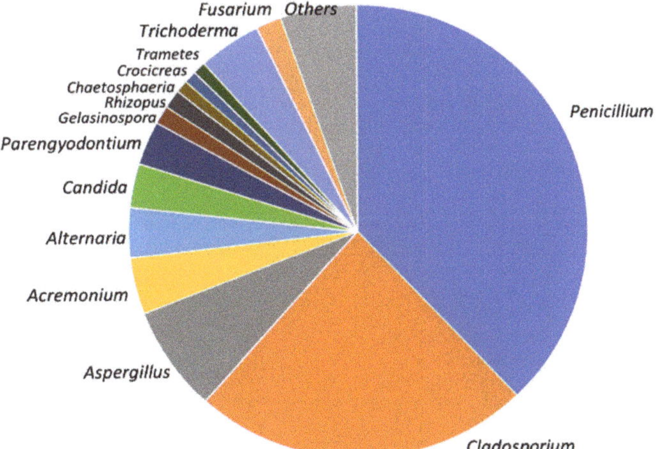

Figure 2. Relative abundance of culturable fungal genera isolated from *Agelas oroides*.

Table 2. Fungal taxa found in association with *Agelas oroides*.

Taxon	Number of Isolated Strains	Genes, in Addition to ITS, Used for Molecular Identification ("-" Indicates No Additional Genes Were Analyzed)
Penicillium steckii	47	Tub2
Penicillium sp.	12	-
Cladosporium limoniforme	10	ACT
Aspergillus sp.	8	-
Cladosporium halotolerans	8	ACT
Cladosporium sp.	8	-
Cladosporium sphaerospermum	8	ACT
Penicillium brevicompactum	8	Tub2
Acremonium sclerotigenum	7	RPB2
Candida sp.	7	-
Parengyodontium album	7	Tub2
Cladosporium ramotenellum	6	ACT
Alternaria alternata	5	GPD
Aspergillus niger	5	Cam
Aspergillus flavus	4	Cam
Penicillium chrysogenum	4	Tub2
Penicillium citrinum	4	Tub2
Alternaria sp.	3	-
Aspergillus tubingensis	3	Cam
Cladosporium aciculare	3	ACT
Cladosporium perangustum	3	ACT
Gelasinospora sp.	3	-
Rhizopus sp.	3	-
Trichoderma sp.	3	-
Acremonium sp.	2	-
Aspergillus protuberus	2	Cam
Chaetosphaeria sp.	2	-
Cladosporium dominicanum	2	ACT
Cladosporium tenellum	2	ACT
Crocicreas coronatum	2	D1/D2
Fusarium acuminatum	2	TEF1
Penicillium adametzioides	2	Tub2
Penicillium capsulatum	2	Tub2
Trametes sp.	2	-
Trichoderma atroviride	2	TEF1
Trichoderma orientale	2	TEF1
Arthrographis kalrae	1	D1/D2
Aspergillus fumigatiaffinis	1	Cam
Aspergillus insulicola	1	benA
Aspergillus sydowii	1	Cam
Aspergillus novoparasiticus	1	Cam
Chaetomium longiciliata	1	Tub2
Cladosporium aggregatocicatricatum	1	ACT
Cladosporium angustisporum	1	ACT
Cladosporium longicatenatum	1	ACT
Cystidiodontia sp.	1	-
Exophiala sp.	1	-
Fusarium brachygibbosum	1	TEF1
Fusarium equiseti	1	TEF1
Microascus sp.	1	-
Monocillium sp.	1	-

Table 2. *Cont.*

Taxon	Number of Isolated Strains	Genes, in Addition to ITS, Used for Molecular Identification ("-" Indicates No Additional Genes Were Analyzed)
Nigrospora osmanthi	1	Tub2
Penicillium astrolabium	1	benA
Penicillium coffeae	1	Tub2
Penicillium digitatum	1	Tub2
Penicillium simile	1	benA
Penicillium sizovae	1	Tub2
Penicillium wotroi	1	Tub2
Pichia guilliermondii	1	RPB2
Pichia sp.	1	-
Plectosphaerella sp.	1	-
Rhodotorula sp.	1	-
Sarocladium bacillisporum	1	D1/D2
Talaromyces funiculosus	1	Tub2
Trichoderma atrobrunneum	1	TEF1
Trichoderma gamsii	1	TEF1
Trichoderma guizhouense	1	TEF1
Zygosporium masonii	1	Tub2, D1/D2
Zygosporium pseudogibbum	1	Tub2

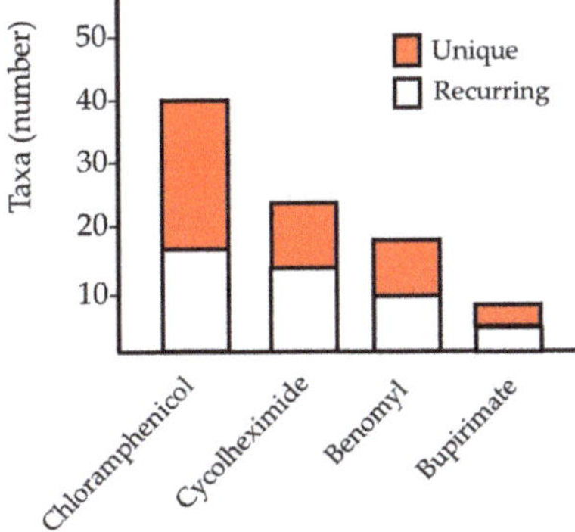

Figure 3. The number of unique and recurring fungal taxa observed on PDA medium with different amendments.

3.2. Members of the *Agelas oroides* Mycobiome Exhibit a Variety of Fungus-Fungus Interactions

A variety of possible interactions can be anticipated to occur between fungi occupying the same niche. To identify some of these potential interactions, a dual culture approach was used. When *Alternaria alternata* and *Parengyodontium album* were cultured together, the colony area of the two fungi was 55% and 44%, respectively, of that measured of the two species when grown in monoculture under the same conditions (12 days at 25 °C). Morphological changes were also evident in the *A. alternata* colonies. Not only was the colony size smaller and not circular, but the hyphae's color was lighter, suggesting less melanin was produced. When *A. alternata* was cultured with *Zygosporium masonii*, which is a much less reoccurring member of the culturable *A. oroides* mycobiome, the growth rate of the former was suppressed by about 60% while that of *Z. masonii* was not significantly affected (Figure 4). In contrast to when cultured in the presence of *P. album*, no marked changes in colony morphology were observed. We also examined the outcome of co-culturing *P. album* and *Z. masonii*. In this case it appears that the two species have inverse effects on each other. While growth of *P. album* in the dual culture decreased by about 40%, that of *Z. masonii* increased by almost two-fold when cultured in the presence of *P. album*. Taken together, it appears that diverse potential interactions can occur between members of the *A. oroides* mycobiome, as is evident even in this relatively simple case, comprised of an interaction matrix of based on co-culturing only two species out of three.

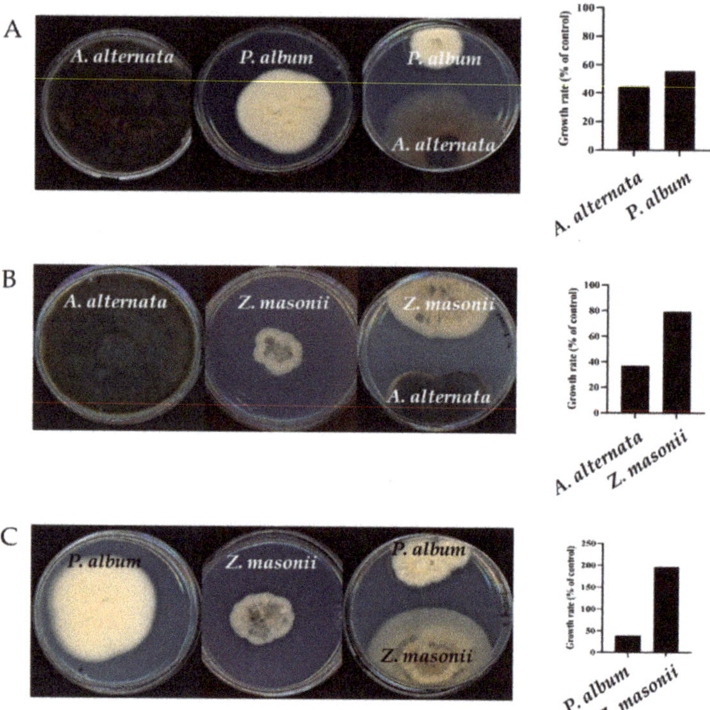

Figure 4. Growth of the *A. oroides* mycobiome constituents *Parengyodontium album*, *Alternaria alternata and Zygosporium masonii* in single and dual cultures. Pictures and measurements were carried out 12, 9 and 18 days after inoculation, for panels (**A–C**), respectively. Within each panel, differences in growth rates are significantly different ($p \leq 0.01$).

3.3. Sponge Holo-Biont Constituents Can Support Growth of *Parengyodontium album*

To determine whether fungi isolated from *A. oroides* have the potential to grow and reproduce under nutritional conditions prevalent in the sponge environment, we cultured five of the most reoccurring species (*Alternaria alternata, Cladosporium halotolerans, Parengyodontium album, Penicillium steckii* and *Zygosporium masonii*) on seawater agar. We were able to clearly observe all stages of their anamorphic life cycle on this low nutrient, high salinity medium. The sensitivity of three of the mentioned species, *C. halotolerans*, *P. album* and *P. steckii* to various temperatures (18–31 °C) which are typical of both mesophotic and shallow depths of the Eastern Mediterranean Sea, was also examined. The optimal temperature for *P. steckii* growth was 18 °C, while *C. halotolerans* and *P. album* exhibited faster growth at moderate temperatures (25 °C and 28 °C). All three fungi exhibited negligible growth at 31 °C and no observable radial growth at 34 °C.

To determine whether the sponge can serve as a sufficient source of nutrients to support the growth of a mycobiome constituent, we cultured *P. album*, known to be a high producer of extracellular proteases [36], on sterile, intact, *A. oroides* collagen-rich tissue fragments. Discs of cultures grown on sea water agar were placed on either autoclaved sponge cubes or cubes that had been depleted (by applying mechanical pressure) of a significant part of their natural dissolved and water-suspended contents. After seven days of incubation at 18°C, growth was clearly seen on the autoclaved sponge fragments while no observable growth was seen on a similar fragment that was also subjected to pressing (Figure 5A,B). 71 days post inoculation, fungal growth was still observed on the autoclaved (yet otherwise unprocessed) sponge fragment, while no observable growth was detected on the pressed sponge sample (Figure 5C,D). To evaluate if the lack of growth could be due to the possible nutrient-depleted state of the pressed sponge sample, 200 µL of growth medium (PDB) were added to one side of a pressed sponge fragment, one week post-inoculation. Eight days after adding this additional source of nutrients, slight, but clearly observable, fungal growth was evident (Figure 5E). Based on these results, we concluded that while sponge collagen was not sufficient to support growth of *P. album*, the sponge holo-biont contains a sufficient source of nutrients to support growth of this *A. oroides*-associated fungus.

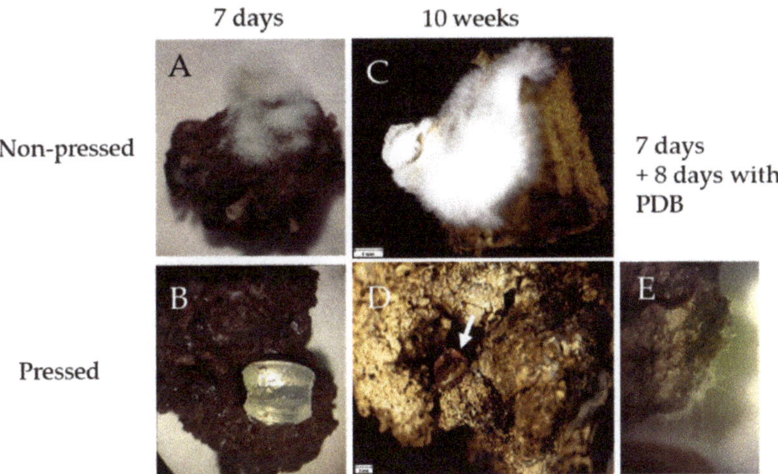

Figure 5. Growth of *Parengyodontium album* on *Agelas oroides*. Mycelial plugs of the fungus on non-pressed (**A**) and pressed (**B**) fragments of *A. oroides*, seven days post inoculation and 10 weeks post inoculation (**C**,**D**) (Plug is marked with arrow). Growth of the *P. album* on a pressed fragment of *A. oroides* 15 days post inoculation (**E**). Seven days after inoculation, potato dextrose broth (PDB) was added to the edge of the sponge.

3.4. *Transfer of Agelas oroides from Mesophotic to Shallow Water Confers Changes in the Mycobiome*

To examine the possible effects of environmental changes on the mycobiome of *A. oroides*, a transplantation experiment was conducted. First, two specimens (I, J) were collected from the mesophotic depth and small samples were cut and used to identify the culturable mycobiome, as part of the experiment described above. The remainder of specimens I and J was placed in in an open seawater system for two weeks and subsequently transferred to a location at 10 m depth, off Mikhmoret beach (Figures 1 and 6). Two weeks later, three fragments from specimens I and J, exhibiting either apparently healthy, unhealthy or dead morphological characteristics, were collected.

Figure 6. Transplanted *A. oroides*. The sponge was isolated from its mesophotic habitat, secured to a base and transferred to a protected surrounding in shallow water. Concrete base length is 18 cm.

Two months post transplantation, one specimen had died and was taken for fungal strain isolation and identification. The other survived until mid-July (four months after transplantation). Strains were isolated from specimens I and J on four occasions: initially when they were mesophotic, later when they were apparently healthy in the shallow water, then when they became sick and later dead. The 30 isolated strains from these sponges were comprised of eight genera (Figure 7). The results indicated that the differences in strain abundance were due to the different sea depth source, and not to the health state of the sponge. Among the fungi isolated from sponge samples at the shallow depth, *Aspergillus* (two strains) and *Penicillium* spp. (one strain) were found in both apparently unhealthy, as well as dead, sponge specimens. *Trichoderma* (six strains) and *Fusarium* spp. (six strains) were isolated from apparently healthy, as well as dead, samples obtained from the same depth. The *Fusarium* spp. colonies (Table 2) were the only fungi isolated here that were not found in any of the mesophotic samples throughout this study. Conversely, one of the most common species isolated from sponges obtained from the mesophotic depth, *P. stekii* (47 strains), was not found in any of the transplanted sponge samples. As one of the most distinct differences between the mesophotic and shallow depths is the temperature, we hypothesized that the higher temperatures prevalent in the shallow water may have affected *P. stekii*. We therefore measured growth of *P. stekii* isolated from the mesophotic sponge samples at different temperatures. Indeed, in spite of the fungus' ability to grow well at temperatures ranging from 18–28 °C, hardly any growth was observed at 31 °C and above (Figure 8), suggesting a possible link between temperature and the presence of *P. stekii* in these sponge samples.

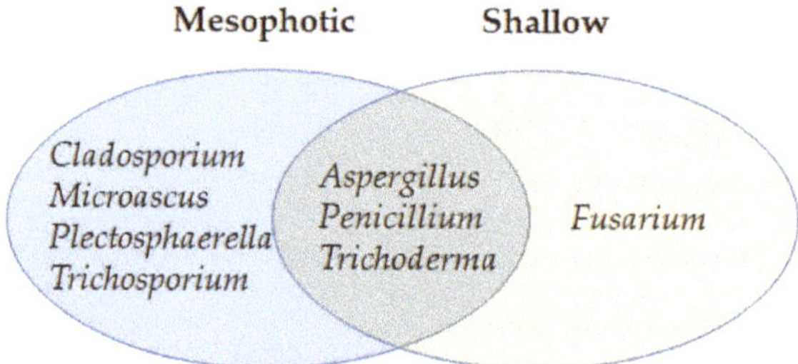

Figure 7. Presence of eight culturable fungal genera in specimens of *Agelas oroides* obtained from mesophotic depth (blue) and transplanted to shallow (10 m) depth.

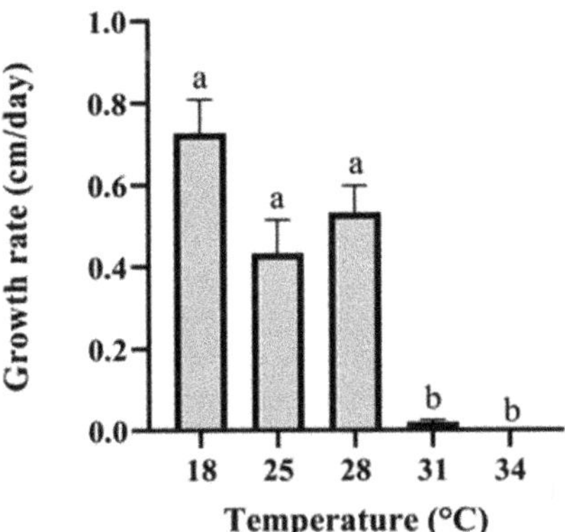

Figure 8. Temperature-dependent growth rate of *Penicillium stekii* isolated from mesophotic *Agelas oroides*. Bars indicate standard error. Different letters above the columns designate the values presented are significantly different ($p \leq 0.05$).

4. Discussion

Fungi have been repeatedly cultured from a wide variety of sessile marine animals [18]. There has been at least one report of the isolation of a sponge-associated fungus, from *Acanthella cavernosa,* collected from a mesophotic depth [37]. In the case of *Agelas* sp., fungi have been, so far, isolated or identified by DDGE analysis from specimens obtained from shallow depths [33,38,39]. In order to determine whether fungi also thrive in *A. oroides* present in mesophotic depths, we sampled mesophotic sponges and analyzed for the presence of culturable fungi. Overall, we isolated 296 fungal colonies from ten *A. oroides* specimens (Table 2). The genera of 237 sponge-derived isolates were determined on the basis of their internal transcribed rRNA gene spacer. Among the cultivated strains, 96% (230 taxa) were identified as Ascomycota and found to be comprised of 30 genera. Paz et al. [17], who studied the fungal community in another Mediterranean sponge, *Ircinia variabilis,* albeit at a shallow depth, also found the presence of representatives of the same four predominant orders: Capnodiales, Eurotiales, Hypocreales and Pleosporales. The high prevalence of Ascomycota found here is in line with previous studies, using either culture-dependent or metagenomic-based approaches [13,40–42]. Fungicide amendments to the isolation media proved advantageous in obtaining a higher diversity of fungal taxa from the sponge host, as has been the case with other sponge species [17,43].

Members of three genera were repeatedly found associated with samples of *A. oroides* (including after host transplantation to shallow water): *Aspergillus, Penicillium* and *Trichoderma.* We therefore concluded that, at least within the geographical area analyzed, these comprise the core culturable mycobiome of *A. oroides*. Nonetheless, these genera are ubiquitous in marine environments and thus caution should be used prior to assigning attributes of specificity to the observed associations. In the case of *Trichoderma* spp., this is the first case of identifying a member of this genus in association with *A. oroides*, even though strains of *Trichoderma*, including a new species, *T. beinartii*, [44], have been found in *Iricinia variabilis* (formerly *Psammocinia* sp.), collected from shallow water along the coast of Israel [45]. In addition to their omnipresent nature in association with marine life forms, many strains belonging to the three mentioned genera have been shown to produce bioactive secondary metabolites [11], including in Mediterranean sponges [10,46–48]. It is likely that this is also the case with some of the strains isolated from the mesophotic *A. oroides* (and perhaps involved in some of the antagonist fungal–fungal interactions observed in this study). The fact that at least some representatives of the three common genera, *Aspergillus, Penicillium* and *Trichoderma* (Figure 7) have been shown to exhibit myco-parasitic capabilities [17,49–51] may be indicative of the potential presence of such interactions among the sponge mycobiont constituents. At least in the case of sponge-derived *Trichoderma* spp. some of the strains isolated from *I. variabilis* have been shown to have myco-parasitic capabilities against other fungi isolated from the same sponge [17]. In this study, *Fusarium* spp. were found only in sponges that had been transplanted to shallow water. This genus is extremely diverse [52] and has been shown to be ubiquitous in terrestrial and marine habitats. The ecological roles of the species identified here have yet to be determined, yet one possibility is that these function as post-transplantation acquired saprobes that feed on decaying sponge tissue. Many *Aspergillus* species are known as animal pathogens [53–55] and strains of some of those species (e.g., *A. sydowii*, *A. fumigatiaffinis*, *A. novoparasiticus* and *A. flavus*) were isolated during this study. While, to the best of our knowledge, fungal pathogens of marine sponges have yet to be described, the possibility of a sponge being a symptomless vector of a fungal pathogen has been previously suggested [18,56] and it is possible that, in this case too, *A. oroides* may harbor potential fungal pathogens of sponges or other marine animals. *Penicillium* spp. were also among the most abundant fungal species isolated (85 strains). More than half (47 strains) were identified as *P. steckii*. This species has been repeatedly found in the marine environment and in association with a variety of organisms, including tunicates, molluscs, fish, algae and sponges [57,58]. While terrestrial *P. steckii* has been reported to favor higher temperatures (optimal growth at 30 °C; [59]), here, we have found that the sponge-associated. *P. steckii* exhibits maximal growth at 18 °C–28 °C, which are the prevalent temperatures at the mesophotic depth from where the host was obtained (Figure 8). This observed preference of lower temperatures may have, along with other factors, contributed to the fact that this species was not isolated from the sponges that had undergone transplantation to the shallow habitat. It also implies adaptation of this specific strain's life to the mesophotic habitat, and raises the question of which adaptations would be advantageous to fungi residing in this ecological niche. Interestingly, the fungus did not last long in the sponges when they were transplanted to shallower (10 m) depth. The hosting sponge, however, survived for several months and exhibited characteristic signs of vitality (e.g., ectosome intact, open oscula, and pumping water), including the rebuilding of their body wall and detaching of the necrotic parts which were a result of the transplantation process. However, once ambient temperatures exceeded 28°C indications of stress were evident and, subsequently, all the hosts died [29]. To what extent the observed differences in the sponge mycobiome are cause or consequence has yet to be determined, but at least in the case of *P. stekii*, temperature may well have had an effect on its presence. Differences in the mycobiome at different depths can be expected. In fact, the stability of core mycobiome constituents on the one hand, and changes in non-core components on the other, even along a much smaller difference in depths, has been observed in another sessile marine animal—*Acropora lorites*, even without transplantation [60]. Seven strains of *P. album* (formerly *Engyodontium album*), noted for its production of proteases and cytotoxic secondary metabolites [61–63], were isolated from the different *A. oroides* samples. To assess the possible relations between *A. oroides* and *P. album*, a sterilized sponge fragment was inoculated with the fungus. Two months post-inoculation the fungus covered most of the fragment. The fact that *P. album* can grow at 18 °C and has the ability to utilize the sponge meso-hyl (inner section) as a sole source of nutrients, supports the possibility that, within *A. oroides*, *P. album* potentially has sufficient nutrients to support growth in the mesophotic niche. However, to date, there have been no reports on the presence of hyphae developing in sponges, and this study is no exception.

Typically, sponges are considered to harbor a higher complexity of microbial diversity than corals, but that diversity is more stable [64]. While the nature of fungal survival, growth and reproduction within the living sponge remains to be elucidated, the fact that fungi are so prevalent in marine sponges suggests that they may interact while occupying this niche, as part of the nature of maintaining a stable mycobiome. Furthermore, analysis of sponge microbiomes among *Ircinia* spp. specimens exhibiting different growth forms indicate the presence of stable associations between host sponges and their microbiomes, and perhaps that they potentially contribute to ecological divergence among *Ircinia* species [65]. It is also likely that interactions between mycobiome constituents are part of the dynamics of microbiome stability. Even though only a limited number of fungal strains were analyzed here, a range of interactions from growth suppression to enhanced colony growth was easily detected (Figure 4). Whether or not such interactions occur within the highly complex holo-biont, remains to be determined, it is likely that the physical proximity of multiple microorganisms is accompanied by a plethora of potential outcomes, some of which were observed here. These would also likely include interactions with bacteria and the host itself.

The results obtained in this study indicate the presence of a diverse mycobiome, which includes core members, that can even be cultured from small sponge fragments. It is highly probable that a significant number of additional taxa are part of the sponge mycobiome (core, as well as transient community members). Some are expected to be difficult to culture under the conditions used here, and some may even be obligate biotrophs (on the sponge itself or other members of the holo-biont). As the current consensus on the definition of marine fungi is based on ecological, rather than a taxonomical, basis [18], we have considered all the culturable fungi described here as part of the sponge holo-biont, regardless of the nature of their symbiosis. Considering that the sponge is a filter-feeder, small particles (~0.1 μm), including fungal spores, can be readily found in the sponge. Kumala et al. [66] described osculum dynamics and its effect on filtering rate, as part of the mechanism by which sponges can control their filter feeding parts and clean them. Even given the sponges' impressive capabilities to digest or extrude particles, fungal propagules can be easily isolated from sponges, indicative of their persistence. As the fungi described here are, for the most part aerobic, availability of oxygen may be a crucial factor to their survival and proliferation within the holo-biont. The ability of the sponge to modulate aerobic and anaerobic metabolism via regulation of its water flow may be one of the factors that determines the fate of its mycobiome [67,68]. It is tempting to speculate that the extent of fungal strain sensitivity to anoxia may play an important role in special and temporal fungal presence and activity within the sponge. Coupling sponge sectioning along genomic and transcriptomic analyses may provide some additional answers to the points raised here and may be instrumental in linking mycobiome residency with mycobiome function within marine sponges.

Author Contributions: Conceptualization, M.I. and O.Y.; investigation, E.B.-D.C., M.I. and O.Y.; resources, M.I. and O.Y.; writing—original draft preparation, E.B.-D.C.; writing—review and editing, M.I. and O.Y.; supervision, M.I. and O.Y.; funding acquisition, M.I. and O.Y. All authors have read and agreed to the published version of the manuscript.

Funding: This work was funded by the Israel Science Foundation (OY); Ministry of Energy (MI); and Israel Nature and Parks Authority (MI).

Data Availability Statement: Sequence data is available at NCBI.

Acknowledgments: We thank Tal Idan for collecting the sponges, and EcoOcean for enabling the sponge collection.

Conflicts of Interest: The authors declare no conflict of interest.

References

1. Van Soest, R.W.M.; Boury-Esnault, N.; Hooper, J.N.A.; Rützler, K.; de Voogd, N.J.; Alvarez, B.; Hajdu, E.; Pisera, A.B.; Manconi, R.; Schönberg, C.; et al. World Porifera Database. Available online: www.marinespecies.org/porifera (accessed on 8 March 2021).
2. Hickman, C.P.; Roberts, L.S.; Keen, S.L.; Larson, A.; L'Anson, H.; Eisenhour, D.J. *Integrated Principles of Zoology*; McGraw-Hill: New York, NY, USA, 2008; pp. 246–259.
3. Pita, L.; Rix, L.; Slaby, B.M.; Franke, A.; Hentschel, U. The Sponge Holobiont in a Changing Ocean: From Microbes to Ecosystems. *Microbiome* **2018**, *6*, 46. [CrossRef]
4. Beazley, L.I.; Kenchington, E.L.; Murillo, F.J.; Sacau, M.D.M. Deep-Sea Sponge Grounds Enhance Diversity and Abundance of Epibenthic Megafauna in the Northwest Atlantic. *ICES J. Mar. Sci.* **2013**, *70*, 1471–1490. [CrossRef]
5. Gerovasileiou, V.; Chintiroglou, C.C.; Konstantinou, D.; Voultsiadou, E. Sponges as "Living Hotels" in Mediterranean Marine Caves. *Sci. Mar.* **2016**, *80*, 279–289. [CrossRef]

6. Lesser, M.P. Oxidative Stress in Marine Environments: Biochemistry and Physiological Ecology. *Annu. Rev. Physiol.* **2006**, *68*, 253–278. [CrossRef] [PubMed]
7. Tsoukatou, M.; Hellio, C.; Vagias, C.; Harvala, C.; Roussis, V. Chemical Defense and Antifouling Activity of Three Mediterranean Sponges. *Z. Naturforsch.* **2014**, *57*, 161–171. [CrossRef] [PubMed]
8. Taylor, M.; Hill, R.; Piel, J.; Thacker, R.; Hentschel, U. Soaking it Up: The Complex Lives of Marine Sponges and Their Microbial Associates. *ISME J.* **2007**, *1*, 187–190. [CrossRef]
9. Taylor, M.; Radax, R.; Steger, D.; Wagner, M. Sponge-Associated Microorganisms: Evolution, Ecology and Biotechnological Potential. *Microbiol. Mol. Biol. Rev.* **2007**, *71*, 295–347. [CrossRef]
10. Cohen, E.; Koch, L.; Myint, T.K.; Rahamim, Y.; Aluma, Y.; Ilan, M.; Yarden, O.; Carmeli, S. Novel Terpenoids of the Fungus *Aspergillus insuetus* Isolated from the Mediterranean Sponge *Psammocinia* sp. Collected along the Coast of Israel. *Bioorg. Med. Chem.* **2011**, *19*, 6587–6593. [CrossRef] [PubMed]
11. Cheng, M.M.; Tang, X.L.; Sun, Y.T.; Song, D.Y.; Cheng, Y.J.; Liu, H.; Li, P.L.; Li, G.Q. Biological and Chemical Diversity of Marine Sponge-Derived Microorganisms over the Last Two Decades from 1998 to 2017. *Molecules* **2020**, *25*, 853. [CrossRef]
12. Schmitt, S.; Tsai, P.; Bell, J.; Fromont, J.; Ilan, M.; Lindquist, N.; Perez, T.; Rodrigo, A.; Schupp, P.J.; Vacelet, J.; et al. Assessing the Complex Sponge Microbiota: Core, Variable and Species-Specific. *ISME J.* **2012**, *6*, 564–576. [CrossRef] [PubMed]
13. Amend, A.; Burgaud, G.; Cunliffe, M.; Edgcomb, V.P.; Ettinger, C.L.; Gutiérrez, M.H.; Heitman, J.; Hom, E.F.Y.; Ianiri, G.; Jones, A.C.; et al. Fungi in the Marine Environment: Open Questions and Unsolved Problems. *MBio* **2019**, *10*, e01189-18. [CrossRef]
14. Nguyen, M.T.H.D.; Thomas, T. Diversity, Host-Specificity and Stability of Sponge-Associated Fungal Communities of Co-Occurring Sponges. *PeerJ* **2018**, *6*, e4965. [CrossRef]
15. Gao, Z.; Li, B.; Zheng, C.; Wang, G. Molecular Detection of Fungal Communities in the Hawaiian Marine Sponges *Suberites zeteki* and *Mycale armata*. *Appl. Environ. Microbiol.* **2008**, *74*, 6091–6101. [CrossRef] [PubMed]
16. Naim, M.A.; Smidt, H.; Sipkema, D. Fungi Found in Mediterranean and North Sea Sponges: How Specific are They? *PeerJ* **2017**, *5*, e3722. [CrossRef] [PubMed]
17. Paz, Z.; Komon-Zelazowska, M.; Druzhinina, I.S.; Aveskamp, M.M.; Shnaiderman, A.; Aluma, Y.; Carmeli, S.; Ilan, M.; Yarden, O. Diversity and Potential Antifungal Properties of Fungi Associated with a Mediterranean Sponge. *Fungal Divers.* **2010**, *42*, 17–26. [CrossRef]
18. Yarden, O. Fungal Association with Sessile Marine Invertebrates. *Front. Microbiol.* **2014**, *5*, 1–6. [CrossRef]
19. Easson, C.G.; Chaves-Fonnegra, A.; Thacker, R.W.; Lopez, J.V. Host Population Genetics and Biogeography Structure the Microbiome of the Sponge *Cliona Delitrix*. *Ecol Evol.* **2020**, *10*, 2007–2020. [CrossRef]
20. Griffiths, S.M.; Antwis, R.E.; Lenzi, L.; Lucaci, A.; Behringer, D.C.; Butler, M.J.; Preziosi, R.F. Host Genetics and Geography Influence Microbiome Composition in the Sponge *Ircinia Campana*. *J. Anim. Ecol.* **2019**, *88*, 1684–1695. [CrossRef]
21. Ferretti, C.; Marengo, B.; De Ciucis, C.; Nitti, M.; Pronzato, M.; Marinari, U.; Pronzato, R.; Manconi, R.; Domenicotti, C. Effects of *Agelas oroides* and *Petrosia ficiformis* Crude Extracts on Human Neuroblastoma Cell Survival. *Int. J. Oncol.* **2007**, *30*, 161–169. [CrossRef] [PubMed]
22. Ferretti, C.; Vacca, S.; De Ciucis, C.; Marengo, B.; Duckworth, A.R.; Manconi, R.; Pronzato, R.; Domenicotti, C. Growth Dynamics and Bioactivity Variation of the Mediterranean Demosponge *Agelas oroides* (Agelasida, Agelasidae) and *Petrosia ficiformis* (Haplosclerida, Petrosiidae). *Mar. Ecol.* **2009**, *30*, 327–336. [CrossRef]
23. Idan, T.; Shefer, S.; Feldstein, T.; Yahel, R.; Huchon, D.; Ilan, M. Shedding Light on an East-Mediterranean Mesophotic Sponge Ground Community and the Regional Sponge Fauna. *Medit. Mar. Sci.* **2018**, *19*, 84–106. [CrossRef]
24. Gonzalez-Zapata, F.L.; Bongaerts, P.; Ramírez-Portilla, C.; Adu-Oppong, B.; Walljasper, G.; Reyes, A.; Sanchez, J.A. Holobiont Diversity in a Reef-Building Coral Over its Entire Depth Range in the Mesophotic Zone. *Front. Mar. Sci.* **2018**, *5*, 29. [CrossRef]
25. Kahng, S.E.; Copus, J.M.; Wagner, D. Recent Advances in the Ecology of Mesophotic Coral Ecosystems (MCEs). *Curr. Opin. Environ. Sustain.* **2014**, *7*, 72–81. [CrossRef]
26. Lesser, M.P.; Slattery, M.; Leichter, J.J. Ecology of Mesophotic Coral Reefs. *J. Exp. Mar. Biol. Ecol.* **2009**, *375*, 1–8. [CrossRef]
27. Bongaerts, P.; Riginos, C.; Brunner, R.; Englebert, N.; Smith, S.R.; Hoegh-Guldberg, O. Deep Reefs Are Not Universal Refuges: Reseeding Potential Varies Among Coral Species. *Sci. Adv.* **2017**, *3*, e1602373. [CrossRef]
28. Tsurnamal, M. Studies on the Porifera of the Mediterranean Littoral of Israel. Ph.D. Thesis, The Hebrew University of Jerusalem, Jerusalem, Israel, 1968.
29. Idan, T.; Goren, L.; Shefer, S.; Ilan, M. Sponges in a Changing Climate: Survival of *Agelas oroides* in a Warming Mediterranean Sea. *Front. Mar. Sci.* **2020**, *7*, 1–11. [CrossRef]
30. Gloeckner, V.; Wehrl, M.; Moitinho-Silva, L.; Gernert, C.; Schupp, P.; Pawlik, J.R.; Lindquist, N.L.; Erpenbeck, D.; Wörheide, G.; Hentschel, U. The HMA-LMA Dichotomy Revisited: An Electron Microscopical Survey of 56 Sponge Species. *Biol. Bull.* **2005**, *277*, 78–88. [CrossRef] [PubMed]
31. Poppell, E.; Weisz, J.; Spicer, L.; Massaro, A.; Hill, A.; Hill, M. Sponge Heterotrophic Capacity and Bacterial Community Structure in High- and Low-Microbial Abundance Sponges. *Mar. Ecol.* **2013**, *35*, 414–424. [CrossRef]
32. Abdel-Wahab, N.M.; Scharf, S.; Özkaya, F.C.; Kurtán, T.; Mándi, A.; Fouad, M.A.; Kamel, M.S.; Müller, W.E.G.; Kalscheuer, R.; Lin, W.; et al. Induction of Secondary Metabolites from the Marine-Derived Fungus *Aspergillus Versicolor* through Co-Cultivation with *Bacillus subtilis*. *Planta Med.* **2019**, *85*, 503–512. [CrossRef]

33. Frank, M.; Özkaya, F.C.; Müller, W.E.G.; Hamacher, A.; Kassack, M.U.; Lin, W.H.; Liu, Z.; Proksch, P. Cryptic Secondary Metabolites from the Sponge-Associated Fungus *Aspergillus Ochraceus*. *Mar. Drugs.* **2019**, *17*, 99. [CrossRef]
34. Bell, T.G.; Kramvis, A. Fragment Merger: An Online Tool to Merge Overlapping Long Sequence Fragments. *Viruses* **2013**, *5*, 824–833. [CrossRef]
35. Schindelin, J.; Arganda-Carreras, I.; Frise, E.; Kaynig, V.; Longair, M.; Pietzsch, T.; Preibisch, S.; Rueden, C.; Saalfeld, S.; Schmid, B.; et al. Fiji: An Open-Source Platform for Biological-Image Analysis. *Nat. Methods* **2012**, *9*, 676–682. [CrossRef]
36. Mandal, S.; Banerjee, D. Proteases from Endophytic Fungi with Potential Industrial Applications. In *Recent Advancement in White Biotechnology through Fungi. Fungal Biology*; Yadav, A., Mishra, S., Singh, S., Gupta, A., Eds.; Springer: Cham, Switzerland, 2019; pp. 319–359. ISBN 978-3-030-10480-1.
37. Le Goeff, G.; Lopes, P.; Arcile, G.; Vlachou, P.; Van Elslande, E.; Retailleau, P.; Gallard, J.F.; Weis, M.; Benayahu, Y.; Fokialakis, N.; et al. Impact of the Cultivation Technique on the Production of Secondary Metabolites by *Chrysosporium Lobatum* TM-237-S5, Isolated from the Sponge *Acanthella Cavernosa*. *Mar. Drugs.* **2019**, *17*, 678. [CrossRef]
38. Frank, M.; Hartmann, R.; Plenker, M.; Mandi, A.; Kurtan, T.; Ozkaya, F.C.; Muller, W.E.G.; Kassack, M.U.; Hamacher, A.; Lin, W.H.; et al. Azaphilones from the Sponge-Associated Fungus *Penicillium Canescens* Strain 4.14.6a. *J. Nat. Prod.* **2019**, *82*, 2159–2166. [CrossRef] [PubMed]
39. Kavruk, E.; Karabey, B.; Temel, H.Y.; Tuna, E.E.H. Phylogenetic Diversity of Mediterranean Sponge Microbiome Using Denaturing Gradient Gel Electrophoresis. *Fres. Environ. Bull.* **2019**, *28*, 10011–10020. [CrossRef]
40. Jones, E.B.G.; Suetrong, S.; Sakayaroj, J.; Bahkali, A.H.; Abdel-Wahab, M.A.; Boekhout, T.; Pang, K.L. Classification of Marine Ascomycota, Basidiomycota, Blastocladiomycota and Chytridiomycota. *Fungal Divers.* **2015**, *73*, 1–72. [CrossRef]
41. Orellana, S.C. Assessment of Fungal Diversity in the Environment Using Metagenomics: A Decade in Review. *Fungal Genom. Biol.* **2013**, *3*, 1–13. [CrossRef]
42. Zhang, T.; Wang, N.F.; Zhang, Y.Q.; Liu, H.Y.; Yu, L.Y. Diversity and Distribution of Fungal Communities in the Marine Sediments of Kongsfjorden, Svalbard (High Arctic). *Sci. Rep.* **2015**, *5*, 1–11. [CrossRef]
43. Passarini, M.R.Z.; Santos, C.; Lima, N.; Berlinck, R.G.S.; Sette, L.D. Filamentous Fungi from the Atlantic Marine Sponge *Dragmacidon Reticulatum*. *Arch. Microbiol.* **2013**, *195*, 99–111. [CrossRef]
44. Du Plessis, I.L.; Druzhinina, I.S.; Atanasova, L.; Yarden, O.; Jacobs, K. The Diversity of *Trichoderma* Species from Soil in South Africa, with Five New Additions. *Mycologia* **2018**, *110*, 559–583. [CrossRef]
45. Gal-Hemed, I.; Atanasova, L.; Komon-Zelazowska, M.; Druzhinina, I.S.; Viterbo, A.; Yarden, O. Marine Isolates of *Trichoderma* spp. as Potential Halotolerant Agents of Biological Control for Arid-Zone Agriculture. *Appl. Environ. Microbiol.* **2011**, *77*, 5100–5109. [CrossRef]
46. Gesner, S.; Cohen, N.; Ilan, M.; Yarden, O.; Carmeli, S. Pandangolide 1a, a New Metabolite of the Sponge-Associated Fungus *Cladosporium sp.* and the Absolute Stereochemistry of Pandangolide 1 and *iso*-Cladospolide B. *J. Nat. Prod.* **2005**, *68*, 1350–1353. [CrossRef] [PubMed]
47. Panizel, I.; Yarden, O.; Ilan, M.; Carmeli, S. Eight Novel Peptaibols from Sponge-Associated *Trichoderma Atroviride*. *Mar. Drugs* **2013**, *11*, 4937–4960. [CrossRef] [PubMed]
48. Koch, L.; Lodin, A.; Herold, I.; Ilan, M.; Carmeli, S.; Yarden, O. Sensitivity of *Neurospora Crassa* to a Marine-Derived *Aspergillus Tubingensis* Anhydride Exhibiting Antifungal Activity that Is Mediated by the MAS1 Protein. *Mar. Drugs* **2014**, *12*, 4713–4731. [CrossRef] [PubMed]
49. Sempere, F.; Santamarina, M.P. Suppression of *Nigrospora Oryzae* (Berk. & Broome) Petch by an Aggressive Mycoparasite and Competitor, *Penicillium Oxalicum* Currie & Thom. *Int. J. Food Microbiol.* **2008**, *122*, 35–43. [CrossRef] [PubMed]
50. Horn, B.W.; Peterson, S.W. Host Specificity of *Eupenicillium Ochrosalmoneum*, *E. Cinnamopurpureum and Two Penicillium Species Associated with the Conidial Heads of Aspergillus. Mycologia* **2008**, *100*, 12–19. [CrossRef]
51. Hu, X.J.; Qin, L.; Roberts, D.P.; Lakshman, D.K.; Gong, Y.M.; Maul, J.E.; Xie, L.H.; Yu, C.B.; Li, Y.S.; Hu, L.; et al. Characterization of Mechanisms Underlying Degradation of Sclerotia of *Sclerotinia Sclerotiorum* by *Aspergillus Aculeatus* Asp-4 Using a Combined qRT-PCR and Proteomic Approach. *BMC Genom.* **2017**, *18*, 674. [CrossRef] [PubMed]
52. O'Donnell, K.; Todd, J.; Ward, T.J.; Robert, V.A.R.G.; Crous, P.W.; Geiser, D.M.; Kang, S. DNA Sequence-Based Identification of Fusarium: Current Status and Future Directions. *Phytoparasitica* **2015**, *43*, 583–595. [CrossRef]
53. Frisvad Jens, C.; Larsen Thomas, O. Extrolites of *Aspergillus Fumigatus* and Other Pathogenic Species in Aspergillus Section Fumigati. *Front. Microbiol.* **2016**, *6*, 1485. [CrossRef] [PubMed]
54. Gonçalves, S.S.; Stchigel, A.M.; Cano, J.F.; Godoy-Martinez, P.C.; Colombo, A.L.; Guarro, J. *Aspergillus Novoparasiticus*: A New Clinical Species of the Section Flavi. *Med. Mycol.* **2012**, *50*, 152–160. [CrossRef]
55. Gourama, H.; Bullerman, L.B. *Aspergillus flavus* and *Aspergillus Parasiticus:* Aflatoxigenic Fungi of Concern in Foods and Feeds. *J. Food Prot.* **1995**, *58*, 1395–1404. [CrossRef] [PubMed]
56. Ein-Gil, N.; Ilan, M.; Carmeli, S.; Smith, G.W.; Pawlik, J.R.; Yarden, O. Presence of *Aspergillus Sydowii*, a Pathogen of Gorgonian Sea Fans in the Marine Sponge *Spongia Obscura*. *ISME J.* **2009**, *3*, 752–755. [CrossRef] [PubMed]
57. Ma, H.G.; Liu, Q.; Zhu, G.L.; Liu, H.S.; Zhu, W.M. Marine Natural Products Sourced from Marine-Derived *Penicillium* Fungi. *J. Asian Nat. Prod. Res.* **2016**, *18*, 92–115. [CrossRef]
58. Nicoletti, R.; Trincone, A. Bioactive Compounds Produced by Strains of *Penicillium* and *Talaromyces* of Marine Origin. *Mar. Drugs.* **2016**, *14*, 37. [CrossRef] [PubMed]

59. Famurewa, O.; Olutiola, P.O. Comparison of Growth and Cellulolytic Enzyme Production in *Aspergillus Chevalieri* and *Penicillium Steckii* from Mouldy Cacao Beans. *Folia Microbiol.* **1991**, *36*, 347–352. [CrossRef] [PubMed]
60. Lifshitz, N.; Hazanov, L.; Fine, M.; Yarden, O. Seasonal Variations in the Culturable Mycobiome of *Acropora Loripes* along a Depth Gradient. *Microorganisms* **2020**, *8*, 1139. [CrossRef]
61. Chellappan, S.; Jasmin, C.; Basheer, S.M.; Elyas, K.K.; Bhat, S.G.; Chandrasekaran, M. Production, Purification and Partial Characterization of a Novel Protease from Marine *Engyodontium Album* BTMFS10 under Solid State Fermentation. *Process. Biochem.* **2006**, *41*, 956–961. [CrossRef]
62. Tsang, C.C.; Chan, J.F.W.; Pong, W.M.; Chen, J.H.K.; Ngan, A.H.Y.; Cheung, M.; Lai, C.K.C.; Tsang, D.N.C.; Lau, S.K.P.; Woo, P.C.Y. Cutaneous Hyalohyphomycosis Due to *Parengyodontium Album* gen. et comb. nov. *Med. Mycol.* **2016**, *54*, 699–713. [CrossRef]
63. Yao, Q.; Wang, J.; Zhang, X.; Nong, X.; Xu, X.; Qi, S. Cytotoxic Polyketides from the Deep Sea Derived Fungus *Engyodontium Album* DFFSCS021. *Mar. Drugs.* **2014**, *12*, 5902–5915. [CrossRef] [PubMed]
64. O'Brien, P.A.; Webster, N.S.; Miller, D.J.; Bourne, D.G. Host-Microbe Coevolution: Applying Evidence from Model Systems to Complex Marine Invertebrate Holobionts. *MBio* **2019**, *10*, 1–14. [CrossRef]
65. Kelly, J.B.; Carlson, D.E.; Low, J.S.; Rice, T.; Thacker, R.W. The Relationship between Microbiomes and Selective Regimes in the Sponge Genus *Ircinia*. *Front. Microbiol.* **2021**, *12*, 1–16. [CrossRef] [PubMed]
66. Kumala, L.; Riisgård, H.U.; Canfield, D.E. Osculum Dynamics and Filtration Activity in Small Single-Osculum Explants of the Demosponge *Halichondria Panicea*. *Mar. Ecol. Prog. Ser.* **2017**, *572*, 117–128. [CrossRef]
67. Hoffmann, F.; Røy, H.; Bayer, K.; Hentschel, U.; Pfannkuchen, M.; Brümmer, F.; de Beer, D. Oxygen Dynamics and Transport in the Mediterranean Sponge *Aplysina Aerophoba*. *Mar. Biol.* **2008**, *153*, 1257–1264. [CrossRef] [PubMed]
68. Lavy, A.; Keren, R.; Yahel, G.; Ilan, M. Intermittent Hypoxia and Prolonged Suboxia Measured in situ in a Marine Sponge. *Front. Mar. Sci.* **2016**, *3*, 1–11. [CrossRef]

MDPI
St. Alban-Anlage 66
4052 Basel
Switzerland
Tel. +41 61 683 77 34
Fax +41 61 302 89 18
www.mdpi.com

Journal of Fungi Editorial Office
E-mail: jof@mdpi.com
www.mdpi.com/journal/jof

www.ingramcontent.com/pod-product-compliance
Lightning Source LLC
LaVergne TN
LVHW070111170726
843515LV00007B/1663

* 9 7 8 3 0 3 6 5 4 7 1 8 3 *